FUTURE TRENDS IN 5G AND 6G

FUTURE TRENDS IN 5G AND 6G

Challenges, Architecture, and Applications

Mangesh M. Ghonge
Ramchandra Sharad Mangrulkar
Pradip M. Jawandhiya
Nitin Goje

CRC Press
Taylor & Francis Group
Boca Raton London New York

CRC Press is an imprint of the
Taylor & Francis Group, an **informa** business

First edition published 2022
by CRC Press
6000 Broken Sound Parkway NW, Suite 300, Boca Raton, FL 33487-2742

and by CRC Press
2 Park Square, Milton Park, Abingdon, Oxon, OX14 4RN

© 2022 Taylor & Francis Group, LLC

CRC Press is an imprint of Taylor & Francis Group, LLC

Reasonable efforts have been made to publish reliable data and information, but the author and publisher cannot assume responsibility for the validity of all materials or the consequences of their use. The authors and publishers have attempted to trace the copyright holders of all material reproduced in this publication and apologize to copyright holders if permission to publish in this form has not been obtained. If any copyright material has not been acknowledged please write and let us know so we may rectify in any future reprint.

Except as permitted under U.S. Copyright Law, no part of this book may be reprinted, reproduced, transmitted, or utilized in any form by any electronic, mechanical, or other means, now known or hereafter invented, including photocopying, microfilming, and recording, or in any information storage or retrieval system, without written permission from the publishers.

For permission to photocopy or use material electronically from this work, access www.copyright.com or contact the Copyright Clearance Center, Inc. (CCC), 222 Rosewood Drive, Danvers, MA 01923, 978-750-8400. For works that are not available on CCC please contact mpkbookspermissions@tandf.co.uk

Trademark notice: Product or corporate names may be trademarks or registered trademarks and are used only for identification and explanation without intent to infringe.

ISBN: 978-1-032-00682-6 (hbk)
ISBN: 978-1-032-00683-3 (pbk)
ISBN: 978-1-003-17515-5 (ebk)

DOI: 10.1201/9781003175155

Typeset in Times
by MPS Limited, Dehradun

Contents

Preface	vii
Editors	ix
Contributors	xi

1 An Organized Study of Congestion Control Approaches in Wireless Sensor Networks   1
Savita Jadhav and Sangeeta Jadhav

2 NR: Architecture, Protocol, Challenges, and Applications   25
Virendra A. Uppalwar and Trupti S. Pandilwar

3 Comprehensive Survey on Device-to-Device Communication for Next-Generation Cellular Technology   63
Priyanka Patil and Vaibhav Hendre

4 Challenges, Opportunities, and Applications of 5G Network   81
Kalyani N. Pampattiwar and Pallavi Chavan

5 Machine Learning and Deep Learning for Intelligent and Smart Applications   95
Reena Thakur and Dheeraj Rane

6 Key Parameters in 5G for Optimized Performance   115
Dhanashree A. Kulkarni and Anju V. Kulkarni

7 Applications of Machine Learning in Wireless Communication: 5G and Beyond   149
Rohini Devnikar and Vaibhav Hendre

8 GREEN-Cloud Computing (G-CC) Data Center and Its Architecture toward Efficient Usage of Energy   163
Devasis Pradhan, K C Priyanka, and Rajeswari

9 SDR Network & Network Function Virtualization for 5G Green Communication (5G-GC)   183
Devasis Pradhan, K C Priyanka, and Rajeswari

10 An Intensive Study of Dual Patch Antennas with Improved Isolation for 5G Mobile Communication Systems   205
T. Prabhu, E. Suganya, J. Ajayan, and P. Satheesh Kumar

11 Design of Improved Quadruple-Mode Bandpass Filter using Cavity Resonator for 5G Mid-Band Applications 219
P. Satheesh Kumar, P. Chitra, and S. Sneha

12 Wavelet Transform for OFDM-IM under Hardware Impairments Performance Enhancement 235
Asma Bouhlel, Anis Sakly, and Salama Said Ikki

13 A Systematic Review of 5G Opportunities, Architecture and Challenges 247
Rishiraj Sengupta, Dhritiraj Sengupta, Digvijay Pandey, Binay Kumar Pandey, Vinay Kumar Nassa, and Pankaj Dadeech

14 The Latest 6G Artificial Intelligence Network Applications 271
K.R. Padma and K.R. Don

15 A Review of Artificial Intelligence Techniques for 6G Communications: Architecture, Security, and Potential Solutions 281
Syed Hauider Abbas, Nazish Siddiqui, and Sanjay Kumar Agarwal

16 Layered Architecture and Issues in 6G 293
N. Krishna Chaitanya, N. V. Lalitha, Gulivindala Suresh, and Mangesh M. Ghonge

17 Artificial Intelligence Techniques for 6G 305
T. Sathis Kumar and N. Kavitha

18 Antenna Array Design for Massive MIMO System in 5G Application 321
Mehaboob Mujawar

Index 327

Preface

Advancements in network, including SD-WAN and intelligent network analytics, have led to an increase in the capacity and capability of networking devices and systems. Scientists and technology designers from across the world are increasingly focusing on these developments as these are considered as the future of networking. The buzzword of today's era, artificial intelligence started growing through networking. The scientific community started looking towards a new paradigm shift and applying artificial intelligence techniques for better understanding the network pattern, helping a machine carry out tasks like humans, increasing the performance of existing network infrastructure with minimal investment, and more. This helps us to keep up with the complexity and scale of next-generation networks. With the evolution of 5G wireless networks, the world is moving into a new era of human interaction. This technological development has brought a revolution in various domains, namely core infrastructure, communication, and virtualization. With advancements in the amount and type of data available for use, it becomes necessary to build an improved security model that can deal with advanced attacking techniques.

This book offers a comprehensive overview of basic communication and networking technologies and exploits recent developments in communication and networking. It focuses on emerging technologies such as 5G, 6G, machine learning, and deep learning solutions for communication and networking, etc. This book also addresses several key issues in energy-efficient systems, cloud computing and virtualization, machine learning, cryptography, and 6G wireless technology and its future.

This volume comprises 18 chapters, highlighting various advancements in communication and networking. Chapter 1 presents the study of congestion control approaches in wireless sensor networks. Chapter 2 describes architecture, protocol, challenges and applications of NR. Chapter 3 presents a comprehensive survey on device-to-device communication for next-generation cellular technology. Chapter 4 describes the 5G technology specifically focused on the challenges, architecture, and applications of 5G. Chapter 5 describes machine learning and deep learning for intelligent and smart applications. Chapter 6 attempts to address the key parameters in 5G for optimized performance. Chapter 7 describes the various applications of machine learning in wireless communication: 5G and beyond. Chapter 8 contains applications of cloud computing (G-CC) data center and its architecture towards efficient usage of energy. Chapter 9 addresses the SDR network and network function virtualization for 5G green communication (5G-GC). Chapter 10 describes an intensive study of dual patch antennas with improved isolation for 5G mobile communication systems. Chapter 11 presents the design of improved quadruple-mode bandpass filter using cavity resonator for 5G mid-band applications. Chapter 12 presents wavelet transform for OFDM-IM under hardware impairments performance enhancement. Chapter 13 discusses the opportunities, architecture and challenges of 5G networks. Chapter 14 presents the latest 6G artificial intelligence network applications. Chapter 15 focuses on architecture, security and potential solutions for 6G communication. Chapter 16 presents the layered architecture and issues with regard to 6G. Chapter 17 describes artificial intelligence techniques for 6G. Finally, Chapter 18 presents the antenna array design for massive MIMO system in 5G application.

We would like to express our deep gratitude to the authors for their contributions. It would not have been possible to reach this proposal submission without their contribution. As editors, we hope this book will encourage further research in 5G and 6G. Special thanks go to our publisher, CRC Press/Taylor and Francis Group.

We hope that this book will present promising ideas and outstanding research contributions supporting further developments in 5G and 6G.

Editorial Team

Editors

Dr. Mangesh M. Ghonge is currently working at Sandip Institute of Technology and Research Center, Nashik, Maharashtra, India. He received his Ph.D. in Computer Science & Engineering from Sant Gadge Baba Amravati University, Amravati, India and an M.Tech degree in Computer Science & Engineering from Rashtrasant Tukadoji Maharaj Nagpur University, Nagpur, India. He authored/coauthored over 60+ published articles in prestigious journals, book chapters, and conference papers. Moreover, Dr. Mangesh Ghonge has authored/edited 8 international books published by recognized publishers such as Springer, IGI Global, CRC Press Taylor & Francis, and Wiley-Scrivener. He has been invited as a resource person for many workshops/FDP. He has organized and chaired many national/international conferences and conducted various workshops. He received a grant from the Ministry of Electronics and Information Technology (MeitY) for organizing a faculty development program. He is the editor-in-chief of International Journal of Research in Advent Technology (IJRAT), E-ISSN 2321-9637. He is also the guest editor for SCIE indexed journal special issue. He worked as a reviewer for Scopus/SCIE Indexed journals. Also, a reviewer in various international journals and for international conferences held by different organizations in India as well as abroad. His 02 patent is published by the Indian Patent office. He has also contributed to the Board of Studies, Computer Science & Engineering of Sandip University, Nashik as a Board Member. Dr. Mangesh Ghonge has 12 years of teaching experience and has guided more than 30+ UG projects and 10 PG scholars. His research interest includes security in wireless networks, artificial intelligence, and blockchain technology.

Dr. Ramchandra Mangrulkar, being a postgraduate from National Institute of Technology, Rourkela, received his Ph.D. in Computer Science and Engineering from SGB Amravati University, Amravati in 2016 and presently is working as an Associate Professor in the Department of Computer Engineering at SVKM's Dwarkadas J. Sanghvi College of Engineering, Mumbai (autonomous college affiliated to the University of Mumbai), Maharashtra, India. Prior to this, he worked as an Associate Professor and Head, Department of Computer Engineering, Bapurao Deshmukh College of Engineering Sevagram, Maharashtra, India. Dr. Ramchandra Mangrulkar has published 48 papers and 11 book chapters with Taylor and Francis, Springer, and IGI Global in the field of interest and also presented significant papers in related conferences. He has also chaired many conferences as a session chair and conducted various workshops on Artificial Intelligence BoT in Education, Network Simulator 2 and LaTeX, and Overleaf. He has also received certification of appreciation from DIG Special Crime Branch, Pune, and Superintendant of Police. He has also received grants in aid of Rs. 3.5 lakhs under Research Promotion Scheme of AICTE, New Delhi for the project "Secured Energy Efficient Routing Protocol for Delay Tolerant Hybrid Network". He is an active member of the Board of Studies in various universities and autonomous institutes in India.

Dr. Pradip M. Jawandhiya has been working as the Principal in Pankaj Laddhad Institute of Technology & Management Studies, Buldana. He has been involved in academics and administration since the past 25 years. He has completed his Diploma in Electronics & Tele Communication in 1990; Bachelor of Engineering (Computer Engineering) in 1993 from Sant Gadge Baba Amravati University, Amravati; Master of Engineering (Computer Science & Engineering) in 2001 from Sant Gadge Baba Amravati University, Amravati; MBA (Human Resource Development) in 2011 from Yashwantrao Chavan Maharashtra Open University, Nashik and Ph.D. (Computer Science & Engineering) in 2012 from Sant Gadge Baba Amravati University, Amravati. He is the Chairman of the Computer Society of India, Amravati Chapter, Amravati, Member of Indian Society for Technical Education, Fellow of IETE,

Fellow of IEI, and several other organizations. He has been invited as a speaker and has delivered a talk on ICT for Rural Development during the 54th Annual Technical Convention of IETE held on September 24 and 25, 2011 at Ahmadabad and many more talks. He has been involved in various academic and research activities. He has attended and served as a reviewer for several national and international seminars, conferences and presented numerous research papers and also published research articles in national and international journals. He is the member of the curriculum development committees of various universities and autonomous colleges and his area of research specialization is Computer Networks, Networks Security, Mobile Ad Hoc Networks, and Security. He is a recognized supervisor (PhD guide) for Computer Science & Engineering as well as for Information Technology in Sant Gadge Baba Amravati University, Amravati. Five Ph.D. research scholars completed Ph.D. under his guidance. He is currently a Member Board of Studies, Computer Science & Engineering, Sant Gadgebaba Amravati University, Amravati.

Dr. Nitin S. Goje is currently working as a professor at Webster University in Tashkent, Uzbekistan. He completed Ph.D. in Computer Science in 2015. He is having 18+ years of experience in teaching, research, and administration with 4 Years in KSA and Erbil. He has published research papers in national/international journals and conferences. He has published books on various topics of Computer Science. He is on the editorial board/reviewer of prestigious international journals. He is an active member of CSI, IAENG, CSTA, and ACM. His research interests include GIS, Image Processing, and Data Mining.

Contributors

Savita Jadhav
Department of E & TC Engineering,
Dr. D. Y. Patil Institute of Technology
Pune, India

Sangeeta Jadhav
Department of IT Engineering, Army Institute of Technology
Dighi Hills Pune, India

Virendra Uppalwar
Senior Engineer (System Software) at Sasken Technologies Limited
Bengaluru India

Trupti Pandilwar
B.E. (Computer Science Engineering), Government College of Engineering, Gondwana University
Gadchiroli

Ms. Priyanka Patil
Assistant Professor at Department of Electronics & Telecommunication, DIT
Pimpri Pune, India.

Prof. Vaibhav Hendre
Professor at Department of Electronics & Telecommunication, G.H. Raisoni College of Engineering & Management
Pune, India.

Kalyani N Pampattiwar
Research Scholar, Department of Computer Engineering, Ramrao Adik Institute of Technology and Assistant Professor, SIES Graduate School of Technology
Nerul, Navi Mumbai, MH, India.

Pallavi Chavan
Department of Information Technology, Ramrao Adik Institute of Technology Nerul
Navi Mumbai MH, India.

Ms. Reena Thakur
Assistant Professor, Department of Computer Science and Engineering, Jhulelal Institute of Technology
Nagpur, India

Dr. Dheeraj Rane
Associate Professor, Medicaps University
Indore, India

Dhanashree Ashish Kulkarni
Dr. D. Y. Patil Institute of Engineering & Technology
Pune, India

Dr. Anju V. Kulkarni
Dr. D. Y. Patil Institute of Engineering & Technology
Pune, India

Rohini Devnikar
Research Scholar at Department of Electronics and Telecommunication Engineering, G H Raisoni College of Engineering & Management
Pune, India

Vaibhav Hendre
Professor at Department of Electronics and Telecommunication Engineering, G H Raisoni College of Engineering & Management
Pune, India

Devasis Pradhan
AIT, Acharya
Bengaluru

Priyanka K C
AIT, Acharya
Bengaluru

Dr. Rajeswari
AIT, Acharya
Bengaluru

T. Prabhu
Department of ECE, SNS College of Technology
Tamil Nadu, India

E. Suganya
Department of ECE, Sri Eshwar College of Engineering
Tamil Nadu, India

J.Ajayan
Department of ECE, SR University
Telangana, India

P. Satheesh Kumar
Department of ECE, Coimbatore Institute of Technology
Tamil Nadu, India

Satheesh Kumar.P
Assistant Professor, Electronics and Communication Engineering Department, Coimbatore Institute of Technology
Coimbatore Tamilnadu, India

P.Chitra
Associate Professor, Electronics and Communication Engineering Department, Coimbatore Institute of Technology
Coimbatore Tamilnadu, India

Sneha.S
PG Scholar, Electronics and Communication Engineering Department, Coimbatore Institute of Technology
Coimbatore Tamilnadu, India

Asma Bouhlel
Laboratory of Electronic and Microelectronic Faculty of Sciences Monastir, University of Monastir
Tunisia

Anis Sakly
Industrial Systems study and Renewable Energy (ESIER), National Engineering School of Monastir, University of Monastir
Tunisia

Salama Said Ikki
Electrical Engineering Department, Faculty of Engineering, Lakehead University, Thunder Bay, ON P7B 5E1
Canada

Rishiraj Sengupta
Network Engineer, Facebook connectivity Deployment, Facebook
London United Kingdom

Dhritiraj Sengupta
Post-Doctoral Fellow, State Key Laboratory of Coastal and Estuarine Research, East China Normal University. 500, DongChuan road
Shanghai, China

Digvijay Pandey
Ph.D. Candidate, IET Lucknow, Lecturer Department of Technical Education Kanpur
India

Binay Kumar Pandey
Assitant Professor, Dept of IT, College of Technology, Govind Ballabh Pant University of Agriculture and Technology
U.K, India

Vinay Kumar Nassa
Principal/Professor, Department of Computer Science Engg, South Point Group of Institutions
Sonepat, India

Dr. Pankaj Dadeech
Computer Science & Engineering, Swami Keshvanand Institute of Technology, Management & Gramothan (SKIT)
Jaipur, India

K.R.Padma
Assistant Professor, Department of Biotechnology Sri Padmavati Mahila Visva Vidyalayam (Women's) University
Tirupati, AP, India

K.R.Don
Reader, Department of Oral Pathology, Saveetha Dental College, Saveetha Institute of Medical and Technical Sciences, Saveetha University
Chennai, Tamil Nadu, India

Syed Hauider Abbas
Integral University
Lucknow, India

Nazish Siddiqui
Integral University
Lucknow, India

Dr. Sanjay Kumar Agarwal
Dolphin PG Institute
Dehradun, India

N.Krishna Chaitanya
Professor in Electronics and Communication Engineering, RSR Engineering College
Kavali, Nellore District
Andhra Pradesh, India

N.V.Lalitha
Assistant Professor in Electronics and Communication Engineering, GMR Institute of Technology, Rajam – 532 127
Andhra Pradesh, India

Gulivindala Suresh
Assistant Professor in Electronics and Communication Engineering, GMR Institute of Technology, Rajam – 532 127
Andhra Pradesh, India

Dr. Mangesh M. Ghonge
Assistant Professor in CSE, Sandip Institute of Technology and Research Center
Nashik, India

T Sathis Kumar
Assistant Professor, Department of Computer Science and Engineering, Saranathan College of Engineering
Tamilnadu, India

Dr Nkavitha
Professor, Department of Information Technology Indra Ganesan College of Engineering, Tiruchirappalli
Tamilnadu, India

Mehaboob Mujawar
Goa College of Engineering
Goa, India

An Organized Study of Congestion Control Approaches in Wireless Sensor Networks

Savita Jadhav[1] and Sangeeta Jadhav[2]
[1]Department of E & TC Engineering, Dr. D. Y. Patil Institute of Technology, Pune
[2]Department of IT Engineering, Army Institute of Technology, Dighi Hills, Pune

1.1 INTRODUCTION

Wireless Sensor Networks (WSNs) find applications in intelligent homes, transportation services, precise agriculture production, environment, habitat surveillance, smart industries, structures, critical military missions, and disaster management [1]. The sensor network supervises and tracks an environment by sensing all the available physical parameters. The various challenges arising during use of WSNs are energy-aware clustering, node deployment, localization, dynamic topology, congestion control, power management, and data aggregation.

Congestion occurs when a packet appearance rate outstrips packet convenience rate [2]. The congestion arises in sensor network due to deterioration of radio link quality, multiple data transmissions over the links, fickle traffic densities, unpredictable and irregular links, and biased data rates. Hence comprehensive analysis of network congestion and local contention is required to attain maximum link usage, to increase network lifetime, to provide fairness among flows, to decrease data loss due to buffer overflow, and to diminish overhead on the network.

Congestion is broadly classified into two types—packet-based and location-based congestions, as shown in Figure 1.1. Packet-based congestion is further divided into two categories. Node-level congestion: This occurs when input load goes beyond the existing capacity, resulting in node buffer overflow. Consequently, the rate of packet service is smaller than the rate of packet arrival, which leads to increase in loss of packet and power wastage. Link-level congestion: This arises when multiple nodes use the same wireless channel, resulting in packet collision. Location-based congestion is classified into three types. First is source congestion: In this case, during the ongoing events the sensor nodes covering the same sensing field sense the event spot simultaneously. All the nodes in that

2 Future Trends in 5G and 6G

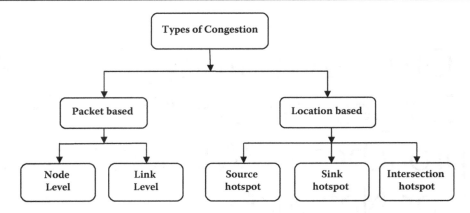

FIGURE 1.1 Classification of congestion.

area try to transmit data to sink and form a traffic hotspot near the event spot. Sink congestion: In this the nodes follow multi-hop and multipath routing which leads to an increase in traffic density at the sink node. Forwarder congestion: The presence of multiple sinks in the network leads to formation of a traffic junction, which leads to increase in traffic at the intersection nodes. The two major factors that can help manage congestion situation are: By adjusting data transfer rate and by regulating network resources. In general, efficient and effectual process should be designed for controlling congestion in WSNs. Further, based on particular applications, the data flow can be event-based, continuous, query-based, and hybrid expressed as follows.

1.1.1 Types of Applications [3]

Event-based applications: Initially the data flow in the network is habitually small but on the occurrence of an event it abruptly increases. The congestion situation arises at the event spot due to increase in data rate.

Continuous sensing applications: The sensor nodes in the network continuously sense the environment and send sensed information to the base station. Periodic sensing is also allowed based on the network load.

Query-based applications: In contrast to event-based application where sensor nodes send information simultaneously on detection of event, in query-based application the sink node sends query messages to source nodes. The source nodes resolve these queries by answering them.

Hybrid application: This application is a combination of the above three techniques. In this periodic sensing, concurrent transmission on triggering of event and answering sink query takes place simultaneously.

1.1.2 Types of Congestion

Congestion mechanism is broadly classified into two types as packet-based and location-based congestion as shown in Figure 1.1.

Packet-based: It is further divided into two types as shown in Figure 1.2.

Node-level congestion: The node buffer overflows on exceeding its capacity as in this case, the rate of packet service is less than the rate of packet arrival. The node level congestion is shown in

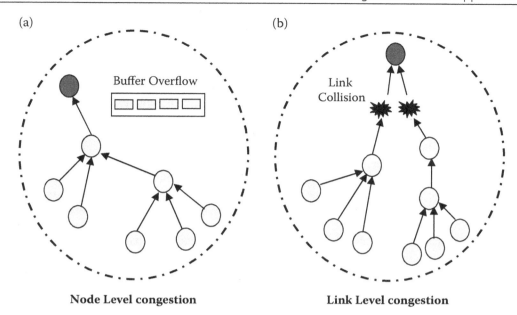

FIGURE 1.2 (a) Node level congestion (b) Link level congestion.

Figure 1.2(a). This congestion ends in packet loss. Packet loss leads to decrease network performance, decrease throughput and causes energy wastage, which directly affects the network lifetime [4].

Link-level congestion: In this technique the nodes use the same wireless link for multi-hop and multipath routing as shown in Figure 1.2(b). Competition for assessing link, packet collision, and bit error rate are the factors that cause congestion in wireless link. Link congestion results in decrease in link level performance and channel bandwidth.

Location-based: This method is further divided into three categories [5].

Source hotspot: The source nodes covering the same sensing field create a hotspot in that area on occurrence of event as all source nodes start simultaneous data transmission towards the sink node. This situation can be overcome by effective recourse utilization, desynchronizing source nodes, and by removing redundant information.

Sink hotspot: The data get aggregated toward the sink node as source nodes follow multipath and multi-hop routing scheme. This concurrent data transmission towards sink creates a hotspot near the sink. The congestion situation can be addressed by deploying multiple sink nodes and by using mobile sinks in the network.

Intersection hotspot: The presence of multiple sinks creates hotspot in the area around the junction of data flows. By using alternate detour paths, by controlling recourses and by using rate regulation at source nodes this situation can be addressed effectively.

The main objectives of this chapter are to provide an impression of the associated study and emphasize the most significant and rationalized congestion control strategies; to highlight diverse tactics in design and upgrade congestion state by covering different approaches, advantages, limitations; to collect, classify, and analyze the key congestion mitigation protocols; to focus the congestion detection, congestion notification, and congestion control schemes; to converse the key challenges in handling congestion state. The chapter is organized as follows: The related work of congestion control is reviewed in section 2. Section 3 presents an overview of related issues in congestion control for WSN. Section 4 describes the performance evaluation methods and analysis. Section 5 winds up this chapter and highlights potential research directions.

1.2 RELATED REVIEW STUDY

A widespread survey of most important congestion control mechanisms is presented in WSN. The chapter concludes by giving directions for future researches [6]. The authors' extant reviews of latest congestion control techniques and the protocols surveyed are analyzed based on three parameters as power consumption, throughput, and delay only [7]. The analysis and categorization of the innovative protocols is provided in [8]. In this survey the review of protocols is split into two categories as resource control and traffic control techniques. Further, the data rate control protocols are classified as reactive protocols and preventive protocols. Comparison is based on congestion detection, notification, control technique, control pattern, application type, loss recovery, evaluation type, and evaluation parameters. The authors present classification, comparison of algorithms, protocols, and mechanisms of congestion control and avoidance in WSNs [2,9,10]. An extensive survey of various congestion mitigation mechanisms in WSNs are presented in [11]. These protocols are classified based on the congestion notification, control pattern, loss recovery, fairness, energy conservation, cross layer/generic, and congestion detection technique. Comparison is based on only performance metrics. In [12], the authors provide a detailed impression and implementation of centralized and distributed congestion alleviation protocols in WSN. Table 1.1 enlists the details of existing survey papers of congestion control techniques in WSN. The congestion control is broadly categorized into two approaches. Most of the survey papers are based on classical approach. The classical approach contains traffic rate regulation, recourse control, queue assisted, priority aware, and hybrid techniques. The soft computing-based approach is divided into fuzzy logic, machine learning, game theory, and swarm intelligence-based techniques. This chapter focuses on the classical approach of congestion control.

1.3 CONGESTION MITIGATION PHASES IN WSN

Congestion control algorithms are classified into three phases. In congestion spotting the most frequently used parameter is measurement of queue length. Every node has a specific queue length; a limit is fixed and congestion is signaled immediately after exceeding of queue length. The congestion situation is notified to upstream nodes either implicitly or explicitly in the notification stage. Finally, congestion is mitigated by using different techniques as shown in Figure 1.3.

1.3.1 Congestion Awareness Phase

Congestion sensing denotes the probable event identification which may lead to congestion occurrence in the network. Various protocols help detect congestion using different combinations of the parameters as stated below [6,8].

Buffer occupancy: Sensor nodes store data packets in buffer before routing it over the radio link. A congestion alarming situation arises when buffer capacity exceeds the predefined threshold value [19].

Channel load: The detection of congestion data packet load on the wireless link can be obtained by using two methods. The first method is channel busyness ratio which is the ratio of time for successful data transmission including collision time to the total time. The second method is congestion degree which is the ratio of packet service time to the packet inter-arrival time [20].

Combination of buffer occupancy and channel load: The above two techniques are combined together in this method. In case of an increase in packet collision and after several unsuccessful MAC retransmissions, data packets are removed from the link. Consequently, the decrease in buffer occupancy

TABLE 1.1 The survey of existing papers based on congestion control in wireless sensor network

REF.	CLASSIFICATION BASED ON
[1]	Classical approach: MAC layer and cross layer techniques
[2]	Classical approach: Reliable data transport technique, congestion control, and avoidance
[8]	Classical approach: Traffic and resource control technique
[13]	Classical and soft computing techniques
[3]	Classical and soft computing techniques: i) Rate, ii) buffer, iii) hybrid or rate and buffer, iv) priority, v) cluster, and vi) multipath routing
[4]	Classical approach: Traffic and resource control technique
[6]	Classical approach: Traffic, resource, priority-aware, queue-assisted technique
[5]	Swarm Intelligence techniques
[14]	Classical approach: Routing schemes aimed with congestion control with centralized strategy, dedicated congestion control schemes with distributed strategy
[9]	Classical approach: Traffic and resource control technique
[10]	Classical approach: Routing protocols to mitigate and control congestion
[7]	Classical approach: Traffic-based, resource-based, hybrid
[15]	Classical approach: Traffic control, resource control, hybrid technique
[11]	Classical approach: Traffic, resource, priority-aware, hybrid technique; soft computing: Swarm intelligence, fuzzy logic, machine learning, game theory based
[16]	Machine learning techniques
[12]	Classical approach: Congestion control, detection, and reliable data transmission techniques
[17]	Classical and soft computing: Centralized and distributed congestion control techniques
[18]	In classical approach: Traffic, resource, priority-aware, hybrid techniques; in soft computing approach: Swarm intelligence, fuzzy logic, machine learning, game theory based techniques

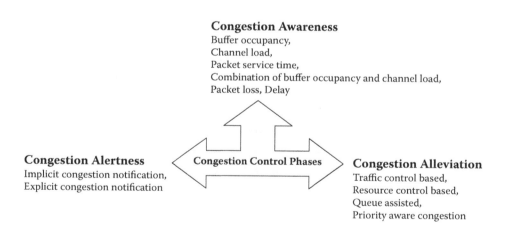

FIGURE 1.3 Three different phases in congestion processes.

due to these drops may mean the absence of congestion when only buffer occupancy is used for congestion detection. Therefore, a hybrid approach is presented using queue length and channel load as congestion detection [21].

Packet service time: Packet service time covers the time taken by packets in wait state, collision resolution state, and packet transmission time at MAC layer [22].

Packet loss: Packet loss is measured at the sender or receiver node. At sender node it is measured by enabling the use of ACKs and at the receiver node it is measured by assigning sequence numbers to data packets.

Delay: It is one hop delay which is the time taken for transmitting the packet from sender node to receiver node [23].

1.3.2 Congestion Alertness Phase

It is further divided in two types. In the first case, congestion situation is notified by piggybacking data packets and by sending acknowledgments whereas extra overhead packets are sent in second case.

Implicit congestion notification: Congestion information in this method is sent by seating CN bit in the header of data packet, by sending acknowledgment bits and by using RTS/CTS handshaking procedure. The advantage of this method is that it reduces overhead on the network [24].

Explicit congestion notification: This method sends congestion information by sending special control packets that contain congestion status. However, the use of these extra packets increases overhead in the network [19].

1.3.3 Congestion Alleviation Phase

The in-depth survey of cutting-edge congestion mitigation protocols in WSNs is presented in this section.

1.3.3.1 Traffic Rate Control-Based Congestion Control

This strategy evaluates the congestion situation by varying the data rate at overcrowded nodes [14]. The data rate is reduced by techniques shown in Figure 1.4. Traditional methods for rate adjustment include drop trail, active queue management, random early detection, congestion window update (CWND, RWND), and AIMD technique. The advantages of traffic rate control methods are that these are less complicated, cost less, and the node level congestion is controlled. The disadvantages of this method is that only the source node controls congestion and reducing the data rate may lead to loss of valuable data. The summary of traffic control-based protocols is provided in Table 1.2. Different traffic control-based protocols are explained as follows.

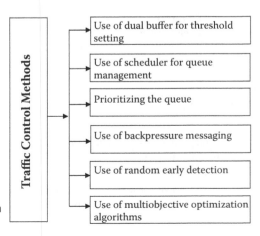

FIGURE 1.4 Different traffic control-based congestion control methods.

TABLE 1.2 Summary of traffic control based protocols

PROTOCOL NAME	CD	CC	CN	COMPARED WITH	MERITS	DEMERITS
AWF (RBCC) [25] (2019)	Queue length	Traffic optimization	Implicit	CDTMRLB,PPI, FBCC	Energy efficient transmission	—
CcEbH [26] (2017)	Queue length	congestion avoidance method	Explicit	DAIPaS	Successfully overcome network congestion and unstable power utilization	Lack of Fairness
PPI [27] (2018)	—	—	—	FCFS	Priority allocation improves congestion free routing	High average delay
Improvement in AFRC [19] (2017)	Buffer occupancy	Buffer occupancy + rate adjustment	Explicit	Advancement in AFRC	Enhances life of entire network	Increase in overhead to maintain mobile node, security challenges
ALACCP [24] (2017)	Buffer occupancy	Rate control	Implicit	TADR, HCCP	Improvement in network performance	Reliability & accuracy of event detection
GASVM [21] (2017)	Buffer occupancy ratio + congestion degree	Rate control	Implicit	CODA, ECODA, DRR	Decrease energy consumed, Packet loss,e2e delay; Improve throughput & network lifetime	Required excess memory for classification, involves high complexity
ACSRO [20] (2018)	Queue length	Rate adjustment + queue length	Implicit	SS, CS, ORA	Congestion is compacted & performance of WSN improved	Large power usage by network results in poor network efficiency

AWF
Adaptive weight firefly (AWF) consists of three stages of network design in which node clustering is done using Greedy approach and hybrid K-means-based algorithms in the first stage [25]. To compute the performance of the system the leftover energy of nodes and number of rebroadcasts are remarkable facets. The congestion status is identified by monitoring the packet flow from source nodes to sink in the system. The queue length is used for detecting congestion. To route data packets in network window size and queue size are used. For notifying the congestion status acknowledgment messages are sent on backward path. No reduction in packet rate if the positive acknowledgment is received or else its uses AWF algorithm to optimize the packet rate. In the final stage, data packets are routed by using Ant Colony Optimization (ACO) algorithm as it helps maximize the throughput.

CcEbH
CcEbH aims at providing both congestion control and energy-balanced hierarchy [26]. The hierarchical topology is a three-layered network. The first layer is for same hierarchical nodes, second layer is for upstream nodes, and the last layer is for downstream nodes. When the downstream node is congested, other lower hierarchy neighbor nodes are used for data forwarding. Queue length, forwarding, and receiving rates are used for congestion detection. The energy expenditure of the bottom-most level hierarchy nodes is stabilized by selecting the high energy nodes.

PPI
PPI guarantees congestion control using packet priority intimation [27]. Each packet consists of a PPI bit to eliminate congestion. The reduction of delay in information communication via selecting higher priority packets is the main function of this protocol. This technique consists of a congestion aware routing protocol based on AODV routing scheme. The three principal ideas of PPI protocol are amending bandwidth utilization, allotting schedule, and assigning priority. This reduces the overhead and overhearing in the network. The main challenge is the networks mobility and data transmission.

PSO-Based Routing Protocol
It is the procedure for prevention of high-quality information loss [28]. This protocol is dissemination of the PPI [27] protocol. This protocol works in five different sections: 1) Wireless body sensor networks, 2) master collector device, 3) scheduler, 4) communication system, and 5) physician's bay and base station. The major objectives are: 1) To design multiple access controllers—the main function of this controller is to decide packet priority and to minimize network traffic, 2) to design a scheduler—the scheduler decides the priority-based data transmission, and 3) to optimize path—design of routing protocol using particle swarm optimization (PSO) for alleviating congestion and minimizing communication latency.

AFRC
Available forward road capacity (AFRC) introduces mobile nodes in the area where nodes are more power-hungry [19]. By adjusting the reporting rate (RR) of the mobile node AFRC minimizes the congestion and network workload. In AFRC network lifetime and energy consumption is evaluated for static and moveable nodes. The mobile node moves around the sink node and forwards the information from congested area. The sensor nodes forward the sensed data to the sink node through mobile nodes. The comparison of power consumption of nodes takes place time to time. The location updates of mobile nodes are redirected to sink nodes if any node in the network fall under its energy speedily. The sink node is moved towards such location. The traffic of hotspot location is diverted using the mobile node. The parameters used for obtaining results are location, speed, and number of mobile nodes.

ALACCP
ALACCP addresses the congestion problem by an adaptive load-aware congestion control protocol [24]. This technique dynamically allocates the data rates of congested nodes. The ALACCP works in three

stages. 1) Congestion discovery phase: This is obtained by managing buffer length and forecasting traffic rate. 2) Data rate adjustment phase: Initially packets in buffer are arranged in descending order using hop count value, this processes repeats itself for every newly arrived packets. 3) Adaptive load-aware path adjustment phase: Adopts two stages as separation and merging of transmission path.

GA-SVM

GA-SVM controls the congestion situation by altering the transmission rate of the current node [21]. Support Vector Machines (SVMs) employ a multi-classification technique to improve network throughput. Genetic algorithm tune the SVM parameters. It decreases the power consumed, data loss, and delay. In substantial traffic condition it improves throughput and network lifetime. The MAC layer link status of downstream node is forwarded to all the nodes. The current node regulates data diffusion rate by means of traffic loading of downstream node. Congestion status is informed to upstream node by sending an Awareness Packet (AP). The retransmission values are decided by buffer occupancy ratio, congestion grade, and data rate.

ACSRO

ACSRO regulates the information rate on the forwarding path to alleviate congestion [20]. The evaluation of stake rate of child node depends on the fitness function. Every child node follows the data provision rate of parent node. The sensing node numbers, priority ID, network bandwidth, and data transmission rate are considered for defining fitness function. After completing one iteration fitness function results with new share rate.

DRCDC

DRCDC minimizes sensor network congestion [29]. The mitigation of congestion in DRCDC scheme takes place by ingeniously decreasing the data transfer rate that carries smaller information contents. At the same time it reduces the alteration rate of information in the system. The DRCDC basically works on spatial and temporal information entropy theory. 1) Congestion alleviation by spatial mode: Information collected by sensor nodes is spatially correlated. The matrix completion method collects missing data from space. 2) Congestion control by temporal mode: This method minimizes data transfer rate of time slots that carry less information.

Water Wave Optimization Algorithm

It formulates an objective function based on network throughput, remaining energy, and data harm rate of the nodes in congested area [30]. Water wave optimization algorithm gives optimum results on application of objective function. This technique is based on the shallow water wave models. The movement of wave is from deep water to slender water which maximizes its wave height and minimizes its wavelength and vice versa. This protocol considers three different processes on the waves as propagation, refraction, and breaking. The water wave algorithm is applied to detect and reduce congestion in WSN.

OFES

OFES presents an exponential smoothing prediction for congestion mitigation [31]. The OFES provides a path determination architecture which is divided into three stages: 1) Establishment of routing path in hierarchical pattern, 2) balancing energy by deviating path, and 3) exponential smoothing for congestion divination. A fuzzy logic system is designed based on hop number, residual power, queue tenancy, and data diffusion rate, which estimates the suitable weights for routing path discovery. Bat algorithm enhances weight over the membership functions.

Congestion Control Protocol

It presents a congestion control mechanism to predict the link quality for various movement stages using HMM model [32]. This system consists of three stages including prediction of link quality, development of congestion free path and rate variation to limit the congestion. This method works on the forecast of movement pattern of nodes and the data flow rate at movement stage to decide the congestion control strategy. The network QoS is improved in view of packet transfer ratio and latency.

Congestion Control Predictor Model
It presents a congestion control predictor model [33] for WSN. The network power utilization, data loss rate, and packet priority are the parameters used in network design. The shortest path selection algorithm inhibits congestion and energy usage.

1.3.3.2 Resource Control-Based Congestion Control

The resources in the network are managed by different techniques as shown in Figure 1.5. It includes enhancement of communications-related protocols—design of routing and MAC protocols—by detecting nonfunctional and energy wasteful activities, reducing over-listening to other nodes transmission, forwarding data in multiple paths, escalating network resources, using uncongested routes, using mobile nodes, turning on ideal nodes, selecting pertinent next hop node, and by incorporating appropriate sleep scheduling [15]. This method is functional when the traffic rate control method fails to accomplish an application. In comparison with rate control technique this method provides balanced energy utilization, high throughput at the cost of higher delay. Different resource control-based protocols are explained as follows. The comparison of various resource control-based protocols is presented in Table 1.3.

CCR
The foremost role of CCR is to instigate a congestion-aware clustering and routing protocol [34]. This protocol illustrates low overhead by performing the setup phase once in the first round. This setup phase segregates the network area into levels and sectors. All the clusters are homogenous. The network load is distributed by dividing the role of cluster head between primary cluster head (PCH) and the secondary cluster head (SCH). The CCR provides stability by using SCH to help PCH. The main task of PCH is data transmission and node accumulation in network.

Protocol for Controlling Congestion
A hierarchical tree and grid structure is designed for creation of preliminary network topology [35]. After that it runs Prim's algorithm to discover pertinent neighboring nodes. A source control algorithm regulates congestion by channeling traffic through an alternative uncongested route from the hotspot location to the sink node. The selection of an alternate route has four steps as maneuvering topology, erecting a hierarchical tree, engendering detour route, and supervising nodes that are lacking energy. An

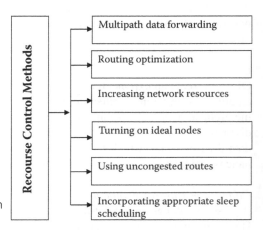

FIGURE 1.5 Different resource control-based congestion control methods.

TABLE 1.3 Comparison of resource control-based protocols

PROTOCOL NAME	CD	CC	CN	COMPARED WITH	MERITS	DEMERITS
CCR [34] (2019)	Buffer occupancy	Resource control	—	LEACH, MHT-LEACH, IMHT-LEACH	Decrease in overflow of data	By transmitting the undesirable data the energy consumption is high
Protocol for CC in WSN [35] (2017)	Queue model	Using uncongested route	Implicit	D3, DPCC, GMCAR	Reduces energy consumption, improved throughput	High delay under heavy traffic load
CDTMRLB [23] (2017)	Buffer occupancy + relative success rate + distance	Rate adjustment + resource control	Implicit	CDTMRLB, CBCC, CAM	Reduced packet drop & energy consumption & improves PDR	Feedback procedure increases network overhead
DACC [36] (2017)	Buffer threshold (dual threshold level queue)	Rate adjustment	Implicit	DACC & noncongestion mechanism	Tested and verified real-time testbed	Energy factor not considered
CAOR [22] (2017)	Packet inter arrival rate + packet service rate	Routing protocol	Implicit	—	Increases life reduces partitioning in network	Simulation and implementation
Water wave algorithm [37] (2017)	—	Clustering	—	CODA	Improved QoS	Very less parameters are used for performance checking
PSOGSA [38] (2018)	Queue length	Rate control	Implicit	Cuckoo search (CS), ACS	QoS improved	Large memory is required to store best solution value. Need improvement in load balancing problem
FCC [39] (2019)	Queuing delay	Resource control (multipath routing)	—	CCOR	Reduces packet loss, power utilization, maximizes network performance	Hop count increases which results in more energy consumption
CCP [40] (2018)	Net packet flow rate	Alternate path selection	Implicit	CRPL-OFO	Improvement in packet loss rate & throughput	Does not take into account energy consumption
ERRP [41] (2018)	Queue length	Resource control		AODV	Reduction in data drop rate improves throughput	Security and power utilization

active queue management (AQM) technique transfers data with wavering priorities. This algorithm attempts to reduce congestion by choosing nodes of the same level and substituting them.

CDTMRLB
CDTMRLB depicts congestion prevention along with power wastage challenge in WSN. If distance between nodes is used for selecting the routes, then it could be frequent utilization of similar paths for routing. Due to this, the node closer to BS will die out of energy soon. There exists consistent energy wastage when routes are chosen randomly, but it leads to reduced data success rate [23]. In this protocol the distance between source and sink node, relative success rate (RSR) value and queue tenure of node are used for selecting routes. A utility function is defined based on these three parameters and employed to all adjacent nodes. For packet forwarding the source node selects the highest utility valued node as its next hop node.

DACC
DACC provides a novel algorithm to defeat the confines of FIFO-based sensors in the network [36]. The DACC operates on two different systems: 1) Works at the base station: At preliminary phase senses congestion ingeniously, 2) works at sensor node: Energetically modifies the duty cycle on the basis of packet marking field. In DACC the congestion state of gateways is reduced by the following three steps. In the first step, DACC embraces two set aside fields in the beacon frame. The first is gateway flag for alerting congestion status of base station and the second is sensor flag for organizing critical and non-critical data. In the second step, it uses the dual marking buffer threshold technique to detect congestion. In the final step, it works on traffic control procedures that minimize the insistent data transmissions among the source and sink nodes by the means of modifying the duty cycle.

CAOR
CAOR is an amalgamation of opportunistic routing, congestion mitigation and forecast sleep time [22]. The left over energy of nodes, distance between nodes, node count, flow rate and wireless channel superiority are measures for computing cost function of relay node. Congestion detection is based on packet inter-arrival time, which is the duration of time that passes since the reception of packet until the next packet appears at the MAC layer and packet service time is the time required by node to transfer data. This is the addition of time in which packet resides in queue, time for collision perseverance, and time to effectively send out the last bit. When data flow rate of a specific node exceeds the predefined threshold value it executes rate adjustment algorithm that governs data transfer rate from network layer to MAC layer. Furthermore, inclusion of sleep scheduling system minimizes energy intake and transmission delay whereas it optimizes the network lifespan and packet distribution ratio.

Water Wave Algorithm
It devises an objective function in virtue of elements like remaining energy, throughput, hop distance, and retransmission number [37]. The optimization being implemented by studying diverse biologically inspired computational intelligence techniques. The clustering operation is performed using firefly algorithm. A fitness function for clustering is built on the values of residual energy and the distance between the nodes. Each node in the cluster shares its information related to queue size, the number of packets lost and residual energy with every other node in the cluster. The number of retransmissions and the distance between nodes are used to route the packets by calculating the optimum route. The Improved Bat Algorithm which is based on the echolocation of bats, firefly algorithm that depends on the attractiveness of the firefly, water wave algorithm which works on the dissemination, deflection, and violation of the waves, PSO, and ACO stand for formulation of objective function to find the optimum solution.

Hybrid Multi-Objective Optimization Algorithm (PSOGSA)
PSOGSA provides multi-objective optimization algorithm to control the congestion problem using PSO and Gravitational Search Algorithm (GSA) [38]. PSOGSA is proposed to optimize the data transmission

rate from all child nodes to the parent node. The power utilization of a node is measured for finding fitness function. For optimization of data transmission rate the priority-based approach is used. The different arrival rates, channel bandwidth, and node energy are inputs of the optimizer.

FCC

A fast congestion control procedure initiates routing with a hybrid optimization algorithm [39]. To select the appropriate next hop node a multi-input time on task optimization algorithm is used with minimal queuing delay in the first phase. The input is collected from three factors: 1) Time elapsed in event waiting, 2) received signal strength (RSS value), and 3) mobility in distinct time intermissions. To make energy effectual path discovery from source to sink an altered GSA is applied in the second phase.

CCP

A game theory approach devises a parent-change process that selects next hop node to alleviate congestion in the network [40]. Formulation of objective function regulates the rank of nodes. The first two bits in rank frame notify the congestion status. The option field consists of sending node's children information, IP addresses, and sender nodes flow rate. An expected rank is calculated based on the objective formulation and the rank field of the DIO sender. First of all, the node tests the CN bit by observing DIO packet of parent node. The clear status of CN bit indicates no congestion and the child node follows normal method of rank calculation. The setting of CN bit indicates congestion on parent node. The linking of child and parent node employs game theory approach for deciding new pair. The redirection of traffic flow to detour path takes place through parent-change process. The nodes refresh pairs by greatest advantage like less hop count and minimum buffer occupancy.

HBCC

The HBCC demonstrates the contention-based hop-by-hop bidirectional congestion reduction algorithm. It applies queue length for congestion detection. The congestion status is split into four classes by noticing queue length of present node and succeeding node. These four classes are 0–0, 0–1, 1–0, and 1–1 (0 for congestion off, 1 for congestion on status) [42]. The HBCC regulates the contention window of the present node when at least one of the two nodes is congested. Thus, it swaps current node priority to access the link. Further, it introduces three novel congestion reduction techniques, such as hop-by-hop receiving-based congestion control (HRCC), hop-by-hop sending-based congestion control (HSCC), and hop-by-hop priority congestion control (HPCC). The HBCC algorithm increases the throughput and reduces loss ratio.

ERRP

ERRP introduces biologically inspired efficient and reliable routing protocol algorithm that controls congestion by sending packets on new pathway [41]. Three different approaches discover congestion-free route. In the first approach path establishment takes place through AODV protocol. Other alternative routes are obtained by biologically inspired techniques. The Artificial Bee Colony (ABC) algorithm works in five phases: Nodes are deployed in the initialization phase, employ bees find their neighboring nodes in second phase, random parents are selected in the crossover phase, onlooker bee is fourth phase, and in the end the scout bee phase. After passing all the stages the optimum route is decided. Result shows the comparison between AODV and ABC algorithms. The largest buffer capacity nodes transmit data from source to destination. Next optimization technique for routing path selection is cuckoo search (CS) algorithm. CS mimics cuckoo species for searching food. At the end, three optimization techniques compared for path selection. ERRP concludes that largest buffer size nodes get selected for establishment of route.

Mobile CC

A mobile congestion control algorithm presents two variants. The first variant uses mobile nodes that establish crucial paths towards sink. The second variant creates completely disjoint paths to the sink by using mobile nodes [43]. It transfers a Congestion Message (CM) to the sink on detection of congestion on node. All the information needed from the sink for extenuating the congestion has been contained in this message. This information incorporates: The congested nodes identification, time required for packet acceptance and transmission per sample, congestion detection status, and neighbor table in sequence. The neighbor table includes NodeIDs, hop number, accepted packet count, and availability flag. On reception of CM message by the sink it calculates the position to place a mobile node. This mobile node provides alternate routes to the sink that reduces congestion. The sink delivers information to the mobile nodes mention position, NodeIDs of sender node, and the next hop NodeID. Upon receiving instructions from the sink, the mobile nodes switch off the radio, approach new location, and turn ON the radio. At the end, the mobile node setups communication with the new node.

ACC-CSMA
It is a lightweight distributed MAC protocol to adjust the conjunction of high and low priority wireless devices in cognitive WSN [44]. The ACC-CSMA forces the accessible spectrum resources, whereas assure rigorous QoS constraints. The channel congestion level is sensed by the system on the basis of local information without exchanging any message. It decides whether all devices autonomously function on a channel or depart in saturation. This procedure vigorously adjusts the congestion status. Furthermore, it obtains optimized number of low priority devices. This technique increases the link utilization without surrendering latency desires of the primary devices. It avoids signaling and computational overheads. Also it can be realized over present wireless protocols with no hardware alteration. ACC-CSMA provides positive insight into less complex distributed adaptive MAC method in cognitive WSNs.

RDML
This protocol is a data compilation approach by vibrant traffic flows for WSNs [45]. The novelty is information diffusion radius which is further integrated as *r*. Many excess energy nodes apply an information diffusion radius as *mr* where *m* is greater than one. The system performance is improved in three stages. 1) The data transferred by nodes in hotspot region is significantly minimized as these nodes can transfer data directly to sink. 2) By decreasing the time required in data aggregation. 3) In third stage, the remaining energy of nodes is wholly consumed for boosting the radius of information diffusion which results in enhancement of energy consumption rate.

1.3.3.3 Queue-Assisted Congestion Control

The congestion state is attempted via queue length of the nodes. At the end, apply simple rate adjustment techniques like active queue management, random early detection, and additive increase multipartite decrease (AIMD) techniques for maintaining the queue length of the nodes at the lowest feasible level [17]. The comparison of queue-assisted protocols is shown in Table 1.4.

Fuzzy CC Based on AQM
A novel active queue management scheme is presented to establish packet loss probability [46]. This procedure is divided in three stages. The first step integrates an innovative RED active queue management (AQM) along with fuzzy proportional integral derivative (FuzzyPID) system for sensing congestion. It sets ICN bit for notifying congestion status implicitly. In the last step, it regulates data transfer rate by applying a fuzzy controller for congestion control. The protocol governs three types of traffics flows—high, intermediate, and low priority. The high priority packets are intended for high priority information which requires small latency, middle priority is for transporting information that necessitates intermediate latency, and low priority is assigned to normal traffic.

DRED-FDNNPID
To control the target buffer queue and sending rate of each node a Dynamic Random Early Detection (DRED) with Fuzzy Proportional Integral Derivative (FuzzyPID) controller is designed [47]. In this

TABLE 1.4 Comparison of queue-assisted protocols.

PROTOCOL NAME	CD	CC	CN	COMPARED WITH	MERITS	DEMERITS
Fuzzy CC based on AQM [46] (2017)	Queue length	Queue-assisted rate adjustment	Implicit	OCMP, PCCP, CCF	Reduction in data loss and delay	Not suitable to monitor mobile patents
DESD + Fuzzy PID [47] (2018)	Queue length	Active queue management + traffic flow control	Implicit	DRED-FPID	QoS and energy consumption improved	Needs improvement in network traffic diversion scheme
EDSMAC [48] (2018)	Queue length	Active queue management	Implicit	IH-MAC	Reduction in size of control packet, overhearing, overall latency	–

technique, a DRED is improved by integrating service differentiation mechanism to differentiate high and low priority traffic. This differentiation is based on the weighted load metric, which services the input traffic according to its priority. The high priority traffic is buffered in a separate queue with low buffer size and low priority traffic is controlled by using DRED algorithm. The fuzzy inference system is enhanced by applying Deep Neural Network (DNN) optimization algorithm known as FDNN. This improves the DREDFDNNPID with self-adaptation to enhance the performance and optimize the average queuing delay.

MAQD
This routing protocol establishes on a neuro-fuzzy inference system for MWSN. MAQD ponder four kinds of alertness: The longevity of sensors, data diffusion latency, network cost, and shortest routing path [49]. A fuzzy inference system chooses suitable routing path based on one of the awareness techniques. In MAQD, Sink Node (SN) accumulates data from source nodes on application of a request message that indicates its alertness type. They instantly create information table and retransmit the query message upon reception of SN's request by query node. All query nodes respond data to SN by means of predetermined path. MAQD works on three stages: 1) Path detection stage, 2) respond stage, and 3) path maintenance stage.

PPSPC with CICADA
This protocol demonstrated the study of WBAN in the healthcare field. It is the fusion of PPSPC and CICADA protocols. It provides protected, reliable, and power-competent data transfer in wireless body area sensor networks [50]. This protocol is subdivided into five stages. The first stage is development of network in which the complete network is partitioned among clusters. The second stage is implementation of hybrid integration of PPSPC with CICADA algorithm. The third stage is creation of protocol. The information concerning data is represented in terms of tables, figures, and reports. Maximization of energy is the next stage and the last stage is performance evaluation by using metrics resembling throughput, packet delivery ratio, and latency in data transmission.

EDS-MAC
It intends an EDS-MAC protocol for versatile traffic in WSNs. This procedure comprises two phases [48]. In the initial phase VSSFFA is projected to form clusters by observing energy of cluster heads. The

16 Future Trends in 5G and 6G

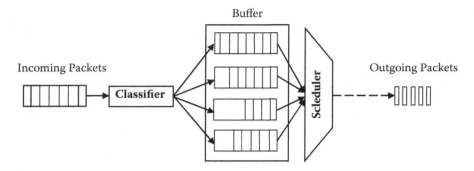

FIGURE 1.6 Priority-based congestion control model.

VSSFFA decreases the cost of positioning head nodes in a cluster. The formation of objective function is based on the enduring energy, intra-cluster distance, node degree, and possible cluster head count. Data transmission process diminishes the latency and control overhead.

1.3.3.4 Priority-Based Congestion Control

This technique makes use of prioritized MAC to furnish the congested nodes with prioritized channel admittance. Incoming packets are classified into different priorities by the classifier. The congestion is supervised through application of diverse priorities in congestion state [18]. Also different schedulers are used for queue management as shown in Figure 1.6. Various priority-based protocols are compared as shown in Table 1.5.

FACC
The FACC and Multipath routing protocols are new means that facilitate steadfast delivery of data [51]. The PCCP algorithm allocates precedence to particular nodes. Scheduling presents a proficient scheduling rate to avoid the packet loss. Through regulating the scheduling rate it avert congestion and advance throughput, reliability in the network. A main route and detour route is determined; after main

TABLE 1.5 Comparison of priority-based protocols

PROTOCOL NAME	CD	CC	CN	COMPARED WITH	MERITS	DEMERITS
FACC [51] (2017)	Queue length	Multipath routing + priority-based scheduling	Implicit	PCCP	Reduction in data loss, delay	Energy factor not considered
PFRC [52] (2017)	Queue size	Rate control + priority based	Implicit	DDRC, FRC	Improved throughput, loss, and delay parameter	–
Priority-based queuing and transmission rate [53] (2017)	Queue length	Traffic length + queue length	Implicit	CSMA/CA	Improved energy consumption	–
FPBCC [54] (2018)	Queue length	Active queue management + rate control	Implicit	PCCP, PHTCCP	QoS improved	Energy consumption not considered

route falls short it gets detour route and sends data. At the end, it presents the particulars of multiple path discovery and maintenance.

PFRC
PFRC presents congestion control algorithm with fair rate controlling and traffic class priority [52]. Depending on traffic priority, queue length and the number of connections to a sensor node the fair allocation of the bandwidth is done in PFRC. Total hierarchy of the network along with real and nonreal time traffic patterns are regarded in the design of PFRC. All traffic patterns allocate diverse priorities.

Priority-Based Queuing and Transmission Rate Control
It presents a fuzzy logic technique to improve power utilization and to decrease packet loss. The operation of fuzzy logic in CSMA/CA technique is accomplished by assigning the buffer length and the data rate to all nodes [53]. A priority level is appended to nodes to facilitate access of wireless link. The backoff exponent (BE) value is dynamically adjusted by observing buffer capacity and data transmission rate of each node. This protocol categorizes nodes into three groups as low, medium, and high priority. Low priority cluster nodes enhance their BE cost to grant other nodes the access of link as they do not have any crucial data that should be sent instantly. The high priority nodes must transmit their data earliest by the means of minimizing BE cost. Medium priority nodes choose a random BE cost.

FPBCC
FPBCC intend random early detection and active queue management-based techniques for specifying congestion in a network [54]. A fuzzy logic system that operates on two inputs and one output is applied to reorganize the maximum drop probability (maxp) of the RED algorithm. The minimum threshold (minth) and maximum threshold (maxth) evaluates the congestion status of all sensor nodes. The parent node regulates data by taking help of fuzzy logical controller (FLC).

1.4 PERFORMANCE METRICS

For evaluation of the performance of congestion control algorithms, different performance metrics are used as listed below. Table 1.6 shows the analysis of different performance metrics used for protocols discussed in section 3. The √ in the column indicates that metric is used to measure the performance.

Network delay: This metric measures the average end-to-end delay of data packet transmission. It is the time taken by the packet to travel from source node to destination node. This delay also takes into account the queuing and the propagation delay of the packets [25,26].

Network throughput: It measures the number of packets received at the destination per unit time. It is usually measured in bits per second (bps) [23,36].

Energy consumption: It is the sum of the energy used by all the nodes in the network. The node energy includes sum of energy utilized for packet transmission, reception, and for idle listening [48,53].

Network lifetime: Network lifetime is measured by basic parameters like the time until half of the nodes die. It is also calculated as the time until message loss rate exceeds a given threshold or by application directed metrics, such as when the network stops providing the application with the desired information about the environment [37,39].

Packet Delivery Ratio (PDR): Packet delivery fraction is the ratio of the number of data packets successfully delivered to the destination nodes and the number of data packets produced by source nodes [25,27].

TABLE 1.6 Performance parameter analyses of congestion control protocols

PROTOCOL NAME	AVERAGE END-TO-END DELAY	PACKET DELIVERY RATIO	THROUGHPUT	ENERGY EFFICIENCY	ENERGY CONSUMPTION	LIFETIME OF THE NETWORK	RELIABILITY	PACKET DROP RATE	AVERAGE RESIDUAL ENERGY	FAIRNESS INDEX
AWF (RBCC) [25]	✓	✓	✓	✓	✓		✓			
CcEbH [26]	✓	✓	✓					✓	✓	
PPI [27]	✓	✓	✓	✓				✓		
Improvement in AFRC [19]		✓			✓					✓
ALACCP [24]		✓		✓						✓
GASVM [21]			✓		✓			✓		
ACSRO [20]	✓		✓		✓			✓		
CCR [34]			✓		✓			✓		
Protocol for CC in WSN [35]		✓								
CDTMRLB [23]	✓	✓	✓		✓					
DACC [36]		✓	✓							
Water wave algorithm [37]		✓	✓			✓				
PSOGSA [38]	✓		✓		✓	✓				
FCC [39]			✓		✓			✓		
CCP [40]			✓		✓			✓		
ERRP [41]	✓		✓		✓			✓		
Fuzzy CC based on AQM [46]	✓		✓		✓					
DESD+ Fuzzy PID [47]	✓		✓		✓			✓		
EDSMAC [48]	✓		✓		✓					
FACC [51]	✓	✓	✓							
PFRC [52]	✓		✓					✓		
Priority base queuing and transmission rate [53]	✓		✓		✓					
FPBCC [54]	✓							✓		

Fairness: It is an indication of fair allocation of network resources such as bandwidth among nodes in the network. The fairness means the sink node receives equal number of packets from each node [19,24].

Packet drop rate: It is the ratio of the total number of lost packets to the total number of sent packets in the network [38].

1.5 CONCLUSIONS

Wireless Sensor Networks (WSNs) face many issues due to resource-constrained nature of the nodes. The network congestion is one of the main challenges that takes place when packet entrance rate surpasses the packet service rate. The chapter compares important techniques on the basis of congestion detection, notification, and alleviation. The analysis of this study specifies that network congestion leads to data loss, excess power dissipation, increased delays, and throughput minimization. The papers selected for this revision are beyond 2017. The protocols discussed in this chapter are headed into four groups as traffic rate, resource management-based, queue-assisted, and priority-based. These techniques are application-specific in nature and have their own merits and demerits. Thus it concludes that in case of heavy traffic conditions, using resource control algorithms is more relevant than data rate alteration. As these algorithms appreciably improve network performance due to uniform energy utilization. The limitations of using recourse control method are increased overhead and delay. On the other hand, traffic control methods are less complicated, less costly, and by using this method node level congestion is controlled with low delay. The limitation of using traffic control method is that it has high energy consumption, low-throughput, and valuable data may be lost. Hence, to control congestion, futuristic congestion control approaches should be explored based on computational intelligence and machine learning techniques. Nowadays, many techniques exist but still require further investigation for use under different applications.

Future directions: The comparative analysis presented in this chapter helps discover innovative exigent, stimulating ideas, issues, and tricky questions on congestion control. Interested future researchers should address various unsolved issues addressed by this chapter. This study analyzes the protocols based on congestion phases along with merits and demerits. It also provides an expeditious outline of foremost congestion control techniques in WSNs.

Real time environment: Further innovative tactics and approaches of examination and verification based on real-life examples should be applied for future studies.

Multipath routing: The predominant contemporary practices shore up single-path routing; multipath data forwarding strategies should be applied to enhance reliability and power efficiency.

Dynamic network: In the current scenario, most of the networks are static; dynamic networks and mobile agent techniques should be used to mitigate congestion control and improve network performance.

Security: As security is momentous predicament in multiple approaches, future researchers should take synergy among congestion control and security for future studies.

Application knowledge: For future studies, the impression of enlightening multimedia applications, mobile WSNs, wearable sensor technology, underground WSNs, and underwater acoustic sensor networks is the arrival of newer WSN generations which raise further challenges and issues in delineation and development of pertinent congestion control protocols.

Hybrid protocols: Hybrid protocols should be designed for future studies as current strategies are based on data rate alteration and stream diversion. For improving reliability, scalability, interrelation between distinct layers; cross-layer approaches can be used.

Future studies should majorly focus on implementation of multi-objective optimization, computational intelligence, soft computing techniques for congestion control as numerous WSNs implementations are hybrid and event-based in nature.

LIST OF ABBREVIATIONS

WSN: Wireless Sensor Networks
MAC: Medium Access Control
RDC: Radio Duty Cycle
CWND: Congestion Window
RWND: Receiver Window
AIMD: Additive-Increase/Multiplicative Decrease
AWF: Adaptive Weight Firefly Optimization
CcEbH: Congestion Control and Energy-Balanced Hierarchy
PPI: Packet Priority Intimation
AFRC: Adaptive Flow Rate Control
ALACCP: Adaptive Load-aware Congestion Control Protocol
GA-SVM: Genetic Algorithm-Support Vector Machines
ACSRO: Adaptive Cuckoo Search-based Optimal Rate Adjustment
DRCDC: Differentiated Rate Control Data Collection
OFES: Optimized Fuzzy Logic-based Congestion Control
CCR: Congestion-Aware Clustering and Routing protocol
CDTMRLB: Congestion Detection Technique for Multipath Routing and Load Balancing
DACC: Dynamic Agile Congestion Control
CAOR: Congestion-Aware Opportunistic Routing
FCC: Fast Congestion Control
CCP: Congestion Control Protocol
HBCC: Hop-by-Hop Bidirectional Congestion Control algorithm
ERRP: Efficient and Reliable Routing Protocol
Mobile CC: Mobile Congestion Control
ACC-CSMA: CSMA-Based Adaptive and Distributed Congestion Control
RDML: Reducing Delay and Maximizing Lifetime
DRED-FDNNPID: Dynamic Random Early Detection (DRED) with Fuzzy Proportional Integral Derivative (FuzzyPID) controller
RED: Random Early Detection
AQM: Active Queue Management
MAQD: Multi-Aware Query Driven
PPSPC: Privacy Preserving Scalar Product for Computation Protocol
CICADA: Cascading Information Retrieval by Controlling Access with Distributed Slot Assignment Protocol
EDSMAC: Energy Efficient Dynamic Scheduling Hybrid MAC
VSSFFA: Variable Step Size Firefly Algorithm
EDSMAC: Energy Efficient Dynamic Scheduling Hybrid MAC
PCCP: Priority-Based Congestion Control Protocol
PFRC: Priority-Based Fairness Rate Control
CSMA/CA: Carrier Sense Multiple Access with Collision Avoidance
FPBCC: Fuzzy Priority-Based Congestion Control
FLC: Fuzzy Logical Controller

REFERENCES

[1]. J. Zhao, L. Wang, S. Li, X. Liu, Z. Yuan, and Z. Gao, "A Survey of Congestion Control Mechanisms in Wireless Sensor Networks", in *2010 Sixth Int. Conf. Intell. Inf. Hiding and Multimedia Signal Process.*, IEEE, pp. 719–722, 2010.

[2]. C. Sergiou, P. Antoniou, and V. Vassiliou, "A Comprehensive Survey of Congestion Control Protocols in Wireless Sensor Networks", *IEEE Commun. Surv. Tutor.*, vol. 16, no. 4, pp. 1839–1859, 2014.

[3]. N. Thrimoorthy and Dr. T. Anuradha, "A Review on Congestion Control Mechanisms in Wireless Sensor Networks", *Int. J. Eng.* ISSN: 2248-9622, vol. 4, no. 11(Version 2), pp. 54–59, 2014.

[4]. P. Budhwar, B. Sharma, and Dr. T. C. Aseri, "Congestion Detection and Avoidance Based Transport Layer Protocols for Wireless Sensor Networks", *Int. J. Eng. Sci. Invention Res. Dev.* e-ISSN: 2278-067X, p-ISSN: 2278-800X, vol. 10, no. 5, pp. 56–69, 2014.

[5]. S. Sendra, L. Parra, J. Lloret, and S. Khan, "Systems and Algorithms for Wireless Sensor Networks Based on Animal and Natural Behavior", *Int. J. Distrib. Sens. Netw.* Article ID 625972, pp. 1–19, 2015. doi: 10.1155/2015/625972

[6]. A. Ghaffari, "Congestion Control Mechanisms in Wireless Sensor Networks: A Survey", *J. Netw. Comput. Appl.*, vol. 52, pp. 101–115, 2015.

[7]. M. A. Jan, S. R. U. Jan, M. Alam, A. Akhunzada, and I. U. Rahman, "A Comprehensive Analysis of Congestion Control Protocols in Wireless Sensor Networks", *Mob. Netw. Appl.*, vol. 23, no. 3, pp. 456–468, 2018.

[8]. M. A. Kafi, D. Djenouri, J. Ben-Othman, and N. Badache, "Congestion Control Protocols in Wireless Sensor Networks: A Survey", *IEEE Commun. Surv. Tutor.*, vol. 16, no. 3, pp. 1369–1390, 2014.

[9]. N. Thrimoorthy and Dr. T. Anuradha, "Congestion Detection Approaches in Wireless Sensor Networks: A Comparative Study", *Int. J. Eng. Res.*, vol. 12, no. 3, pp. 59–63, 2016.

[10]. M. Zeeshan, F. Khan, and S. R. Jan, "Congestion Detection and Mitigation Protocols for Wireless Sensor Networks", *Int. J. Sci. Res. Comput. Sci. Eng. Inf. Technol.*, vol. 1, no. 1, 2016.

[11]. S. S. Shamsabad Farahani, "Congestion Control Approaches Applied to Wireless Sensor Networks: A Survey", *J. Electr. Comput. Eng. Innovations*, vol. 6, no. 2, 2018.

[12]. B. Nawaz and K. Mahmood, "Congestion Control Techniques in Wsns: A Review", *Int. J. Adv. Comput. Sci. Appl.*, vol. 10, no. 4, 2019.

[13]. Sunitha G. P., Dilip Kumar S. M., and Vijay Kumar B. P., "Classical and Soft Computing Based Congestion Control Protocols in Wsns: A Survey And Comparison", in *IJCA Proc. Nat. Conf. Recent Advances Inf. Technol. NCRAIT(2)*, pp. 1–8, 2014.

[14]. S. A. Shah, B. Nazir, and I. A. Khan, "Congestion Control Algorithms in Wireless Sensor Networks: Trends and Opportunities", *J. King Saud Univ. - Comput. Inf. Sci.*, vol. 29, no. 3, pp. 236–245, 2017.

[15]. H. A. A., Al-Kashoash, H., Kharrufa, Y., Al-Nidawi, and A. H. Kemp, "Congestion Control in Wireless Sensor and 6lowpan Networks: Toward the Internet of Things", *Wirel. Netw.*, vol. 25, no. 8, pp. 4493–4522, 2019.

[16]. D. P. Kumar, T. Amgoth, and C. S. R. Annavarapu, "Machine Learning Algorithms for Wireless Sensor Networks: A Survey", *Inf Fusion*, vol. 49, pp. 1–25, 2019.

[17]. A. Bohloulzadeh and M. Rajaei, "A Survey on Congestion Control Protocols in Wireless Sensor Networks", *Int. J. Wirel. Inf. Netw.*, vol. 27, no. 3, pp. 1–20, 2020.

[18]. D. Pandey and V. Kushwaha "An Exploratory Study of Congestion Control Techniques in Wireless Sensor Networks", *Comput. Commun.*, vol. 157, pp. 257–283, 2020.

[19]. S. B. Tambe and S. S. Gajre, "Novel Strategy for Fairness-aware Congestion Control and Power Consumption Speed with Mobile Node in Wireless Sensor Networks", *Smart Trends in Systems, Security and Sustainability, Lecture Notes in Networks and Systems, Springer*, vol. 18, 2017.

[20]. V. Narawade and U. D. Kolekar, "ACSRO: Adaptive Cuckoo Search Based Rate Adjustment for Optimized Congestion Avoidance and Control in Wireless Sensor Networks", *Alex. Eng. J.*, vol. 57, pp. 131–145, 2018.

[21]. M. Gholipour, A. T. Haghighat, and M. R. Meybodi, "Hop-by-hop Congestion Avoidance in Wireless Sensor Networks Based on Genetic Support Vector Machine", *Neuro Computing*, vol. 223, pp. 63–76, 2017.

[22]. M. Shelke, A. Malhotra, and P. N. Mahalle, "Congestion-aware Opportunistic Routing Protocol in Wireless Sensor Networks", *Springer Nature Singapore Pte Ltd., Smart Computing and Informatics, Smart Innovation, Systems and Technologies* 77, doi: 10.1007/978-981-10-5544-7_7

[23]. A. M. Ahmed and R. Paulus, "Congestion Detection Technique for Multipath Routing and Load Balancing in WSN", *Wirel. Netw.*, vol. 23, pp. 881–888, 2017.

[24]. T. S. Chen, C. H. Kuo, and Z. X. Wu, "Adaptive Load-aware Congestion Control Protocol for Wireless Sensor Networks", *Wirel. Pers. Commun.*, vol. 97, pp. 3483–3502, 2017.

[25]. V. Srivastava, S. Tripathi, K. Singh, and L. H. Son, "Energy Efficient Optimized Rate Based Congestion Control Routing in Wireless Sensor Network", *J. Ambient. Intell. Humaniz. Comput., Springer*, vol. 11, pp. 1325–1338, 2019.

[26]. W. Chen, Y. Niu, and Y. Zou, "Congestion Control and Energy-balanced Scheme Based on the Hierarchy for WSNs", *IET. Wirel. Sens. Syst.*, vol. 7, no. 1, pp. 1–8, 2017.

[27]. M. P. Shelke, A. Malhotra, and P. Mahalle, "A Packet Priority Intimation Based Data Transmission for Congestion Free Traffic Management in Wireless Sensor Networks", *Comput. Electr. Eng.*, vol. 64, pp. 248–261, 2017.

[28]. M. P. Shelke, A. Malhotra, and P. Mahalle, "PSO-Based Congestion Free Critical Data Transmission in Health Monitoring System," *2018 IEEE Punecon, Pune, India*, pp. 1–8, 2018. doi: 10.1109/PUNECON.2018.8745433

[29]. J. Tan, W. Liu, T. Wang, S. Zhang, A. Liu, M. Xie, M. Ma, and M. Zhao, "An Efficient Information Maximization Based Adaptive Congestion Control Scheme in Wireless Sensor Network", *Special Section on Artificial Intelligence for Physical-Layer Wireless Communications, IEEE Access*, vol. 7, pp. 64878–64896, 2019.

[30]. M. S. Manshahia, "Water Wave Optimization Algorithm Based Congestion Control and Quality of Service Improvement in Wireless Sensor Networks", *Trans. Netw. Commun.*, vol. 5, no. 4, 2017.

[31]. P. Aimtongkham, T. G. Nguyen, and C. So-In, "Congestion Control and Prediction Schemes using Fuzzy Logic System with Adaptive Membership Function in Wireless Sensor Networks", *Hindawi. Wirel. Commun. Mob. Comput.*, vol. 2018, Article ID 6421717, 2018.

[32]. N. Thrimoorthy and Dr. T. Anuradha, "Congestion Control in Wireless Sensor Network Based on Prediction Modeling with HMM", *A UGC Recommended Journal, IJCSC*, vol. 9, no. 1, pp. 73–79, 2018.

[33]. N. T. Panah, R. Javidan, and M. Rafie Kharazmi, "A New Predictive Model for Congestion Control in Wireless Sensor Networks", *J. Eng. Sci. Technol.*, vol. 12, no. 6, pp. 1601–1616, 2017.

[34]. M. Farsi, M. Badawy, M. Moustafa, H. Arafat Ali, and Y. Abdulazeem, "A Congestion-aware Clustering and Routing (ccr) Protocol for Mitigating Congestion in WSN," *IEEE Access*, vol. 7, pp. 105402–105419, 2019.

[35]. H. D. Nikokheslat and A. Ghaffari, "Protocol for Controlling Congestion in Wireless Sensor Networks", *Wirel. Pers. Commun.*, vol. 95, pp. 3233–3251, 2017.

[36]. M. J. A. Jude and V. C. Dinesh, "DACC: Dynamic Agile Congestion Control Scheme for Effective Multiple Traffic Wireless Sensor Networks," in International Conference on Wireless Communications, Signal Processing and Networking (WiSPNET), Chennai, IEEE, pp. 1329–1333, 2017.

[37]. M. S. Manshahia, M. Dave, and S. B. Singh, "Computational Intelligence for Congestion Control and Quality of Service Improvement in Wireless Sensor Networks", *Trans. Mach. Learn. Artif. Intell.*, vol. 5, no. 6. 2017.

[38]. K. Singh, K. Singh, L. H. Son, and A. Aziz, "Congestion Control in Wireless Sensor Networks by Hybrid Multi-Objective Optimization Algorithm", *Comput. Netw.*, vol. 138, no. 19, pp. 90–107, 2018.

[39]. C. J. Raman and V. James, "FCC: Fast Congestion Control Scheme for Wireless Sensor Networks using Hybrid Optimal Routing Algorithm", *Clust. Comput.*, vol. 22, pp. 12701–12711, 2019.

[40]. C. Ma, "A Congestion Control Protocol for Wireless Sensor Networks", Springer Nature Switzerland AG 2018 X. Chen et al. (Eds.): *CSoNet 2018*, LNCS 11280, 2018, pp. 356–367. doi: 10.1007/978-3-030-04648-4_30

[41]. M. Kaur and A. Malik, "An Efficient and Reliable Routing Protocol Using Bio-Inspired Techniques for Congestion Control in WSN", in 2018 4th International Conference on Computing Sciences (ICCS), Jalandhar, IEEE, pp. 15–22, 2018.

[42]. J. Wang, X. Yang, Y. Liu, and Z. Qian, "A Contention-Based Hop-By-Hop Bidirectional Congestion Control Algorithm for Ad-Hoc Networks", *Sensors*, vol. 19, no. 16, 2019.

[43]. A. Nicolaou, N. Temene, C. Sergiou, C. Georgiou, and V. Vassiliou, "Utilizing Mobile Nodes for Congestion Control in Wireless Sensor Networks", *2019 15th Int. Conf. Distrib. Comp. Sensor Syst.*, 2019. doi:10.1109/DCOSS.2019.00047

[44]. S. Zhuo, H. Shokri-Ghadikolaei, C. Fischione, and Z. Wang, "Online Congestion Measurement And Control In Cognitive Wireless Sensor Networks", *IEEE Access*, vol. 7, pp. 137704- 137719, 2019.

[45]. Q. Wang, W. Liu, T. Wang, M. Zhao, X. Li, M. Xie, M. Ma, G. Zhang, and A. Liu, "Reducing Delay and Maximizing Lifetime for Wireless Sensor Networks with Dynamic Traffic Patterns", *IEEE Access*, vol. 7, pp. 70212- 70236, 2019.

[46]. A. A. Rezaee and F. Pasandideh, "A Fuzzy Congestion Control Protocol Based on Active Queue Management in Wireless Sensor Networks with Medical Applications", *Wirel. Pers. Commun.*, vol. 22, no. 9, 2018., doi: 10.1007/s11277-017-4896-6

[47]. Monisha V. and Ranganayaki T., "A Service Differentiation Aware Dynamic Random Early Detection and Optimized Fuzzy Proportional Integral Derivative for Active Queue Management Congestion Control in Mobile Wireless Sensor Network", *Int. J. Comput. Sci. Netw.,* vol. 16, no. 7, July 2018.

[48]. V. Sundararaj, S. Muthukumar, and R. S. Kumar, "An Optimal Cluster Formation Based Energy Efficient Dynamic Scheduling Hybrid Mac Protocol for Heavy Traffic Load in Wireless Sensor Networks", *Comput. Secur.,* vol. 77, pp. 277–288, 2018.

[49]. A. I. Saleha, K. M. Abo-Al-Ezb, and A. A. Abdullahc, "A Multi-Aware Query Driven (MAQD) Routing Protocol for Mobile Wireless Sensor Networks Based on Neuro-fuzzy Inference", *J. Netw. Comput. Appl.,* vol. 88, pp. 72–98, 2017.

[50]. M. Mishra, S. Mishra, B. K. Mishra, and P. Choudhury, "Analysis of Power Aware Protocols and Standards for Critical e-Health Applications", in Springer International Publishing AG, Internet of Things and Big Data Technologies for Next Generation Healthcare, Studies in Big Data 23, 2017.

[51]. S. Iniya Shree, M. Karthiga, and C. Mariyammal, "Improving Congestion Control in WSN By Multipath Routing with Priority Based Scheduling", in International Conference on Inventive Systems and Control (ICISC-2017), 978-1-5090-4715-4/17/$31.00, IEEE, 2017.

[52]. S. K. Swain and P. K. Nanda, "Priority Based Fairness Rate Control in Wireless Sensor Networks", presented at the IEEE WiSPNET 2017 conference 978-1-5090-4442-9/17/$31.00, 2017.

[53]. I. Bouazzi, J. Bhar, and M. Atri, "Priority-Based Queuing and Transmission Rate Management using a Fuzzy Logic Controller in WSNs", *ICT Express,* 2017, doi: 10.1016/j.icte.2017.02.0012405-9595/c.

[54]. F. Pasandideh and A. A. Rezaee, "A Fuzzy Priority Based Congestion Control Scheme in Wireless Body Area Networks", *Int. J. Wirel. Mobile Comput.,* vol. 14, no. 1, 2018.

NR: Architecture, Protocol, Challenges, and Applications

Virendra A. Uppalwar[1] and Trupti S. Pandilwar[2]

[1]B.E., M.Tech (Electronics & Comm.), Senior Engineer - System Software, Sasken Technologies Limited, Bengaluru, India
[2]B.E. (Computer Science Engineering), Government College of Engineering Chandrapur, Gondawana University

2.1 INTRODUCTION

The achievement of wireless communication technology relies on the coverage capacity of the network because the coverage is the only parameter of wireless communication that defines the ecosystem and the system capacity. The expeditious research and development in the present-day system, in terms of new research, development of new use cases, and solutions for future challenges will shape the future technology.

The development of the new era of wireless technology 5G or New Radio (NR), already showcases the opportunity in the areas of IoT, V2X, M2M, AI, VR, Robotics, and more, and a variety of use cases from communication background endure their long-term growth and benefits for humans beings in the future. The 5G is expected to set a new revolution in and outside of the globe. The mobile industry is expecting that 71% of the world's population will be using a mobile phones, and 5G will cover about 65% of the world's population by 2025.

The 5G has the ability to support high range of spectrum use for multi-purpose services and will provide an efficient spectrum management capability using ultra-dense network (UDN) as compared with the existing wireless communication networks. The HO procedures in UDN are a challenging task because 5G is expected to interact with the existing network and also with new use cases at the same time. 5G has been upgraded over the air procedures for easy HO performance and to perform the services with improvisation. Basically, the UDN is a heterogeneous wireless access network, which is an infrastructure of a combination of small cells (Pico, Femto, Macro). It provides new use cases for services like V2X, IoT, and M2M. One of the most crucial tasks for 5G is to provide services to high-scale 5G-driven applications. The 5G architecture comprises a combination of cloud computing, SDN, and NFV for new business opportunities also. And it is expected to address the issues associated with the OPEX and CAPEX as well.

In this chapter our aim is to cover the conceptual introduction of 5G. We discuss the introduction, architecture, protocol, challenges, and applications of this future technology. We focus more on the architecture, protocol, and challenges related to 5G as compared to its applications. We think that there is no edge for 5G applications and to elaborate all in this small chapter is next to impossible. We focus on NG-RAN, 5G system architecture, network slicing, and network deployment options of 5G, specifically with regard to the architecture. We explain the NSA and SA briefly just for the ease of explanation. We focus more on the protocols in this chapter. We also elaborated the 5G protocol stack from UE and gNB point of view. This part mainly focuses on functional architecture, procedures, expected services from upper and lower layers, and general description for each protocol. The 5G system has Physical, MAC, RLC, PDCP, SDCP, RRC, NAS protocols, which are discussed. These protocols are further divided into Layer 1, Layer 2, and Layer 3. We discuss this in the protocol section further.

Similar to the other wireless networks, 5G also has an assemblage of challenges related to every part of its system. Based on the study, we discuss two important challenges of 5G, the first one is building complex and dense network and the second is security issues and potential targets in 5G networks. We have added subtopics in both challenges to explain them more systematically. We discuss the future threats in the network and available solutions to overcome these challenges. In the final portion of the chapter, we discuss the use cases of 5G. Yes, we do agree that mentioned use cases are not sufficient to explain the application of 5G but we have skipped some part for self-study. We understand that few technical pages are not adequate to explain an advanced technology like 5G, but we try our best to explain this technology in the chapter in a simplified manner, such that a person from nontechnical background will also find it easy to understand 5G technology [1,2].

2.2 ARCHITECTURE

During upgradation of the previous generation networks, more emphasis was given to the end users. Every working node of the system was focused only to provide optimum services to the end user. In the era of 4G and 4G advance, some new aspects come into the picture, which were not just focused on end user but also found new opportunities in the wireless world. In the last cellular network development, process designers offered increased data rates, full IP services, and better connectivity and we made use of this network system, i.e., 4G/LTE.

In 4G, users are experiencing advanced services like VoLTE, which provide full voice call satisfaction to the users.

The architecture is a sequence of multiple nodes and in the case of LTE, it is likely to be a UE, eNB, S-GW, P-GW, MME, HSS, or PCRF. Individual nodes are a combination of dominant software and hardware and are integrated into exclusive devices. These nodes are responsible for the formation of wireless networks.

5G or New Radio is a quantum leap in wireless communications. 5G introduced a vertical pattern of use cases which improves the multiplicity of the architecture. It is not just designed to provide communication services for end clients, but it is also looking for new clients in other services like V2X, M2M, IoT, AI, healthcare, manufacturing and, more. Such multifariousness created a demand for an extended software world, rather than "stiffened" hardware devices. Because of a wide variety of service providing ability, it is challenging for operators to deploy the infrastructure for edge-level services. 5G architecture has been designed such that it runs on software to provide services. SDN, NFV, and cloud computing provide backend support to make a 5G network more software-based as compared to any existing wireless networks [3,4].

2.2.1 New-Generation Radio Access Network

A 5G architecture is a combination of 5G access networks (5G-AN) and 5G core networks (5G-CN). The 5G access points are the working design of the new-generation radio access network (NG-RAN). To communicate with each other it uses an interface "Xn". It is capable of smartly collaborating with other 3GPP and non-3GPP access points. The TCP/IP transport layer helps 5G architecture to form a hybrid network to support different smart access technologies. The transport layer supports different QoS to meet users' requirements (Figure 2.1).

The most important point in NG-RAN is gNB, and is similar to eNB in LTE. The gNB architecture is a combination of a central unit (gNB-CU) and gNB-DU which can include one or more than one gNB structures. The "F1" interface connects the gNB-CU and gNB-DU units. The gNB-CU is further divided into gNB-CU-CP and gNB-CU-UP to support the control plane and user plane protocol stack. NG-RAN communicates with 5G core network over the "NG" interface. Basically, an interface is a communication link between any two nodes.

This architecture provides an opportunity for UE to hook up with external data networks and the PDU session mechanism plays a crucial role in providing connectivity. The PDU session is a sequence of next-generation (NG) tunnels. We can consider it as a "Bit Pipe", which subsequently joins the UE with core network, and also extends UE connection to an external data network. The core network is responsible to create, maintain, and end tunnels and radio bearers. This procedure is much needed for UE state and user mobility. The PDU session handling in 5G is similar to LTE, where it transfers the user plane data. The 5G QoS parameter and supported user data units are different as compared to LTE. The 5G QoS model follows the new QoS flow structure, this provides the finest QoS differentiation. In 5G, distinct QoS flow may reside in a single PDU session.

New-Generation RAN (NG-RAN) handles the establishment, maintenance, and release procedure of PDU sessions over the radio interface. It encounters physical impairments (e.g., power reduction, interference, fading, session multiplexing(scheduling), inter-gNB handovers). On the other hand, NR Core Network (CN) leads the other remaining procedures of PDU session. However, we should be watchful of the other processes that are not similar to radio access [5–7].

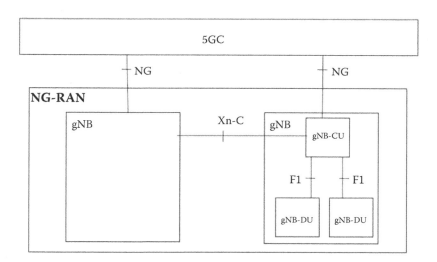

FIGURE 2.1 New Generation Radio Access Network.

2.2.2 Introduction to 5G System Architecture

I. Access and Mobility Function

The AMF is responsible for handling 5GC signaling during communication with UE. UE connects with AMF over the N1 interface. The AMF procedures are similar to the MME in LTE network. AMF allows users to perform NAS signaling procedures with the core network and perform other NAS-level procedures. Using ciphering and integrity protection creates secure communication for UE with core NW. AMF is responsible to perform mobility management (5GMM) and session management (5GSM) procedures. In addition, the authentication, registration management, and connection management procedures are also managed by the AMF (Figure 2.2).

II. Session Management Function

The SMF is responsible for handling the control part of PDU session. SMF sets up NG tunnels. It takes help from DHCP server to allocate IP addresses. A single UE can be associated with multiple SMFs. The SMF procedures are identical to session-related procedures managed by MME, SGW-C, and PGW-C in LTE. SMF takes care of IP-related functionality of 5G, such as IP address allocation. SMF shares the QoS parameters and user policy details to RAN via the AMF node.

III. Policy Control Function

The PCF is an interconnected node responsible for handling the policy rule for control plane function. It carries network slicing details, and roaming information. PCF is similar to PCRF node in LTE.

IV. User Plane Function

The UPF manages NG-U tunnel and services like traffic policy enforcement, QoS, anchoring for handover. A UE can be associated with multiple UPFs. The UPF is an amalgamation of the SGW and PGW from LTE. UPF is responsible for handling the data packet service, packet inspection, and QoS handling.

V. Authentication Server Function

The ASF implements the EAP authentication server and stores keys for operation. The ASF is identical to Home Subscriber Server (HSS) in LTE.

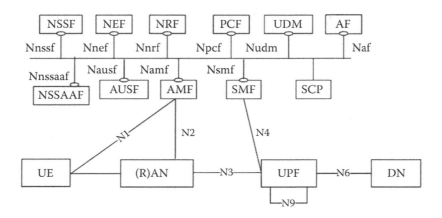

FIGURE 2.2 5G System architecture.

VI. Unified Data Management

The UDM carries UE-related information like credentials and identifiers. The UDM performs the operation to generate AKA credentials, which is useful during NAS procedures. It takes care of user identification, access authorization, and subscription management procedures.

VII. Application Function

The AF corresponds to an AS that can interact with other control plane NFs. It handles the application that influences traffic routing of the data and accessing NEF. AF interacts with the policy framework of the network operator for policy control.

VIII. NF Repository Function

It verifies the service registration and discovery function. It manages the NF profile and feasible NF instances of different users.

IX. Network Exposure Function

The NEF starts a procedure for exposing capabilities and events.

X. Network Slice Selection Function

The NSSF handles the selection of network slice instances to deliver it to the user. It regulates the allowed NSSAI. It quests AMF set to be used to serve the user [8,9].

2.2.3 Network Slicing in 5G

Telcom experts are anticipating great opportunities in the network slicing for 5G. This technology provides capacity to create numerous virtual networks on shared infrastructure in cloud. It has the capacity to share a space for other service providers to set up their applications and services to take on various specialized services such as AR, VR, V2X, online games, etc. (Figure 2.3).

"Network Slicing" has been designed to deliver the vertical service demand in 5G infrastructure. It is a block of virtual resources over the physical infrastructure provided by the operator, simply a logical network. This network has the capability to serve various types of services to clients. It is a malleable

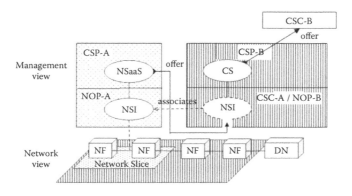

FIGURE 2.3 Network slicing in 5G.

model used to distribute resources; we called it the network-as-a-service (NaaS) model. It has the intelligence to fulfill client's demands by performing customizations in network slices for diversified and convoluted 5G use cases. The network slicing implementation depends on the SDN, NFV, and cloud computing fundamentals, a combination of this automation provides a very promising service-based architecture to network operators. However, this virtual system also has some security concerns. We will discuss the security of the network in the challenges section [10,11].

2.2.4 5G Network Deployment Option

There are multiple network deployment options available in 5G to upgrade from existing architecture to service-based 5G architecture. This evolution is based on technical, strategic, and economic factors. The network deployment option will be very crucial for different types of services. Based on the service requirement, user will be shifted to multiple network deployment options in the future. This will create new challenges in building and deployment of 5G network. In this chapter, we discuss the 5G non-standalone and standalone architecture (Figure 2.4).

I. 5G Non-Standalone

Initially, 5G starts with NSA architecture as it is difficult and challenging to deploy a fully 5G dependable network at a time. All operators are not in a condition to afford the deployment of a 5G network. So, in the initial phase of 5G, operators decided to shift on the 5G with NSA deployment option. Both operators and customers will be happy with the deployment of NSA. The non-standalone architecture consists of a 5G radio and a 4G core part. If we try to define it in a one-line statement, then it will look like,

NSA = 5G Radio + 4G Core

In this deployment, UE depends on 5G radio for user plane and on 4G for control plane support. Now, mobile network operators realize the real deployment tractions in 5G. It is possible that some operators will believe in the existing LTE network to provide a 5G coverage. Because 5G coexists with LTE and other 3GPP2 technologies, it entirely depends on the technical capacity of an existing network. Operators now have to manage the existing 4G architecture efficiently to deploy 5G nodes. It is the operator's duty to ensure better 5G services without major issues. This combinational deployment of 5G over the existing 4G network is introduced as NSA type 5G deployment.

NSA mode is the best way to handle congestion in LTE NW or to deliver improved data throughput, because it allows operators to take advantage of their existing network assets rather than deploying an ultimately new end-to-end 5G network.

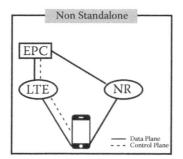
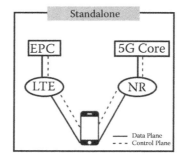

FIGURE 2.4 5G Deployment option.

TABLE 2.1 Differences in NSA and SA.

TECHNOLOGY	NSA	SA
RAT	In NSA mode, UE is capable of dealing with both LTE and 5G for services	In SA mode UE is only capable of dealing with 5G
Plane Support	LTE provides CP capability while 5G supports UP capability	In the case of SA mode, 5G will be used both in U-Plane and C-Plane
Paging	UE receives the paging information through LTE	UE receives the paging information through 5G

II. 5G Standalone

It consists of a radio part and a core part from 5G architecture. If we want to define this in a one-line statement, then it looks like,

SA = 5G Radio + 5G Core

In standalone 5G deployment, UE can access standalone 5G carrier only, and there is no mutual understanding with LTE carrier. 5G SA is a good option to showcase industry 4.0 services. It will share better alignment to support the 5G use cases and decipher the potential of next-gen mobile technology. SA allows an operator to achieve M2M communications or ultra-reliable low latency IoT services. It also provides the network slicing functionality [12,13].

Differentiation of NSA and SA architecture is based on the RATs, plane support, and paging channel information (Table 2.1).

2.3 PROTOCOL STACK MODEL OF UE AND CORE NETWORK

We can differentiate 5G protocol into two parts based on the procedures and services. 5G comprises a control plane protocol stack and a user plane protocol stack. In these stack, the control plane is used to take care of the user connection management, QoS policies, and authentication of the user, whereas user plane deals with data traffic forwarding.

2.3.1 5G Control Plane

The NR control plane has NAS and RRC sublayer. The control plane is responsible for creating communication between UE and NW. NAS layer which have two sublayers MM and SM is bound to create a communication between UE and core network components like AMF (for MM) and SMF (for SM). Another protocol of the NR control plane is RRC (Radio Resource Control), which is responsible for communication between UE and gNB. We discuss 5G-NAS and 5G-RRC protocols in the control plane. The following figure represents the control plane structure of the 5G/New Radio (Figure 2.5).

2.3.1.1 5G Non-Access Control Protocol

The 5G-NAS is the most decisive protocol in the stack. It creates communication between UE and the core network (AMF and SMF). The 5G NAS protocol consists of two sublayers, 5G mobility

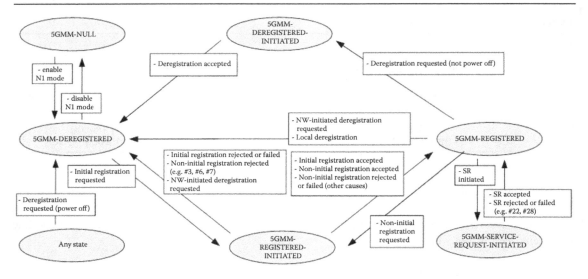

FIGURE 2.5 Control plane protocol stack.

management (5GMM) and 5G session management (5GSM). The 5GMM is responsible for identification of users, creates a secure way of communication and handles the mobility of users. It also handles the RRC connection establishment and release procedures. We discuss below the procedure, state and services of an NAS sublayer.

2.3.1.1.1 5GMM Sublayer Introduction
32..1.1.1 5GMM Procedures There are three important 5GMM procedures mentioned in 3GPP TS 24.501; the separation of these three procedures is based on the way they are initiated.

I. 5GMM Common Procedures

It is must for the UE to enter in 5GMM-CONNECTED mode to initialize 5GMM common procedures. The procedures belonging to this type are:

i. 5GMM common services initiated by the network:
   o. Network-initiated NAS transport
   o. Primary authentication and key agreement procedure
   o. Security mode control
   o. Generic UE configuration update
   o. Identification

I. 5GMM common services initiated by the UE:
   o. UE-initiated NAS transport
   o. MM status: Upon receiving the MM protocol data, if error condition encounters during a process, the status report will trigger the UE to update NW.

II. 5GMM specific procedures

Only a single UE-initiated procedure can run at a time. It is specific for each of the access network(s) that the UE is camping on. The procedures belonging to this type are:

- Registration: Only UE can trigger a registration request to register over the NW for 5GMM context or services. UE can update the location/parameter(s) using this procedure.
- De-registration: To release a 5GMM context or services, UE or NW can initiate a de-registration procedure.
- eCall inactivity: The UE that configures with eCall mode only can follow this procedure, UE follows this procedure to release a 5GMM context or services.

III. 5GMM connection management procedures

These CM procedures are initiated by the UE to create a secure connection with NW. UE demands for the resource reservation to send data or services using these procedures.

The connection management procedures are:

- Service requests
- Paging service
- Notification service

2.3.1.1.1.2 5GMM States   5GMM is one of the most complicated and challenging procedures in the development of 5G. The decision mechanism during the 5GMM procedures only depend on the 5GMM states. 5GMM states guide a UE for the next step. We explain below the 5GMM main states of UE and NW, but do not cover all transition states in this chapter.

I. 5GMM main states in the UE: (Figure 2.6)

Table 2.2 explains the 5GMM main state in UE.

II. 5GMM main states in the network: (Figure 2.7)

The following table describes the 5GMM main states in the network (Table 2.3).

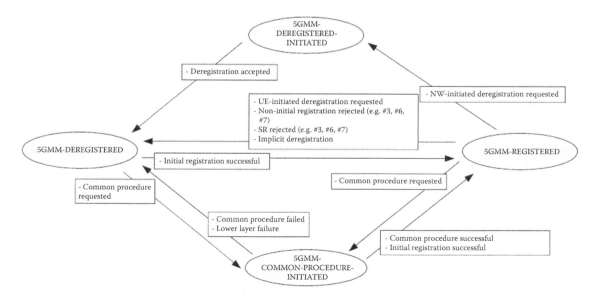

FIGURE 2.6   5GMM main states in the UE.

TABLE 2.2 State and description of 5GMM main state in UE.

STATE	DESCRIPTION
5GMM-NULL	In this state, 5G services are disabled in the UE. UE shall not perform mobility management functions in this state.
5GMM-DEREGISTERED	The UE is not attainable by the NW in this state. There is no 5GMM context established between UE and NW. The location of UE is also unknown to NW, so UE performs an initial registration procedure to obtain 5GMM context.
5GMM-REGISTERED-INITIATED	When UE starts the initial registration procedure or the non-initial registration procedure, it enters the 5GMM-REGISTERED-INITIATED state and waits for the NW response.
5GMM-REGISTERED	During the 5GMM-REGISTERED state UE is able to establish the 5GMM context. More than one protocol data unit (PDU) session may be established in this state.
5GMM-DEREGISTERED-INITIATED	When UE starts the de-registration procedure then UE may enter this state, which means the UE requests to release 5GMM context and waits for NW response.
5GMM-SERVICE-REQUEST-INITIATED	UE comes into this state only when it triggers a service request procedure and is waiting for NW's response.

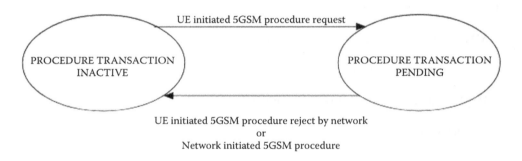

FIGURE 2.7 5GMM main states in the network.

TABLE 2.3 State and description of 5GMM main state in network.

STATE	DESCRIPTION
5GMM-DEREGISTERED	There is no 5GMM context establishment for UE in this state or the 5GMM context is marked as deregistered. In this state NW may answer the initial registration procedure or to a de-registration procedure triggered by UE.
5GMM-COMMON-PROCEDURE-INITIATED	When NW starts a common 5GMM procedure then the network enters the 5GMM-COMMON-PROCEDURE-INITIATED state and expecting feedback from the UE.
5GMM-REGISTERED	The 5GMM context has been established for NW in this state. More than one protocol data unit (PDU) session may be established in this state.
5GMM-DEREGISTERED-INITIATED	When NW starts the de-registration procedure then NW may enter this state. In this state, NW is expecting feedback from the UE.

2.3.1.1.2 5GSM Sublayer Introduction

The 5GSM sublayer is responsible for creating the session between the UE and the SMF, which means it supports PDU session handling in the UE and core NW. The content of this message is not depicted by AMF during 5GSM procedures. The message part shall include the PDU Session ID during the SM procedure [15].

The 5GSM sublayer is responsible for authentication and authorization. 5GSM can establish, modify, and release PDU sessions. During the HO of 3GPP and non-3GPP technology, it handles the request for HO during a live PDU session.

2.3.1.1.2.1] 5GSM Procedures There are three types of 5GSM procedures described by 3GPP as stated below.

I. Network-initiated procedures or procedures related to PDU sessions
   o. PDU session authentication and authorization
   o. Network-initiated PDU session modification
   o. Network-initiated PDU session release

II. UE-initiated procedures or transaction-related procedures
   o. UE-requested PDU session modification
   o. UE-requested PDU session release
Common procedure for both UE and NW
   5GSM status procedure

2.3.1.1.2.2 5GSM Sublayer States The 3GPP explains the 5GSM sublayer state for UE and network.

I. UE 5GSM states for PDU session handling

Figure 2.8 explains the 5GSM state for PDU handling in UE.

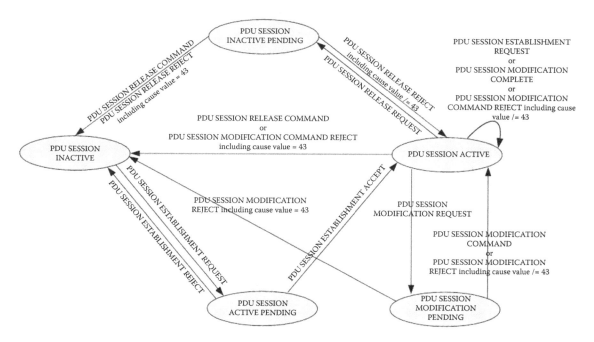

FIGURE 2.8 5GSM sublayer states in the UE.

36 Future Trends in 5G and 6G

The following table describes the 5GSM sublayer states and procedures of UE PDU session handling in 5GSM (Table 2.4).

i. The procedure transaction states in the UE (overview) (Figure 2.9):

The following table describes the procedure transaction states in the UE (Table 2.5).

II. NW 5GSM states for PDU session handling: (Figure 2.10)

The following table describes the 5GSM sublayer states for PDU session handling in NW (Table 2.6).

ii. The procedure transaction states in the network (Figure 2.11):

TABLE 2.4 State and description of 5GSM in UE.

STATE	DESCRIPTION
PDU SESSION INACTIVE	There is no PDU session remaining in this state.
PDU SESSION ACTIVE PENDING	UE triggers a PDU session establishment procedure towards NW and is waiting for a response from the NW.
PDU SESSION ACTIVE	PDU session is live in UE in this state.
PDU SESSION INACTIVE PENDING	In this state, UE is expecting the result of the PDU session release procedure from the NW.
PDU SESSION MODIFICATION PENDING	In this state, the UE has initiated a PDU session modification procedure towards the NW and is waiting for a response from the NW.

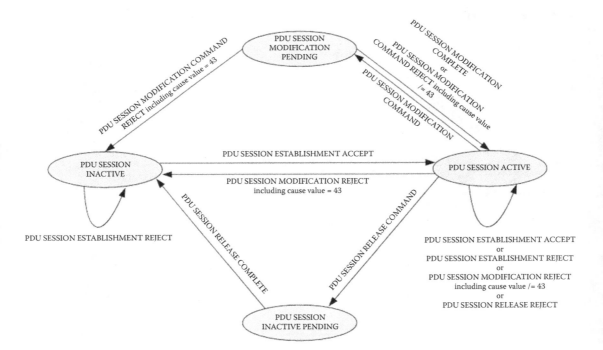

FIGURE 2.9 The procedure transaction states in the UE.

2 • NR:Architecture, Protocol, Challenges   37

TABLE 2.5 Procedure transaction states in the UE.

STATE	DESCRIPTION
PROCEDURE TRANSACTION INACTIVE	UE triggers a 5GSM procedure request towards the NW and waiting for feedback.
PROCEDURE TRANSACTION PENDING	Triggers 5GSM procedure request rejected by NW, after which this state comes into picture.

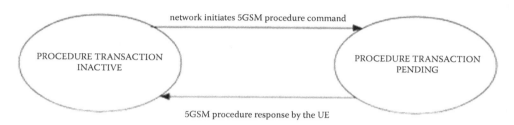

FIGURE 2.10 5GSM sublayer states in the network.

TABLE 2.6 5GSM sublayer states in the network.

STATE	DESCRIPTION
PDU SESSION INACTIVE	There is no PDU session existing in this state.
PDU SESSION ACTIVE	During this state PDU session should be active in the NW.
PDU SESSION INACTIVE PENDING	In this state, NW initiates a PDU session releases procedure towards UE and expects a response from the UE.
PDU SESSION MODIFICATION PENDING	In this state, NW initiates a PDU session modification procedure towards UE and expects a response from the UE.

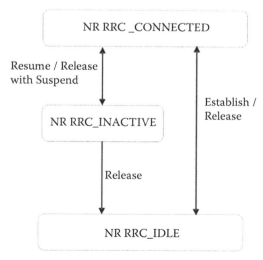

FIGURE 2.11 The procedure transaction states in the network.

TABLE 2.7 Procedure transaction states in the network.

STATE	DESCRIPTION
PROCEDURE TRANSACTION INACTIV	NW initiates 5GSM procedure command
PROCEDURE TRANSACTION PENDING	It is UE's responsibility to share 5GSM procedure response to NW

Table 2.7 describes the procedure transaction states in the network [14].

2.3.1.2 Radio Resource Controller

Radio resource control protocol is described by 3GPP in TS 38.331, where RRC stands between UE and gNB. The network's RRC layer shares the broadcasting system information (MIB, SIB) and paging information with UE. RRC layer establishes, maintains, and releases RRC connection between UE and gNB. It also handles the mobility functionalities of UE, QoS management, and UE measurement report. Radio link failure (RLF) disclosure and restoration of the new radio link between UE and gNB is also one of the important functions of the RRC layer.

2.3.1.2.1 The RRC Services
  I. Upper layer expects the following services from RRC:
   o. RRC layer broadcasts the common control information related to the 5GNAS layer
   o. MT call notification (UE's RRC_IDLE mode notification)
   o. Provides warning to UE's about the Earthquake and Tsunami Warning Service (ETWS) and Commercial Mobile Alert System (CMAS)
   o. It is responsible for the transmission of dedicated control information to a specific UE

  II. The RRC expects the following services from a lower layer:
   o. Integrity protection
   o. Ciphering
   o. Sequential delivery of data

2.3.1.2.2 The RRC Functions
- Broadcasting the system information to UEs. It carries NAS common information, PLMN information, and this type of information is very important for UE during RRC_IDLE and RRC_INACTIVE state procedures
- Network uses broadcast information to share ETWS notification and CMAS notification in an emergency
- The paging information is the main source of the network that tells UE "I have something for you". After receiving this message the UE decodes the content (paging cause) of the paging message and based on paging cause the UE has to initiate the appropriate procedure
- It is responsible for establishment/modification/suspension/resumption/release of RRC connection
- This includes the assignment/modification of UE identity (C-RNTI)
- RRC is responsible for addition/modification/release of carrier aggregation
- The RRC handles the ENDC service for 5G using addition/modification/release of the RRC connection between E-UTRA and NR RAT
- The RRC is responsible for the initial security activation of AS
- RRC handles the handover procedures and measurement reporting mechanism of UE
- RRC recovers the UE from radio link failure

2.3.1.2.3 The RRC States and Procedures
The operation of the RRC layer is led by a state machine. It defines the current state of UE. RRC_Inactive is a new state introduced in the 5G, the other two RRC_Connected and RRC_Idle states are similar to that in LTE in 5G. In 5G the "RRC_Inactive" state is introduced to minimize latency and to

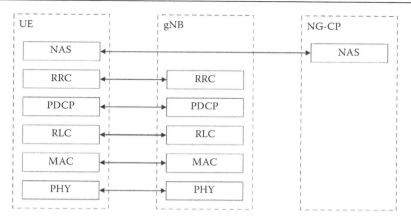

FIGURE 2.12 5G RRC states.

reduce signaling load. We can see the 5G RRC states in the figure below. During the initial power-up procedure, the UE stays in RRC_Idle. UE can enter in RRC_Connected state only after successful initial attach or RRC connection establishment procedure. If there is no procedure performed by UE after the initial power-up, UE can enter into the RRC_Inactive state. UE can enter in RRC_Connected state after resuming its session. When the connection releases from RRC_Connected or RRC_Inactive state, UE can move to RRC_Idle (Figure 2.12).

I. Idle Mode Procedure
   o. UE can perform Public Land Mobile Network (PLMN) selection procedure
   o. UE can receive system information (MIB, SIB)
   o. UE is capable to perform cell selection or cell reselection
   o. UE can receive a paging messages
   o. UE can receive DRX for CN paging

II. Inactive Mode Procedure
   o. UE can receive system information (MIB, SIB)
   o. UE is capable to perform cell selection or cell reselection
   o. UE can receive RAN paging

III. Connected Mode Procedure
   o. If UE is in connected mode means there is a connection in control plane and user plane between UE and Core NW
   o. UE supports CA in NR in this mode and UE can receive paging information [15]

2.3.2 5G User Plane

The NR user plane protocol stack is responsible for the user plane activity, which means all data handling-related aspects are taken care of by the user plane protocol stack. SDAP is a new protocol added in 5G for the alignment between QoS and DRB. The remaining protocols are similar to LTE, which is PDCP, RLC, MAC, and Physical protocols. We discuss the 5G control plane protocols and cover the architecture, functions, and services expected by lower and upper layers (Figure 2.13).

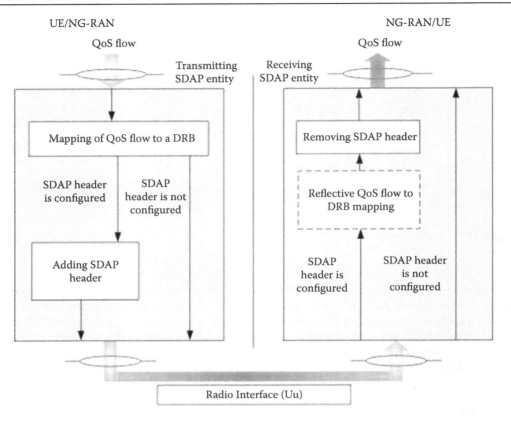

FIGURE 2.13 User plane protocol stack.

2.3.2.1 Service Data Adaptation Protocol

SDAP is a new protocol added in 5G by 3GP for the alignment between QoS flow and DRB. The SDAP sublayer is configured by the RRC layer. The SDAP is responsible for imprinting the QFI in UL and DL packets. The SDAP offers 5GC QoS flows across the air interface. It creates an exact QoS flow within a PDU session to an equivalent DRB (Figure 2.14).

The SDAP will add the QFI on transmitted packets. It ensures that received packets follow the correct forwarding treatment beyond the 5G system traveling. During each PDU Session, SDAP will be configured. This helps in dual connectivity mode, whereby the MCG and SCG both will have a separate SDAP configuration on the device.

I. The upper layer except following services from SDAP:
   o. The SDAP transfers the user plane data to the upper layer

II. The lower layer shares the following services with SDAP:
   o. Lower layers transfer the user plane data to the SDAP layer
   o. SDAP expects the in-order delivery except when out-of-order delivery is configured by RRC [16]

2.3.2.1.1 The SDAP functions
The SDAP supports the following functions:

- It is responsible for the transfer of user plane data

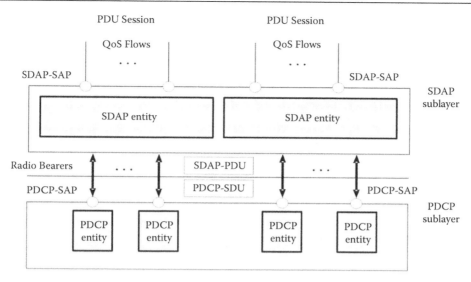

FIGURE 2.14 SDAP sublayer structure view.

- It is responsible for QoS flow and DRB mapping in DL and UL
- It marks the QoS flow ID in DL and UL packets (Figure 2.15)

2.3.2.1.2 The SDAP Procedures
- Based on the RRC requests, the UE performs the establishment of an SDAP entity procedure
- Based on the RRC requests, the UE performs the release of an SDAP entity procedure
- After the reception of an SDAP SDU from the upper layer and lower layer for QoS flow, the transmitting SDAP entity performs the UL and DL data transfer, respectively

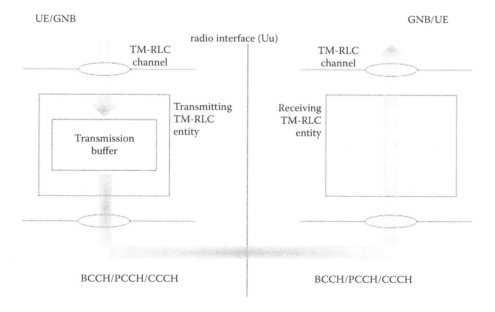

FIGURE 2.15 SDAP layer functional view.

42 Future Trends in 5G and 6G

- In the case of QoS to DRB mapping SDAP responsible for configuration, reflective mapping and DRB release based on the RRC layer information
- The SDAP informs the NAS layer of the RQI and QFI for each received DL SDAP data PDU [16]

2.3.2.2 Packet Data Convergence Protocol

According to TS 38.323, the PDCP is set up by upper layers. The PDCP layer is responsible for mapping the DCCH and DTCH logical channels. We can see in the following figure that the PDCP entity is placed in the PDCP sublayer. Some PDCP entities specify for UE and are associated either with control plane or user plane based on the radio bearer it is carrying the data for.

PDCP maintains the PDCP SN number. PDCP offers header compression and decompression using the ROHC protocol. It provides services such as ciphering and deciphering, integrity protection and integrity verification, timer-based SDU discards, duplicate detection and discarding of SDUs and re-ordering and in-order delivery of SDUs. In the following diagram, PDCP is precisely linked with RLC UM and RLC AM mode only, there is no collaboration with RLC TM mode because there is no path for RLC TM mode with PDCP (Figure 2.16).

I. The upper layer expects the following services from PDCP:
   o. The PDCP layer transfer the user plane and control plane data to the above layers
   o. The PDCP layer handles the header compression mechanism
   o. The PDCP layer is also involved in security using ciphering and integrity protection procedures

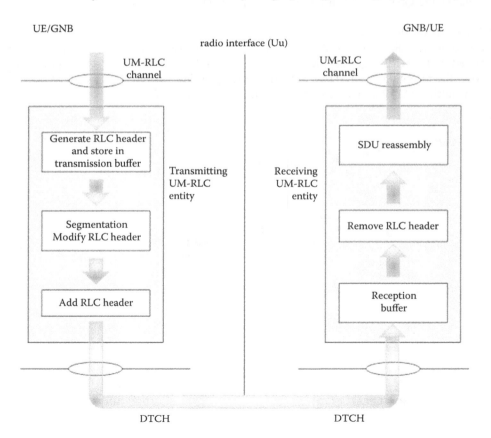

FIGURE 2.16 PDCP layer structure view.

According to 3GPP, the maximum supported size of PDCP SDU is 9000 bytes and for PDCP Control PDU it is 9000 bytes.

   II. The lower layer shares the following services with PDCP:
      o. The RLC acknowledges successful data transfer and sends a sign of positive PDCP PDU reception
      o. The RLC shares the results from any unsuccessful data delivery

2.3.2.2.2 The PDCP Functions
- The PDCP transfers user plane and control plane data (Figure 2.17)
- It is responsible for the maintenance of PDCP SN numbers
- The PDCP handles the header compression and decompression procedure adopting the ROHC technique
- The PDCP handles its security mechanism, i.e., ciphering and deciphering, which applies to both control plane and user plane
- The integrity protection and verification for control plane means for RRC/NAS layer message
- It handles duplicate data packets. If the data packets are not in a sequence then it performs the reordering, and shares the data packets in order with other layers
- It discards the duplicate packets and it has timer-based SDU discarding mechanism

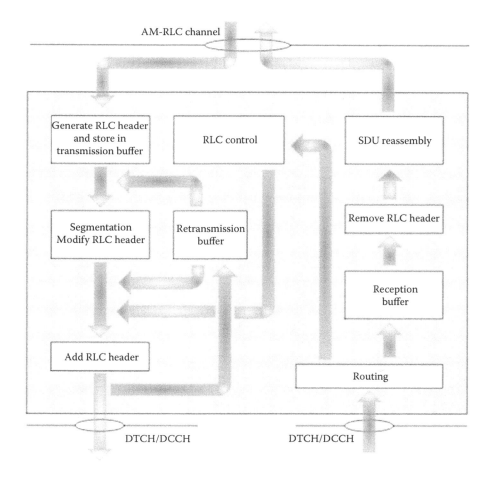

FIGURE 2.17 PDCP layer functional view.

2.3.2.2.1 The PDCP Procedures
- It is responsible for PDCP entity handling, including the entity establishment, reestablishment, and release procedures
- It handles the transmission and reception of data, i.e., data transfer operation
- It is responsible to discard the SDU based on the discard timer expire value
- Status reporting and data recovery at PDCP entity
- PDCP is also responsible for header compression and decompression, ciphering and deciphering, integrity protection, and verification
- It is responsible to activate the activation and deactivation of the PDCP duplication process for SRB and DRB and perform the PDU discard procedure based on duplicate detection [17]

2.3.2.3 Radio Link Control

The upper layer RRC generally depends on the RLC composition. There are three different entities in RLC that perform functions of the RLC sublayer. RLC is structured with UE and gNB. The source of data packets for RLC is its upper layer, and it transmits these data packets to its companion entity on another end via lower layers and vice versa. The RLC layer has TM, UM, and AM operational modes based on the data types. Based on these, an RLC is classified as TM RLC, UM RLC, and AM RLC entities. This distribution is dependent on the mode of data transfer that the RLC entity is configured to provide (Figure 2.18).

2.3.2.3.1 Introduction of RLC Entity
I. TM RLC entity (Figure 2.19):

TM mode means the "*no processing to RLC data*". This mode is only responsible for buffer data. There is no operation performed on data by TM RLC mode. There is no operation in RLC for header addition,

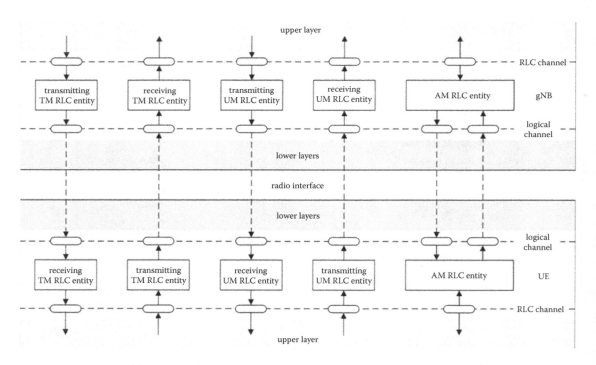

FIGURE 2.18 RLC sublayer model.

2 • NR: Architecture, Protocol, Challenges 45

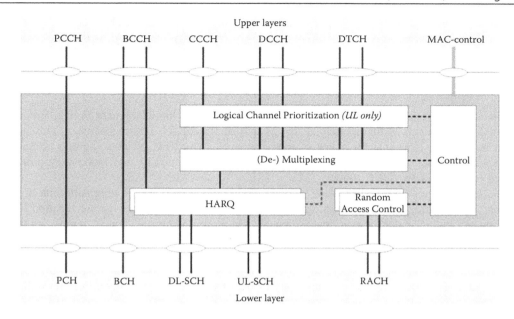

FIGURE 2.19 TM mode entity.

reordering of data, segmentation, and reassembly. That is why RLC input and RLC output are mirror data packets to each other.

Transparent mode is applicable on the transmitting and receiving sides. The logical channels BCCH, CCCH, and PCCH are responsible for the flow of data packets for the TM RLC entity. A TM RLC entity sends and receives RLC data PDU, i.e., TMD PDU.

II. UM RLC entity

There is no acknowledgment mechanism in this mode. In RLC UM entity transmitting side does not require any reception response from the receiving side (Figure 2.20).

In UM mode the transmitting side buffers the data and sticks the RLC header to transmitting data. This entity splits data into small pieces, i.e., "segmentation of data" and based on the segmentation status it modifies and adds the RLC header.

According to TS 38.322, the "concatenation" operation is displaced from the RLC layer and added into the MAC layer. In case of LTE, UM mode is responsible for performing the concatenation process

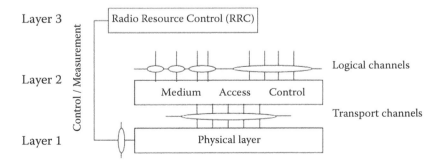

FIGURE 2.20 UM mode entity.

also. The receiving side of UM is responsible for buffering the data and it also helps reorder the unsequenced chunks of data. At receiving point, it removes the RLC header added by the peer transmitting side and reassembles the data.

III. AM RCL entity

In RLC acknowledge mode, the transmitter side requires the acknowledgment from receiving entity in successful and unsuccessful operations. It has a special ACK/NACK mechanism. This acknowledgment mechanism is a must for every transmission cycle. The RLC window, polling bit and ACK/NACK scheduling are introduced in RLC. These three concepts lead to AM mode very prominently compared to the other two RLC modes.

The functionality of AM entity is the same as UM entity except for the retransmission buffer and RLC control procedure. In RLC AM transmission mode, after the segmentation/concatenation there is a mechanism to add the header for every data segment. AM mode creates twin copies of data and transmits one copy to the lower layer, i.e., MAC layer, and sends one copy to the retransmission buffer to store it. If the transmitting entity receives NACK in a response or there is no response from the receiver and for a specific duration. Buffer can use stored in the same copy of data for retransmission (Figure 2.21).

If the RLC gets ACK in response, the copy of data from the retransmission buffer would be discarded. There is no difference in receiver side operation in AM RLC as compared to LTE, but the transmitter side is of NR RLC AM, which does not support concatenation.

I. RLC layer provides the following services to the upper layer protocol:
   o. RLC transfers TM data to upper protocol
   o. RLC transfers UM data to upper protocol
   o. It transfer AM mode data to the upper layer, including ACK of delivery
II. RLC layer expected following services from the lower layer protocol:
   RLC transfers data to the lower layer
   o. RLC expects the notification of a transmission opportunity and size of RLC PDU to be transmitted

2.3.2.3.2 The RLC Functions
- RLC shares upper layer data to lower layer using different RLM mode entities
- In the case of RLC AM data transfer mode it provides the functionality of error correction through ARQ
- RLC UM and AM mode support for segmentation and reassembly procedure
- In the case of AM mode resegmentation of RLC SDU segments and duplicate detection and detection of protocol error is possible
- SDU discard support possible in RLC UM and AM
- In case of failure RLC reestablishment is possible [18]

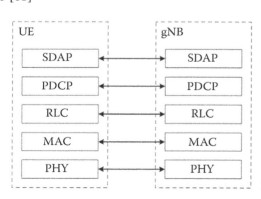

FIGURE 2.21 AM mode entity.

2.3.2.4 Medium Access Control

The Medium Access Control protocol is the L2 protocol. A mapping between logical and transport channels is one of the important works of MAC. It is responsible for logical channel prioritization. Multiplexing and demultiplexing procedures are initiated based on the request from the PHY layer. TB is a shipper between MAC and PHY that carries data. MAC transfer this data over transport channels. It handles scheduling information reporting and error correction through HARQ (Figure 2.22).

The MAC configuration controls the RRC layer. The MAC handles BCH, PCH, DL-SCH, UL-SCH, RACH transport channels.

  I. The upper layer expects the following services from MAC:
     o. MAC shares the data with the upper layer
     o. MAC is responsible for radio resource allocation

  II. The lower layer expects the following services from MAC:
     o. MAC expects the data from the lower layer, i.e., Physical layer
     o. Lower layer shares the HARQ feedback with MAC
     o. Physical layer responsible for scheduling request to MAC
     o. MAC expects the various measurement reports from the lower layer (Channel Quality Indication (CQI))

2.3.2.4.1 The MAC Functions
- MAC handles the mapping between logical and transport channels
- The MAC handles the multiplexing of MAC SDU from logical channels. These SDU are delivered to the physical layer on TB using the transport channel and also handle demultiplexing
- Scheduling information reporting examine by MAC
- Error correction through HARQ
- MAC takes care of logical channel prioritization

There are some important procedures that are handled by MAC. The RACH for UE, DL-SCH data transfer for NW, UL-SCH data scheduling for NW, UL-SCH data scheduling for NW, UL-SCH data transfer for UE, scheduling request, discontinuous reception, semi-persistent scheduling, PCH reception, and BCH reception.

2.3.2.4.2 The MAC Procedures
- The MAC is responsible for the RACH procedure. There are two types of RACH processes in 5G similar to LTE.

  I. Contention-free RACH process

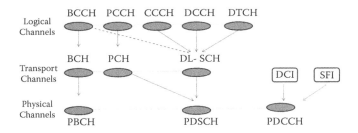

FIGURE 2.22 MAC structure overview.

48 Future Trends in 5G and 6G

II. Contention-based RACH process
- o. The MAC maintains the uplink time alignment based on the RRC configuration parameters
- o. The MAC handles the DL-SCH and UL-SCH data transfer
- o. The MAC layer takes care of PCH reception for paging information and BCH reception for broadcast information
- o. To save battery power in devices MAC has the Discontinuous Reception (DRX) mechanism
- o. The MAC handles the MAC reconfiguration and MAC reset
- o. It is responsible for the bandwidth part operation, beam failure detection, and recovery procedures
- o. One of the most important work is handling MAC CEs in different scenarios [19]

2.3.2.5 Physical Layer

The PHY is the L1 layer in the protocol stack. It has a connection with TX and RX and other hardware in the system. It is the genesis of PHY signals along with the modulation scheme. PHY layer is authoritative for establishing aspects of physical channels and is responsible for defining the UL as well as DL physical channels and frame structure, including physical resources to shared information and different modulation mapping like BPSK, QPSK, etc. (Figure 2.23).

The above figure explains radio interface architecture over layer 1. The PHY layer collaborates with MAC and RRC. It has SAP between different layers/sublayers. The higher layers expect the data transportation from the PHY layer. It uses a transport channel via the MAC sublayer for data transport services to upper layers. Transport channels are mapped between MAC and PHY for communication and the PHY channel provides an air interface between UE and gNB. The PHY carries the control plane as well as the user plane data.

PHY layer handles multiple access schemes based on Orthogonal Frequency Division Multiplexing (OFDM) with a cyclic prefix (CP), this fills the space in the data streams. In the case of uplink, Discrete Fourier Transform-spread-OFDM (DFT-s-OFDM) with a CP is supported in NR physical layer. 5G uses the two basic forms of full duplexing, first is Frequency Division Duplex (FDD) and the second is Time Division Duplex (TDD), both are enabled to support transmission in the paired and unpaired spectrum.

In the case of DL, physical channels have PDSCH, PDCCH, PBCH, and in UL PRACH, PUSCH, PUCCH. The reference signals as well as the primary and secondary synchronization signals are also defined in the PHY layer.

5G PHY layer supports modulation schemes QPSK, 16QAM, 64QAM, and 256QAM in DL and QPSK, 16QAM, 64QAM, and 256QAM for OFDM with a CP and $\pi/2$-BPSK, QPSK, 16QAM, 64QAM, and 256QAM for DFT-s-OFDM with a CP in UL.

Radio characteristics are measured by the UE and NW and transferred to higher layers. It consists of measurements for intra-frequency and inter-frequency handover, inter-RAT handover, timing measurements, and measurements for RRM. These measurement reports are crucial during HO. The PHY has procedures like cell search, power control, UL synchronization, UL timing control, random access-related procedure, HARQ, beam management, and CSI-related procedures to improve the 5G system [20].

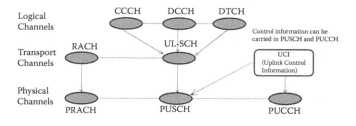

FIGURE 2.23 Physical layer interaction overview with other layers.

2.3.3 5G Channel Mapping

The protocol PHY, MAC, RLC carries the information through a set of channels. There are three types of channels in 5G similar to LTE. It is logical channels, transport channels, and physical channels and it is applicable in UL and DL in both directions. Based on the related condition of information, it traverses through different channels (Figure 2.24).

2.3.3.1 DL Channel Mapping

I. DL Logical Channel
The logical channels stand between RLC and MAC protocol. It defines the characteristics of information. It has broadcast, paging, control, and traffic channels to carry the information.
   o. The BCCH carries broadcast information for UEs, this broadcast information is transferred over the MIB and SIB OTA messages within a cell. This broadcast information carries network-related information, which is very important for UEs to camp on the network
   o. The PCCH carries paging messages. In the case of NSA mode, LTE provides the paging messages to user's devices
   o. The CCCH is used to share control information partnership with RA
   o. The DCCH shares control information to/from a device. This logical channel is very important and helps set user configuration parameters for individual devices
   o. The DTCH shares user information to/from a device. This channel is used for unicast uplink and downlink transmission of user data (Figure 2.25)

II. DL Transport Channel
The transport channels stand between MAC and PHY layer. The transport channels have an interworking mechanism with the logical and physical channels. It defines how the information is transported to the PHY layer.
   o. The BCH maps to the BCCH, it uses a part of MIB information to receive transmission from BCCH
   o. The transport paging channel PCH is used to share the paging information. The PCH supports discontinuous reception (DRX), which helps the device save power

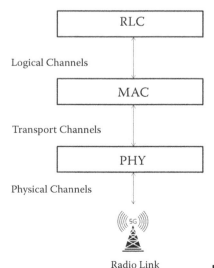

FIGURE 2.24 5G Channel mapping.

o. The BCCH shares the SIB information with the DL-SCH and there is a direct mapping of CCCH, DCCH, DTCH with the DL-SCH channel. In the 5G transport channel, DL-SCH is used for downlink data transmission. It supports DRX and hybrid ARQ

III. DL Physical Channel
The physical channel works in collaboration with the PHY layer. Physical channel characteristics include access protocols, timing and data rates.
o. The broadcast message has its physical channel, i.e., PBCH. It carries the part of SIB information to access the network
o. The PDCCH handles DCI and SFI information. It also carries the scheduling information
o. The PDSCH carries the information of paging messages, delivery of SIB, and RA response messages

2.3.3.2 UL Channel Mapping

I. UL Logical Channel
o. UL channel mapping in 5G is similar to the DL. Uplink logical channels have similar working. CCCH, DCCH and DTCH share the same information with transport channel as they share with DL logical channels (Figure 2.26)

II. UL Transport Channel
o. UL-SCH is a replica of DL-SCH in the uplink, it is used for the transmission of UL information
o. A self-contained process random access channel (RACH) exists for the initial connection at the transport channel. MAC handles the RACH procedure. UE receives initial parameters from the network to initiate a RACH in broadcast information. Random access procedure has a very important role in the protocol stack

III. UL Physical Channel
o. The PRACH supports the RACH in the physical channels
o. The PUSCH is like PDSCH in UL physical channels
o. The UE uses the PUCCH channel to share the feedback of DL TBs with gNB. This feedback informs the gNB bout TB delivery, if is successful or not
o. It is also used by UE to share hybrid-ARQ acknowledgments with gNB

Compared to LTE, 5G NR physical channels have a few feature additions. Physical channel includes a phase tracking reference signal (PTRS) to track phases and time scheduling and a demodulation reference signal, for the UL control channel and the DL broadcast channel [21–23].

2.4 CHALLENGES

2.4.1 Building Complex and Dense Network

5G network is expected to lead the next industrial revolution 4.0, which is the nerve center of interconnectivity, automation, ML, and real-time data mining. This industrial revolution connects the physical world with the digital world for significant developments.

The 5G is the combination of all wireless communication technology. 5G supports IoT, V2X, AI, and M2M services. To provide all-time connectivity, 5G depends upon technologies like WiFi,

Femtocell, Picocell, Microcell, Macrocell, and Fiber optics network. It is very challenging to build a hybrid network to support such advanced communication technology. We discuss the challenges in mobility, resource management, energy management, and spectrum management.

2.4.1.1 Mobility

Deployment of a proper and dense network for this type of heavy digital service is a requirement of the future. The deployment of Femtocell, Picocell, Microcell, Macrocell provides an UDN for better coverage and capacity. The aim is to provide high-speed internet to every corner of the globe. The conventional cellular network generally provides coverage to the large physical area compared to UDN. The UDN network formation consists of thousands of access points (APs) or base stations (BSs) in 1 km^2 compared to the traditional cellular network which consists of three to five BSs in 1 km^2 physical area. This UDN network creates a new issue in the network like more noise and the cell-edge effect where mobility of user and interference are negatively affected [24].

As compared to the conventional cellular network, operators are experiencing new mobility issues in UDN in terms of HO complexity and performance requirements of the network. Here, mobility is a clear scenario where the user selects/reselects the BS (Handover/Handoff), there is no complication in this mobility procedure for the user. The frequency interference, the noise of the signal is limited in earlier networks. But due to new infrastructure requirements, it is important to develop the UDN network using multiple small cells and in this case, the interference may be high. Imbalance in signal transmission power in multi-cell (picocell, femtocell, microcell) infrastructure will also increase the complexity of the HO mechanism.

From a user point of view (UE), scrutinizing the signal quality of camped cells and confirming the availability of neighboring small cells falls under the cell search procedure. This procedure is also responsible to collect the reference signal value (RSRP, RSRQ) and SINR value, which handles the HO decision policy during mobility. The reference signal terminology and values are based on the UE's RAT. The drawback of the reference signal receiving power scheme is that the reference signal power is not always equal to the reference signal quality. In UDN, these schemes do not look fair and effective. 5G CA supports up to 16 contiguous and non-contiguous Component Carriers (CC) to improvise the 5G system to reduce HO procedure duration and improve the QoS [25–27].

2.4.1.2 Resource Management

The new radio system (5GS) has adequately superior performance criteria. The high transmission rate for user data, better spectrum utilization for multi-layer services, and upgraded energy consumption mechanism are some of the important criteria of 5GS. To achieve these super-performance criteria, better resource management mechanism comes into picture in UDN. Two steps, the RACH process and the contention-free RACH, are also important in achieving resource management. Hardware, software upgradation process have the potential for high resource management. But the OPEX and R&D cost is very high. There is huge scope in research of resource management. We can consider the base station deployment strategy, spectrum management, and energy management for resource management in 5G [28].

I. Energy Management

In 5G, energy management is a challenging task. The growing user count and all-time connecting devices in the 5G network will increase the data traffic in the system. In UDN, the user will experience some new types of services like V2X and IoT. This means the UDN is a platform of multi-layer service with various connected devices at a time. It will increase per-user data consumption and we will see a high rise in device battery consumption. In the future, the battery drainage will be more because a user will experience multiple digital services using single devices. In a multi-cell scenario, a cell search procedure during HO will drain the user battery more compared to a conventional cellular network system. Access points (AP), webs of small cells in UDN, will considerably take up more energy to run the system than user devices.

Advanced modulation and coding scheme in 5G has the potential to improve the air interface efficiency. The scheduling procedure in the MAC layer or power control management at BS can be pursued to improve network efficiency. In UDN, not all AP of small cells may have equal user presence in their cells, which means not all APs are being used equally in a single network. So may be some APs will be used in ideal conditions while others are highly utilized. This undistributed load of APs will create a serious energy issue; thus, it is very important in this condition to deploy a smart load management system to save energy. Self-Organizing Networks (SONs) are helpful in resolving this type of issue in UDN. Specific to the UE, a manufacturer can upgrade the battery mechanism for the energy-efficient battery for users. In the future, manufacturers can adopt lithium-sulfur batteries, silicon anode lithium batteries, or cobalt-free lithium-ion batteries [29,30].

II. Spectrum Management

"Spectrum need is directly proportional to the communication need". Spectrum is a rare resource in communication technology. It provides an opportunity for operators to provide a new spectrum to mobile users but requires careful planning and best spectrum sharing policies to achieve the 5G goal. The new digital world depends on the fast-growing internet world, the mobile user count is increasing day-by-day, and mobile users are experiencing the online world. The data consumption per user has drastically increased in the past half decade and will need more bandwidth in the future. Hence spectrum management is an important resource management criterion that impacts system performance of the UDN network. The concept of spectrum management is a technique of smooth spectrum sharing with the users. Operators are facing new issues in spectrum sharing specifically in 5G. Operators are working on introducing improvements in small cell, carrier aggregation, cognitive radio, small cell, high-spectrum access, and M-MIMO, etc.

i. **Small Cell**

The user support capability of 5G is 100 times better than the existing technology, and the deployment of small cells allows the infrastructure to fulfill this requirement. Full-Duplex (FD) helps overcome spectrum issues in small cell infrastructure.

ii. **Carrier Aggregation**

The CA was initially added in LTE and it is a significant part of 5G. Carrier aggregation can utilize spectrum resources more effectively and improve the system performance.

iii. **Cognitive Radio**

CR automatically explores the convenient channels in the spectrum. The transceiver can intelligently search for available channels using cognitive radio. This is a very promising technology to use available resources smartly.

iv. **MIMO**

Advanced MIMO technology in 5G allows the user equipment to transmit and receive multiple signals on the same channel. There are different MIMO combinations available in 5G for better system performance.

In spectrum sharing, we have dynamic spectrum access (DSA) policy, which shares the spectrum to the primary user as well as the secondary user. The primary user is a licensed user and the secondary user is considered an unlicensed user in the network. In spectrum sharing the priority is always given to the primary user and authority to the secondary user is only given whenever it would not create any interference with the primary user during the spectrum sharing or utilization. In dynamic spectrum sharing, cognitive radio and geolocation database are crucial pillars. The conventional network can fulfill the user requirement using spectrum management. But in an ultra-dense network, small cell infrastructure will initiate new challenges for the operator. But still, operators have the option of massive MIMO antennas, advanced modulation, and coding schemes to improve spectrum sharing [31,32].

III. Small-Cell Deployment

In addition to the system performance analysis, small-cell deployment and coworking are the important criteria. Due to limited capabilities and restricted coverage range operators have limited options to work. Dense deployment of small cell APs is a theoretically good plan but practically it will create high interference and will surely decrease the system performance. The random deployment of APs and dynamic on-off APs will raise new challenges in the network system. Small cell deployment is planned to set up in the dense urban area, where the penetration loss, diffraction loss, the fading effect may be observed because of the height of buildings, the height of trees or dense trees, WiFi antennas and high pollution. Random deployment of wireless core networks and dynamic on-off capability initiate unfamiliar challenges. To overcome these challenges, centralized system approach may not be feasible in UDN; however, distributed system for small cell deployment somehow manages this issue [33].

2.4.2 Security Issues and Potential Targets in 5G Networks

Accommodating a wide variety of applications and basic packet transmission at high speed, 5G security is expected to play a crucial role in future communications technology. Linking heavy industries and user-friendly applications over the 5G internet network will create a new software-based communication model, but it is a big threat to user privacy at the same time [34]. According to tech experts, 5G is a secure network but still, it is suffering from cascading effects. The internet data in 5G has decision-making parameter; 5G network depends on the collection and variety of data. Thus, strong security data protection measures are needed for 5G. The security checks and protocols present in earlier generation's networks are not effective anymore, 5G security protocols should be created with strong security checks. The following are the potential targets in 5G networks for security vulnerability.

I. Mobile core and external IP networks

1. Mobile core and external IP networks
2. Access networks
3. User equipment

The principle of 5G architecture is software-based, and this is the drawback of this high standard network in terms of user security, which means the 5G wireless network is sensitive to security foible. In the case of 5G network slicing, the 5G operator will provide diverse services over the virtualized network. Maintaining the security of this virtualized network is a challenging task for vendors, operators, service providers, software developers, users, and all those who are a part of this system. We already discussed above that the 5G NSA architecture combines the use of New Radio and LTE network core. As a result, these networks take over all the security issues of LTE networks from the get-go [19]. Research already indicated vulnerable issues in LTE networks for denial of service (DoS) through diameter exploitation. This means the 5G NSA network will be vulnerable to DoS, too.

A distributed denial-of-service (DDoS) attack is a targeted nasty effort to disarrange the normal data traffic of a server (targeted), resulted in the breakdown of service, or network by a flood of unsecured internet data traffic. In the case of a 5G network, it will be used to increase the signaling amplification or target the Home Subscriber Server (HSS) or other entity to saturate the network. DDoS attacks specifically target external entities of the core network, which work as a database of the users. In the case of GSM, hackers target the HLR/VLR of mobile network [35,36].

II. Heterogeneous Access & Mobility

The heterogenous access network is an intrinsic part of the 5G network. A heterogeneous status provides concurrent access to diverse technologies in 5G. Femtocell, Picocell, Microcell, Macrocell provide heterogeneous capabilities to 5G network to provide industry 4.0 services. The mobility of a user is a test in a limited range network. Dealing with Femto, Pico, Micro, Macrocell during the use of services will create HO issues and also create a space for security threats. The message from proxy AP or network leads to a DoS attack against a UE during HO. Attackers may follow the user footprints using the UE location in specific or in multiple cells. Sometimes, attackers plan a packet scheduling mechanism through a DoS attack to steal the bandwidth of UE.

The attackers interfere the user devices with other malicious devices (physical tampering with equipment), configuration attacks, attack through proxy points over the security protocol, attacks on core network from negotiable entities. Sometimes hackers attack RRM (radio resources and management) to increase HO in a heterogeneous network, and attack the user identity privacy and credential theft from open access nodes. These are some of the possible scenarios of security threats in a heterogeneous communication network [37,38].

III. User Equipments

In 5G, we have a new category of user devices. Users are not stuck with a mobile phone, they have the option of wearable devices and entertainment devices. Many powerful smartphones, tablets, smart-watches, smart goggles, AR/VR devices make use of 5G technology. The smartwatch equipped with a special user identity, global positioning system, camera, audio recorder, and all-time internet connectivity options is very attractive and the best option for daily use. Users can communicate through message, email, call using a smartwatch. That is why users are always under a security threat.

Why the user device is threatened? The answer is a variation of connectivity preferences over multiple technologies and the online presence of more users. 5G devices are not just connected with the 5G network but can connect with WiFi, IMS, Bluetooth, NFV. This inter-connectivity is vulnerable to a security threat. The popularity of unsecured devices is more because of the low cost of devices and this is also a cupcake for hackers. These low-cost devices accept the open market operating systems (OS) and third-party applications, which is also a loophole in the security process. Mobile malware is a personal security threat. The user always prefers the free and unsecured applications downloaded from an untrusted app store, which creates a space for the malware attack. This malware attack targets users to exploit or steal personal information like photos, contact, bank details, etc.

Mobile botnets are a serious UE security threat, an attacker can automatically select a set of UEs and spread a malware attack on this specific set of UE. The negotiable UE can operate remotely using botmaster in such a big malware attack [39]. After considering all security and privacy issues in a cellular network, standardization groups 3GPP, IETF, 3GPP2 and other works accordingly to enhance security and privacy mechanism over the software and hardware to develop a secure network for users.

The STIR (Secure Telephony Identity Revisited) SHAKEN (Secure Handling of Asserted information using toKENs) is one of the promising systems for calling number identification for end users. It follows the common public-key cryptography mechanism, to verify and ensure the calling party's contact number is secure or is it a bot [40].

During the upgrade program of the product, various operational and maintenance mechanisms are introduced. The routine system patch updates during the software upgradation are important to trim the risk of malicious exercise. Software suppliers always do the security risk assessment of their product to develop vulnerable free software and releasing appropriate fixes at a specific period to update the product.

2.5 APPLICATION

The 5G use cases in the real world are not countable, 5G has a large scope from heavy industrial applications to daily life applications. We cannot define 5G applications in one word or few sentences. Smart factories, connected vehicles, smart offices, advanced mining, smart healthcare, smart transport model are some of the use cases of 5G. Robots can operate heavy machinery in the mining and port sector.

The 5G applications will increase real-time control, security, performance and automation. It will create new jobs and will change the service sector. The improvisation in 5G use cases generates new cash flow for the consumers. The 5G is the only technology that is going to bring change in the existing business model of the world. We will consider three major applications for discussion.

2.5.1 Vehicle-to-Everything

V2X system is a definite future of the automobile sector; no one can say a no to a feature like this. The "V" denotes vehicle and "X" stands for "everything" in V2X. "V" always will be constant but the meaning of "X"changes depending on duties. There is a lot of scope in this application like vehicle-to-vehicle (V2V), vehicle-to-machine (V2M), vehicle-to-pedestrian (V2P), and vehicle-to-device (V2D) (Figure 2.27).

V2X system upgrades the overall efficiency of the automobile sector and will reduce human error in the operation. It improves the customer's safety and energy efficiency. For the success of this service, connectivity of vehicle with the "X" is the most important parameter. Pedestrians, bikes, cars, heavy trucks, and all connected "X" should exchange messages with "V" to indicate their actual location, speed of the vehicle, moving direction, and remaining properties with each other to install and maintain the V2X infrastructure. We can consider the following basic V2X communication categories.

   I. Vehicle-to-Machine

The V2M communication system helps create a collective infrastructure of roadside devices. Streetlights, signal lights, CCTV surveillance, smart road signage display, speed detectors, advertising display are some of the essential parts of the transport infrastructure.

The ZigBee, MAC, PHY protocols, and high-speed internet can improve V2M system. The smart collaboration-available technology and high-level protocols of V2M can provide an extraordinary controlled and secure transport infrastructure.

   II. Vehicle-to-Network

The V2N is the most important concept in V2X communication technology. In V2N, the vehicle can access the telecom operator network for connectivity. This telecom network is a combination infrastructure of multiple RAT and small cells. It creates an all-time connectivity bubble for vehicles. The dedicated short-range communication (DSRC) is used to communicate with other vehicles on the road. The combination of DSRC and telecom operator infrastructure grants vehicles the following.

- The vehicle can receive the broadcast information from the network. This broadcast information shares the real-time on-road situation with vehicles including data of accidents, traffic conjunction, weather reports, etc.
- The vehicles can communicate with any other vehicles on road. Vehicles can share their information like geographical coordinates, speed, etc.

This improves the performance of the vehicles and also creates a secure transport system.

III. Vehicle-to-Devices

We can consider V2D as a subset of V2X. V2X allows vehicle to share information with other smart devices like Alexa, smart car screen, tablet, mobiles using the protocols Bluetooth and ZigBee, etc. This allows wearable devices to connect with mobiles and also with vehicles.

IV. Vehicle-to-Grid

The V2D provides a smart collaboration of vehicles with an electric grid. It is a new concept in V2X. Hydrogen Fuel Cell Vehicles (HFCEV), Battery Electric Vehicles (BEV), and Plug-in Hybrid Vehicles (PHEV) need a smart electric grid for operation. The future of vehicles depends on smart grid technology. Tesla, Google, Ola, Uber are developing smart vehicles in line with the anticipated future requirements.

V. Vehicle-to-Vehicle

The V2V allows vehicles to share real-time information. The Dedicated Short-Range Communication (DSRC) allows this short-range communication. The geographical location, speed, and other required information can be exchanged using DSRC. The V2V concept works on a mesh network. Under this, each vehicle is considered as a node that can share, receive and transmit signals. The V2V prevents road accidents and can save human lives.

VI. Vehicle-to-Pedestrian

The V2P is also a new category of V2X communication. In V2V communication pedestrian is also considered as a node. Because pedestrian can take own decision in real-time. LiDAR helps to trigger a collision warning for pedestrians and vehicles. The 360-degree CCTV surveillance on the road also makes sure about road safety. Third-party apps will be an important part of the V2P system. Consider you are using a smart car equipped with V2X technology, which means your location is well known to everything that is in range of approximately 10 m from you. The pedestrian can communicate with your car with his smartwatch, the role of watch here is just to provide a warning to his user while crossing the road. We call this the V2P system. It is a very challenging technology to deploy as well as to use.

VII. Vehicle-to-Infrastructure

The intelligent transport system is a bidirectional communication mechanism of the vehicle and road infrastructure. The road infrastructure consists of surveillance cameras, signboards, traffic signals, smart displays, etc. It is important to share the detailed information of roads with real-time data with a vehicle for a collision-free system. The automated vehicle needs real-time data, and this can only be shared by smart infrastructure.

VIII. Vehicle-to-Cloud

The V2C communication is a crucial system in V2X communication. Because all real data can be shared and stored in the cloud only, it is very difficult for product developers to give space to store real-time information on the device, as this type of data needs huge space for storage. The V2C can analyze the big collection of data in a short time and can improve the decision policy of the nodes and can improve the system. The operating mechanism of V2X is complicated, V2X needs a bunch of diverse technology for successful operations. It needs various types of sensors to sense the distance, objects, and path. It requires all-time internet connectivity to connect with an operator or with other technical goods. The WiFi, Bluetooth, RFID, Laser techniques are also always required for V2X. The merger of all these smart technologies provides a short-range, low latency, and high-reliability vehicle to everything connectivity. This helps to manufacture a smart vehicle with speed limit alert, collision warning, electronic parking, and smart toll payment.

The most important factor of V2X technology is security. The user cannot clean his footprints in this technology. This means the operator will surely be able to collect routine data, and this can prove to be very harmful for users [41–44].

2.5.2 Internet of Things

IoT means Everything connected, Everywhere. The IoT is changing the lifestyle of human beings. It is a technology that will connect every essential (physical object) entity of humans with the internet. The IoT can connect any physical object to the internet. IoT can use the power of the internet to collect and analyze real-time data of the world and can serve the world in a better way. The physical objects may be study tables, water bottles, shoes, doors, kitchen appliances, and more. Wireless technology and embedded system have an important role in the development of IoT.

IoT is a very special technology, physical objects can work with minimum human cooperation. We can operate IoT devices without keyboard commands. IoT device developers can develop applications also, which is the single point of contact for IoT devices. According to the DataProt, in the next decade, operating IoT devices are expected to surpass 25.4 billion devices. Following are some key aspects of IoT infrastructure (Figure 2.28).

I. IoT protocols and communication standards:

IoT devices can communicate with other devices and for communication, it needs a special set of protocols. For short-range communication and low power consumption, IoT devices prefer Bluetooth or Zigbee protocols. The queuing telemetry transport (MQTT) is the lossless, bidirectional protocol that runs over the TCP/IP. It works on a publish-subscribe mechanism that is developed to exchange messages between IoT devices. The role of GSM, LTE, 5G, and WiFi is considerable in the development of the IoT. The IoT infrastructure depends on all-time internet connectivity and telecom operators can fulfill this requirement.

II. IoT data and cloud infrastructure

The real-time data collection of billions of IoT devices is not just few gigabytes or terabytes but is way more. Cloud infrastructure provides a space to collect such zettabytes of data for further evaluation. In IoT, the data is the only source of information and that is why it is very important to develop software that can analyze real-time data in a fraction of seconds and take decisions without human interaction.

For example, real-time data collected from pressure sensors in a water tank could be evaluated by a software in a device and can trigger an alarm to shut valves off. This is a regular example from our daily life, IoT devices now support us to inform the status of water in our household water tank.

III. IoT device management

All IoT devices need to work on authentification, provisioning, configuration, monitoring, and software updates to become productive. This fundamental procedure allows IoT devices to follow security procedures, create own identity, create communication with authenticating the device, and update device software regularly.

IV. Provisioning and authentication

The provisioning process allows user devices to enter the network system. This procedure allows the service provider to provide required services to the users. The authentication process is an integral part of any system, authentication process allows the system to authenticate a user device based on the identity. This is a two-way process, which means the authenticity of both parties verified by each other for security purposes. After the successful authentication procedure, the system allows the user device to enjoy the services.

V. Configuration

In the IoT world, most of the devices work as remote devices and it is essential to provide a configuration capability to such devices to recover from errors, adopt the new changes to update and configure the devices.

VI. Monitoring

Monitoring and diagnostics are a crucial part of the IoT system. The device needs to display the battery consumption, available resource management, update the status of service, track the bugs in the software, which are some of the procedures that fall under the IoT.

VII. Software updates

Software updates and maintenance allow the developer to fine tune a device regularly. The development team upgrades the system through a software update patch that helps resolve the issues in the system. Electronics circuits, embedded systems, data analytics, sensors, laser technology, wireless network, all-time internet connectivity integrate the physical object with IoT and it creates a big threat to cybersecurity. All these technologies work together to create a smart techno system for human beings [45–47].

2.5.3 Machine-to-Machine

M2M is the future of industry 4.0. It is a software-controlled communication between machine-to-machine without human interaction. Machine learning (ML) and artificial intelligence (AI) allow smart communication between systems. 3GPP mentioned M2M as Machine Type Communication (MTC) in their standards. The infrastructure of M2M is a consolidation of cellular network, internet, data processing, system software, RFID, sensor, Bluetooth, and WiFi. M2M mechanism depends on the sensor's data, it collects the real-time data from the machine sensors and transmits data through the cellular network, ethernet, etc.

The history of the M2M communication technology is very old. Telemetry was developed by the Russian army in 1845, which was the first M2M project. The GSM data connectivity has given new wings to this technology in the 1990s but the IoT is responsible for the success of M2M technology (Figure 2.29).

We can consider the following figure as a working mechanism of the M2M technology. There are three primary data collectors a company, a small machine, and a car. These three primary data collectors have a set of subnets (components) inside it like sensor, Bluetooth, RFID, operating chips. These components are responsible for collection of real-time data and share with the primary data collector. Now, collected data will be shared with the cloud via a gateway. Because of limited data storage capacity at primary data collection, it is necessary to store and process data over the cloud.

All the operational commands can trigger after analyzing the data provided by different components. The role of wireless and wired communication is also most important to install M2M infrastructure.

Low mobility, network priority, high reliability, small data transmission, security are some key features of M2M technology. It is also very effective in the healthcare sector. In medical centers or at hospitals, the processes are automated to improve efficiency and safety. Robots are doing complicated surgeries. Telemedicine is already implemented in some places. Following are some industrial applications:

- Automated maintenance of the car or any machine
- Procedure for requesting spare parts automatically based on car analysis
- Data collection for processing by other equipment smartly [48–50]

2.6 SUMMARY

In this chapter, we discussed the introduction, architecture, protocol, challenges, and applications of 5G. Our aim is to introduce you to the 5G technology and not cover every aspects of 5G. We presented a small introduction to 5G. In architecture, we have covered UE, gNB and core network architecture with an introduction to NSA and SA mode. The most important part in 5G to understand is protocols, if someone wants to mug up 5G technology from depth then focus on the study of protocols. In the section on protocols, we explained control plane and user plane protocols. We discussed the NAS, RRC, SDAP, PDCP, RLC, MAC, and Physical protocols. We focused on the architecture, function, procedure, and service of the respective protocols. We discussed in short the procedures because it is not possible to add a description for each procedure in this small chapter. The most important are the challenges and applications, about which we have added very limited information because the scope of both topics is huge. We think this chapter is sufficient to better understand the basic concepts in 5G.

REFERENCES

[1]. A. Y. Ding and M. Janssen. "Opportunities for Applications using 5G Networks", in *Proc. 7th Int. Conf. Telecommun. Remote Sens.*, Barcelona, Spain, 2018. doi: 10.1145/3278161.3278166

[2]. C.-X. Wang et al., "Cellular Architecture and Key Technologies for 5g Wireless Communication Networks",*IEEE Commun. Mag.*, vol. 52, no. 2, pp. 122–130, Feb. 2014.

[3]. R. Baldemair et al., "Evolving Wireless Communications: Addressing the Challenges and Expectations of the Future",*IEEE Veh. Technol. Mag.*, vol. 8, pp. 24–30, Mar. 2013.

[4]. A. Gupta and R.K. Jha, "A Survey of 5G Network: Architecture and Emerging Technologies", *IEEE Access*, vol. 3, pp. 1206–1229, July 2015.

[5]. T. Tran et al., "Mobile Core Network", *Deliverable D4.1*, 5G-PPP 5G-Xcast, Valencia, Spain, Jun. 2018.

[6]. J. Hart et al., "Converged Core Network", *Deliverable D4.2*, Valencia, Spain, 5G-PPP 5G-Xcast Project, Oct. 2018.

[7]. 3GPP, TS 38.410, "NG-RAN; NG General Aspect and Principles (Release 15) V15.2.0", Dec. 2018.

[8]. System Architecture for the 5G System; Stage 2(release 15), Jun. 2018, [online] Available: http://www.3gpp.org/ftp/specs/archive/23_series/23.501/.

[9]. Procedures for the 5G System; Stage 2(release 15), Jun. 2018, [online] Available: http://www.3gpp.org/ftp/specs/archive/23_series/23.502/.

[10]. S. Zhang, "An Overview of Network Slicing for 5G", *IEEE Wirel. Commun.*, vol. 26, no. 3, Jun. 2019.

[11]. R. Trivisonno, X. An, and Q. Wei, "Network Slicing for 5G Systems: A Review from an Architecture and Standardization Perspective", in *IEEE Conf. Standards for Commun. and Netw. (CSCN)*, Helsinki, Finland, pp. 36–41, Sept. 2017.

[12]. A. El Rhayour and T. Mazri, "5G Architecture: Deployment Scenarios and Options", in *2019 Int. Symp. on Adv. Elect. and Commun. Technol. (ISAECT)*, 27–29 Nov. 2019.
[13]. 5G Deployment: Standalone vs. Non-Standalone from the Operator Perspective, *IEEE Commun. Mag.*, vol. 58, no. 11, Nov. 2020.
[14]. 3GPP TS 24.501: 5G; Non-Access-Stratum (NAS) Protocol for 5G System (5GS).
[15]. 3GPP TS 38.331: 5G; NR; Radio Resource Control (RRC).
[16]. 3GPP TS 37.324: LTE; 5G; Evolved Universal Terrestrial Radio Access (E-UTRA) and NR; Service Data Adaptation Protocol (SDAP).
[17]. 3GPP TS 38.323: 5G; NR; Packet Data Convergence Protocol (PDCP).
[18]. 3GPP TS 38.322: 5G; NR; Radio Link Control (RLC) protocol.
[19]. 3GPP TS 38.321: 5G; NR; Medium Access Control (MAC) protocol.
[20]. 3GPP TS 38.201: 5G; NR; Physical Layer; General description.
[21]. NR and NG-RAN Overall Description; Stage 2 (Release 15), 06 2018.
[22]. E. Dahlman, S. Parkvall, and J. Sköld, 5G NR The Next Generation Wireless Access Technology, San Diego: Academic Press, 2018.
[23]. 5G Data Channels: Physical; Transport; & Logical; [online] Available: https://www.electronics-notes.com/articles/connectivity/5g-mobile-wireless-cellular/data-channels-physical-transport-logical
[24]. Ultra Dense Network in 5G by Hao Peng, 2015-05-06.
[25]. T. 23.009. Handover Procedures. 3GPP Technical Report, Apr, 10.0.0, 2011.
[26]. T. 36.839. Mobility Enhancement in Heterogeneous Network. Sep, 11.0.0, 2012.
[27]. C. Yu. "Survey on Algorithms and Strategies for Mobility Enhancement under Heterogeneous Network (HetNet) Deployment Circumstances". Inter-national Journal of Future Generation Communication and Networking, vol. 9, no. 1, pp. 187–198, 2016.
[28]. H. Zhuang, J. Chen, and D. O. Wu, "Joint Access and Backhaul Resource Management for Ultra-dense Networks". 2017, in *IEEE Int. Conf. Commun. (ICC)*; 21–25 May 2017. doi: 10.1109/ICC.2017.7996390
[29]. J. G. Andrews, S. Buzzi, W. Choi, S. V. Hanly, A. Lozano, A. C. K. Soong, et al., "What will 5G be?", *IIEEE J. Sel. Areas Commun.*, vol. 32, no. 6, pp. 1065–1082, June 2014.
[30]. F. Ahmed, A. A. Dowhuszko, and O. Tirkkonen, "Network optimization methods for self-organization of future cellular networks: Models and algorithms", *Self-Organized Mobile Communication Technologies and Techniques for Network Optimization.* IGI Global, pp. 35–65, June 2016.
[31]. "SK Telecom's View of 5g Vision Architecture Technology and Spectrum", *SK-Telecom Seoul South Korea White Paper*, 2014.
[32]. L. Zhang, M. Xiao, G. Wu, M. Alam, Y.-C. Liang, and S. Li, "A Survey of Advanced Techniques for Spectrum Sharing in 5G Networks", *IEEE Wirel. Commun.*, vol. 24, no. 5, pp. 44–51, 2017.
[33]. J. Xu, J. Wang, Y. Zhu, Y. Yang, X. Zheng, S. Wang, et al. "Cooperative Distributed Optimization for the Hyper-dense Small Cell Deployment". *IEEE Comm. Mag.*, vol. 52, no. 5, pp. 61–67, 2014.
[34]. 5G Networks Present New Risks and Security Challenges, By CISOMAG. https://cisomag.eccouncil.org/5g-networks-security-challenges/
[35]. I. Ahmad, T. Kumar, M. Liyanage, J. Okwuibe, M. Ylianttila, andA. Gurtov, "5G Security: Analysis of Threats and Solutions", in , doi: 10.1109/CSCN.2017.8088621.2017 IEEE Conf. Standards Commun. and Netw. (CSCN) Helsinki, Finland
[36]. A. Perrig, J. Stankovic, and D. Wagner, "Security in Wireless Sensor Networks", Commun. ACM, vol. 47, no. 6, pp. 53–57, June 2004.
[37]. W. H. Chin, Z. Fan, and R. Haines, "Emerging Technologies and Research Challenges for 5G Wireless Networks", *IEEE Wirel. Commun.*, vol. 21, no. 2, pp. 106–112, April 2014.
[38]. A. Ghosh, J. Andrews, R. Ratasuk, E. Ratasuk, B. Mondal, P. Xia, et al. "Heterogeneous Cellular Networks: From Theory to Practice", *IEEE Commun. Mag.*, vol. 50, no. 6, pp. 54–64, June 2012.
[39]. M. N. Tehrani, M. Uysal and H. Yanikomeroglu, "Device-to-Device Communication in 5g Cellular Networks: Challenges Solutions and Future Directions", *IEEE Commun. Mag.*, vol. 52, no. 5, pp. 86–92, May 2014.
[40]. Combating Spoofed Robocalls with Caller ID Authentication; [online] Available: https://www.fcc.gov/call-authentication.
[41]. W. Fei-Yue, "Parallel Control and Management for Intelligent Transportation Systems: Concepts, Architectures, and Applications", *IEEE Trans. Intell. Transp. Syst.*, vol. 11, pp. 630–638, 2010.
[42]. P. Alexander, D. Haley, and A. Grant, "Cooperative Intelligent Transport Systems: 5.9-GHz Field Trials", *Proc. IEEE*, vol. 99, pp. 1213–1235, 2011.
[43]. K. Hirose, K. Ishibashi, and Y. Yamao, "Low-Power V2M Communication System with Fast Network Association Capability", in 2015 IEEE 2nd World Forum on Internet of Things (WF-IoT), 14–16 Dec. 2015.

[44]. M. Hasan, S. Mohan, and T. Shimizu, "Securing Vehicle-to-Everything (V2X) Communication Platforms", *IEEE Trans. Intell. Veh.*, vol. 5, no. 4, Dec. 2020.
[46]. https://dataprot.net/statistics/iot-statistics/, "IoT statistics"
[45]. G. A. Akpakwu, B. J. Silva, G. P. Hancke, and A. M. Abu-Mahfouz, "A Survey on 5g Networks for the Internet of Things: Communication Technologies and Challenges", *IEEE Access*, vol. 6, pp. 3619–3647, 2017.
[47]. J. Lin, W. Yu, N. Zhang, X. Yang, H. Zhang, and W. Zhao, "A Survey on Internet of Things: Architecture Enabling Technologies Security Privacy and Applications", *IEEE Internet Things J.*, vol. 4, no. 5, pp. 1125–1142, Oct. 2017.
[48]. I. Cha, Y. Shah, A. U. Schmidt, A. Leicher and M. V. Meyerstein, "Trust in M2M communication", *IEEE Veh. Technol. Mag.*, vol. 4, no. 3, pp. 69–75, Sep. 2009.
[50]. Y. Mehmood et al., "Mobile M2M Communication Architectures Upcoming Challenges Applications and Future Directions", *EURASIP J. Wireless Commun. and Netw.*, no. 1, 2015. doi: 10.1186/s13638-015-0479-y
[49]. A. Osseiran et al., "The Foundation of the Mobile and Wireless Communications System for 2020 and Beyond", in *Proc. IEEE VTC-Spring Wksp.*, 2013.

Comprehensive Survey on Device-to-Device Communication for Next-Generation Cellular Technology

3

Priyanka Patil[1] and Vaibhav Hendre[2]

[1]*Assistant Professor at Department of Electronics & Telecommunication, Dr. D.Y. Patil Institute of Technology, Pimpri, Pune, India*

[2]*Professor at Department of Electronics & Telecommunication, G.H. Raisoni College of Engineering & Management, Pune, India*

3.1 INTRODUCTION

Cellular network is now four generations old. Now, the next-generation cellular technology is expected to fulfill all the requirements in mobile network by providing fast multimedia-rich data exchange, improved data rates, minimum latency, as well as minimum energy consumption of devices. To fulfill these requirements, different technologies are introduced in 5G like D2D, massive multiple input multiple output (MIMO), multi-RAT, small cell, beam forming, millimeter waves (mm W), energy harvesting, green communication, and non-orthogonal multiple access (NOMA) [1]. 5G offers offloading of traffic from cellular networks to peer-to-peer users. For this, it requires the execution of direct links for mobile UEs to minimize the load on base stations (BSs).

5G KEY ENABLING TECHNOLOGIES

Significant advancements with regard to smart phones, wearables, and other data-consuming equipment with improved multimedia applications are projected to exhibit higher spectral efficiency and advanced

TABLE 3.1 λ5G key enabling technologies with performance characteristics

MAJOR FACTORS	PEAK DATA RATE 10 [20] GB/S	DATA RATE (DATA RATE ~10 GB/S)	SPECTRAL EFFICIENCY 3–5X RELATIVE TO 4G	MOBILITY APPROX. 350 KM/H	ENERGY, COST-EFFICIENT	REDUCED LATENCY (2~5 MILLISECONDS END-TO-END LATENCIES)
D2D communication		✓	✓	✓	✓	✓
Mm-wave	✓		✓		✓	
Radio access technology	✓	✓	✓	✓	✓	
Advanced network				✓	✓	✓
Advanced MIMO		✓	✓		✓	
Multiple access			✓			✓
Small cell				✓	✓	

technology in next-generation cellular networks. For better results, primary technologies have been introduced for improving capacity and data rates, such as i) Multi-RAT, ii) massive multiple input and multiple output (MIMO), iii) millimeter-wave (mm-wave), iv) enhanced D2D connectivity, v) small cell, etc. (Table 3.1).

D2D communication is an important technology for upcoming wireless communication applications. It is a modality that allows direct communication, without involving the BS. As compared with a traditional wireless cellular network, in case of minimum distance among connected devices, D2D technology is found to be more energy-saving with enhanced efficiency, throughput, and minimum delay. D2D also helps offloading traffic from cellular networks with high performance. There are many challenges in successfully implementing the D2D technology. This wireless technique presents a new device-centric communication with direct communication between two devices. The challenges like resource management, peer discovery, appropriate mode selection, recourse distribution, and management, security, and mobility management are associated with D2D communication. Many ongoing studies are focused on improving spectral efficiency and interference management [2]. D2D plays an important role in 5G communication to achieve all the requirements of a cellular network user. D2D communication is the most challenging and promising technology that offers a massive user capacity and network overcrowding control, high data rates, and assured QoS [3]. For D2D communication operations, reutilization of spectrum is an important aspect. For D2D communication, cellular spectrum and the unlicensed industrial, scientific, and medical (ISM) spectrum utilization are main aspects with several challenges like interference management and avoidance, QoS and delay, and some more, which are highlighted in [4].

Qualcomm's Flash-LinQ develops the concept of D2D communications [5]. This system consists of i) synchronization between time and frequency resulting from cellular spectrum, ii) device discovery, iii) link management, iv) distributed power, data-rate and link development of cellular network. Underlying physical layer technology like OFDM/OFDMA supports Flash-LinQ. It is a useful system designed for device detection, synchronization, and resource management in D2D communication. The third Generation Partnership Project (3GPP) is also focused on D2D communication as proximity services (Pro-Se) [6,7]. Pr-Se has been discussed in detail in [8]. Currently, 3GPP introduced a new technique for D2D user security and safety networks [9]. For device discovery and D2D communication, the IEEE 802.X group has also worked on the same [10]. In the table below, we compare different short-range transmission techniques with different features (Table 3.2).

TABLE 3.2 Comparison of short-range transmission techniques

FEATURE NAME	D2D	WI-FI DIRECT	NFC	ZIGBEE	BLUETOOTH 4.0	UWB
Standardization	3GPP LTE-Advanced	802.11	ISO 13157	802.15.4	Bluetooth SIG	802.15.3a
Frequency band	Licensed band for LTE-Advanced	2.4 GHz, 5 GHz	13.56 MHz	868/915 MHz, 2.4 GHz	2.4 GHz	3.1–10.6 GHz
Max distance	10–1000 m	200 m	0.2 m	10–100 m	10–100 m	10 m
data rate	1 Gbps	250 Mbps	424 kbps	250 kbps	24 Mbps	480 Mbps
Device discovery	BS coordination	ID broadcast and embed soft access point	Radio-frequency	Broadcasting of ID and coordinator assistant	Manual pairing	Manual pairing
Uniformity of service	Yes	No	No	No	No	No
Application	Public safety, content sharing, local advertising, cellular relay	Content sharing, group gaming, device connection	Contactless payment systems, Bluetooth and Wi-Fi connections	Entertainment and control, monitoring	Object exchange, peripherals connection	Wireless USB, high-definition video, auto radar

To enable D2D communication, various wireless technologies with minimum range are used, such as Bluetooth, Wi-Fi, and LTE. All these technologies generally differ in the speed of data rate, distance between two devices, device discovery, and applications. For example, for Bluetooth, the maximum data rate is 50 Mbps and the distance is approximately 240 m, Wi-Fi direct is capable of 250 Mbps rate with 200 m range, and LTE direct provides 13.5 Mbps data rate and distance of 500 m [11].

Categories of D2D communication: a) Device-and-gateway domain—it is an established connectivity between IoT gateway and users pairs, b) core network—when large number of devices communicate directly, this domain is used to collect total data from D2D pairs, c) applications domain—it supports various applications in IoVT, IoMT, public security, smart homes, etc.

The 3GPP-LTE release 12 supports D2D communication and is an important element in 5G networks. The IEEE 802.11 family supports D2D communication with ad hoc mode, hence is part of the IEEE 802.11 networks [12]. The 5G network tire is divided in two types: a) Macro cell tier—In this, conventional communication is established between the BS and device. b) Device Tire—Only device-centric communication is established. According to the requirement the most suitable mode is selected. It consists of four communication modes.

Mode 1: Device relaying with operator-controlled link establishment (DR-OC)—This mode supports the system in which two devices are in proximity and can send data. All the tasks between the user equipment are done by the BS. Improved rate, minimum power utilization, and load reduction are advantages of mode-1 (Figure 3.1).

Mode 2: Direct D2D communication with operator-controlled link establishment (DC-OC)—In the absence of an active mobile network connection, DUs can select another communication interface with other neighboring devices which are linked to the BSs. Here, two devices communicate directly using the control link operated by the BS (Figure 3.2).

Mode 3: Device relaying with device-controlled link establishment (DR-DC)—In this type, many devices are connected to an active connection device, which is connected to a BS and again it establishes a network with the additional devices. All devices in the established network receive same data (Figure 3.3).

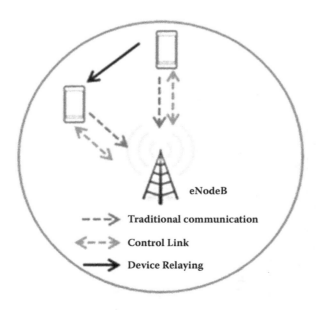

FIGURE 3.1 DR-OC.

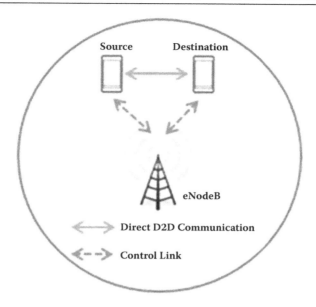

FIGURE 3.2 DC-OC.

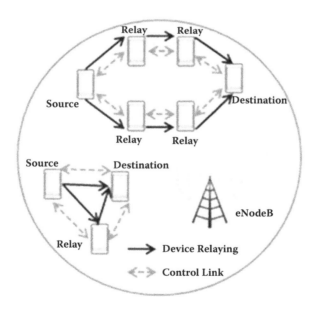

FIGURE 3.3 DR-DC.

Mode 4: Direct D2D communication with device-controlled link establishment (DC-DC)—Here, many mobile users connected to the Wi-Fi offload data connection for communication purposes. All the controls are managed by the BS. Higher data rate, less power consumption, and minimum traffic overload of BSs are offered by Wi-Fi offload (Figure 3.4).

Other Scenarios: In this, the communication can be efficiently established within a proximity [13].

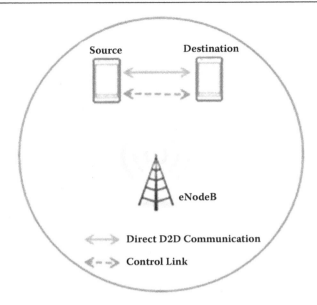

FIGURE 3.4 DC-DC.

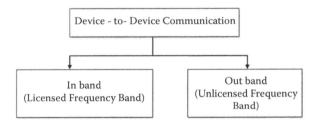

FIGURE 3.5 Types of D2D communication.

3.2 CLASSIFICATION OF BANDS

Inband and Outband Communication [14] (Figure 3.5).

3.2.1 Inband D2D Communication

Here, a licensed frequency band is used. The interference control is possible in inband communication. The licensed spectrum is divided into overlay and underlay communication. In underlay, working of devices is limited under definite cellular network to improve spectrum efficiency with better system output, whereas in overlay unused spectrum is used (Figure 3.6).

3.2.1.1 Underlay

In D2D underlay, cellular network links are shared with cellular user equipment (CUE), which is useful in obtaining better spectral efficiency by controlling the interference between the channels. For this

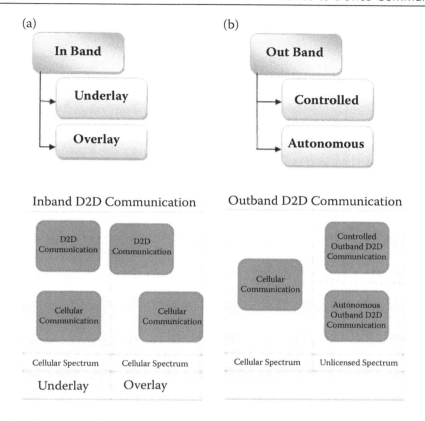

FIGURE 3.6 Classification of inband and outband communications with spectrum allocation.

reason, in underlay communication interference management is executed [15]. In this type, the same band is shared by cellular users and D2D users, which results in improved spectrum efficiency. In underlay, interference for the cellular and D2D users is a major concern. To overcome this interference problem different allocation and power control algorithms have been presented in [16].

3.2.1.2 Overlay

Here, the resources are divided into two bands: a) Cellular user band and b) D2D user band. In an overlay system, the bands used are different, which results in no interference between the devices; however, it reduces the spectrum efficiency.

3.2.2 Outband D2D Communication

In this type, unlicensed spectrum is used for D2D communications. In outband, interference problem does not occur. The categories of outband are enlisted below.

3.2.2.1 Controlled

Cellular network controls D2D users.

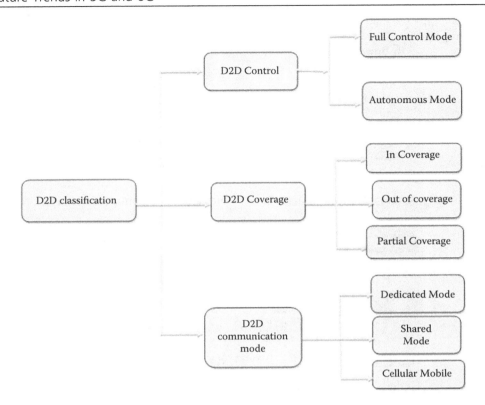

FIGURE 3.7 D2D main classification.

3.2.2.2 Autonomous

Cellular network does not have control over D2D communication (Figure 3.7).

3.3 D2D MAIN CLASSIFICATION

3.3.1 D2D Control

It is divided into two modes: a) Full control mode—Devices are managed by an operator and a cellular network. b) Autonomous mode—Devices themselves are in charge totally.

3.3.2 D2D Coverage

It is divided into thr1ee parts: a) In coverage—Both the devices are present in cellular network coverage, which operates through a licensed band. b) Out of coverage—Here, the devices are out of cellular network coverage, which access an unlicensed band such as scientific medical radio band and industrial band. c) Partial coverage—In this one, the device is in cellular network coverage while the second device is out of cellular coverage.

3.3.3 D2D Communication Mode

It specifies if the D2D users share the resource block with other users such as D2D or cellular users. This mode is again divided into three types: a) Dedicated mode, b) shared mode, c) cellular mode [17].

3.4 INTEGRATED FEATURES

Following are the integrated features of D2D communication (Figure 3.8).

3.4.1 D2D Integrated with Millimeter Wave (mmWave)

Millimeter wave is the most promising technology among the next-generation networks. The operating frequency band ranges from 30 GHz to 300 GHz. In D2D communication, effective utilization of bandwidth is possible. In [18], scheduling mechanism in millimeter wave small cell D2D transmission has been proposed. D2D communication using millimeter wave supports a number of direct concurrent links that improve network capacity with minimum latency and enhance throughput. In this, some challenges like hardware design, additional interface, and interferences occur [19].

3.4.2 Internet of Things (IoT)

Minimum energy consumption, support for interoperability, and big data handling for different applications are the basic requirements of IoT applications. To fulfill all these requirements, D2D is the best emerging technology. For IoT applications, resource allocation and resource optimization are very important aspects in multicell D2D communication, which are explained in [20].

3.4.3 Artificial Intelligence (AI)

AI has the capability to manage resources of massive capacity and reduce interference and nonlinearities in RF components. Also it supports network optimization for quality of service (QoS) [21]. Integrating D2D technology with AI offers more throughput and improved QoS [22,23].

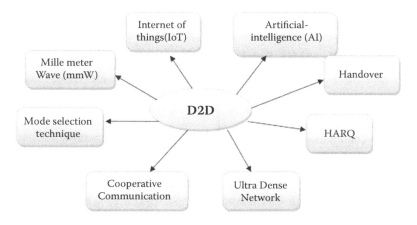

FIGURE 3.8 D2D communication integrated features.

3.4.4 Handover

In D2D communication, devices select neighbor cells at a certain point in time. There are two types of handovers in D2D technology. a) Joint handover—When devices are within proximity they select a joint handover. b) Half handover—When devices are not in proximity they choose half handover [24]. In [25], the method of handover procedure for D2D communication is proposed. User equipment and BS operations with different methods for handover are given in [26].

3.4.5 Hybrid Automatic Repeat Request Operation (HARQ)

In HARQ two techniques are combined to perform a smooth operation in D2D communication. First is automatic repeat request transmission (ARQ). It is used to overcome the packet loss problem within transmission which support to D2D communication for next-generation cellular applications and second is forward error correction. In D2D two type of HARQ exits:

a. Direct HARQ—In this, the D2D receiver sends the acknowledgement directly.
b. Indirect HARQ—Acknowledgement is sent by D2D receiver through eNB [27].

Methods and apparatus used to reduce the transmission error in D2D communication with HARQ are proposed in [28].

3.4.6 Ultra Dense Network (UDN)

Deployment of UDN supports future wireless network and fulfills user demands. Integrating UDN with D2D results in improvement of different important factors like power consumption, energy efficiency, deployment of large small cells, traffic offloading, etc. [29]. Combination of D2D and UDN is a important enabler for latent solution to fulfill next generation mobile network challenges.

3.4.7 Cooperative Communication

InD2D communication, in case the distance between the users is not sufficient for communication [30], to overcome this problem, cooperative communication plays a very important role that improves the performance of D2D communication in data unloading between user equipment (UE).

3.5 RESEARCH CHALLENGES

3.5.1 Peer Discovery

The demand for D2D communication is more in recent trends and technology for the growth of next-generation cellular network. Discovering peers is the most efficient method in network requirement, in which the users discover nearby users with low power consumption. Peer discovery is divided into two categories: a) Restricted discovery—In this type of discovery, for UE detection, prior permission is required which would help maintain user privacy. b) Open discovery—In this type, the UEs are discovered during the time in which they are in proximity of the other users [31] (Table 3.3).

TABLE 3.3 Device discovery methods

SR. NO.	REFERENCE NO.	METHOD OF DISCOVERY
1	[32]	1. Centralized discovery
		2. Distributed discovery
		3. Inband discovery
		4. Outband discovery
2	[33]	1. Centralized
		2. Distributed
3	[34]	1. Sequence-based structure
		2. Message-based structure
4	[35]	1. Power-based device discovery
		2. Distributed
		3. Centralized
5	[36]	1. Direct discovery
		2. Bluetooth discovery
		3. Wi-Fi device
		4. Network supported
		5. Packet- and signature-based
		6. Demand-based
6	[37]	1. Centralized
		2. Distributed
		3. Pilot discovery
		4. Frequency of discovery
7	[38]	1. Direct discovery: Wi-Fi, Bluetooth, sensor network
		2. Network-assisted discovery: Fully controlled, loosely controlled

3.5.2 Resource Allocation (RA)

To improve system performance with regard to different parameters like power consumption, time, and available spectrum in D2D communication, resource allocation is an important factor. In [39], different resource allocation algorithms and methodologies are analyzed and evaluated. Resource allocation is done by device themselves or BS. As per the participation of BS in RA, spectrum allocation and supervision can be carried out by two different allocations: a) Centralized and b) distributed.

    a. Centralized—In centralized type all the allocations and managements are controlled by the BS.
    b. Distributed—All the controls are managed by the user itself. Distributed RA reduces the network difficulty (Table 3.4).

3.5.3 Mode Selection

In D2D cellular network the UEs can communicate with the BS directly, which improves the performance in terms of delay and network throughput. There are many designing challenges such as resource management, network overloading or two user equipment working in the same mode. Mode selection is a very important aspect in D2D communication. A total of four mode selection categories are available [44].

TABLE 3.4 List of resource allocation algorithms

SR. NO.	REFERENCE	ALGORITHM	DESCRIPTION
1	[40]	1. Perfect channel cross-layer optimization resource allocation algorithm	This algorithm focuses on • User access control • User optimal power matching • Selection of user channel multiplexing
		2. Hybrid RA algorithm based on channel probability and statistics	This algorithm supports users with • Access control with outage probability • Power allocation multiplex by D2D users • Optimal channel selection
		3. Optimal resource allocation algorithm based on dichotomy	When the number of users increase, the system module becomes complicated. To overcome this problem with minimum cost and improved efficiency this algorithm is proposed
2	[39]	1. Channel allocation algorithm	It describes channel allocation in original optimization (between D2D users and cellular user)
		2. Power allocation algorithm	In this algorithm, energy efficiency is compared and larger one is applied as final allocation
3	[41]	1. Constrained DA algorithm 2. Coalition formation algorithm	In this algorithm, investigation of recourse allocation problem in downlink D2D communication is performed
5	[42]	1. The coalition formation algorithm for the D2D pairs resource allocation	In this algorithm, the D2D pair makes switch operation in a random order for allocation
6	[43]	1. Lagrangian decomposition-based (LDB) method	Selection of limited number of users to improve energy efficiency
		2. Subcarrier assignment	Analysis of large number of users
		3. Mode switching	It is used to avoid unnecessary calculations of power allocation and subcarriers of entire system

3.5.3.1 Pure Cellular Mode

When the least resources are available and more interference problems are present in a network, the pure cellular mode is selected. D2D users cannot transmit their data in pure cellular mode, which is why it is called as pure cellular mode (Figure 3.9).

3.5.3.2 Partial Cellular Mode

UEs communicate through BS without co-channel spectrum sharing (Figure 3.10).

3.5.3.3 Dedicated Mode

In this mode, dedicated spectrum resources are available for communication between the UEs (Figure 3.11).

3 • Device-to-Device Communication   75

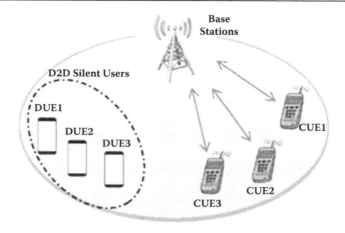

FIGURE 3.9 Pure cellular mode.

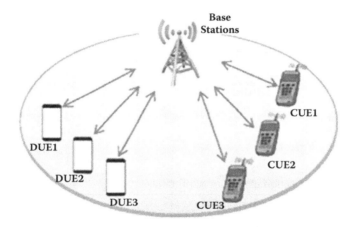

FIGURE 3.10 Partial cellular mode.

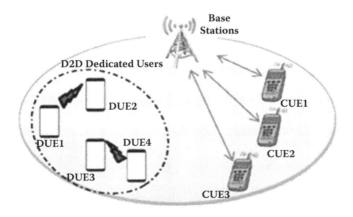

FIGURE 3.11 Dedicated mode.

3.5.3.4 Underlay Mode

Uplink and downlink resources are available for D2D users and cellular users (Figure 3.12) (Table 3.5).

3.5.4 Interference Management (IM)

In D2D technology, IM is one of the significant challenges. If CU and D2D users make use of same cellular resource, it leads to an interference issue. Interference management is subcategorized into three parts (Figure 3.13).

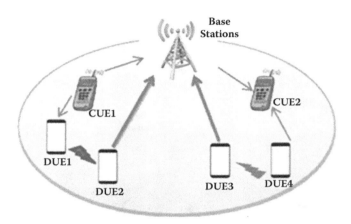

FIGURE 3.12 Underlay mode.

TABLE 3.5 Different mode selection algorithms with details

SR. NO.	REFERENCES	ALGORITHM	DESCRIPTION
1	[45]	Multiuser mode selection algorithm	This algorithm allows the network to select the required mode of operation
2	[46]	1. Optimal mode selection in local route, direct, and relay communications	It offers three different modes of link establishment for multiple pairs of devices
		2. Optimal mode selection (OMS) for direct communications	Reformation of optimization problems in reuse of UL channel of the existing cellular user equipment
		3. OMS for direct and relay D2D	It allows reuse of same UL channels of the existing cellular users
3	[47]	A distributed mode selection using evolutionary game	It defines controlled mode selection in a distribute manner
4	[48]	Distributed coalition formation algorithm	This algorithm is designed to develop a framework to study the joint mode selection and link allocation problems in D2D communication
5	[49]	Heuristic mode selection (MS) algorithm	This algorithm is designed to evaluate the interference and noise power as well as the transmit power, SINR, and throughput

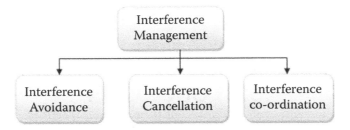

FIGURE 3.13 Interference management categories.

Depending upon the mode of operation, two types of interferences occur in D2D communication. They are intracellular and intercellular interference, also four different interference scenarios exist in D2D communication [50].

1. Interference from DUs to CUs
2. Interference from CUs to DUs
3. Combination of both 1 and 2
4. Interference between D2D pairs (Table 3.6)

3.5.5 Power Control

In cellular networks, the users in proximity create D2D links directly with BSs for data exchange. In this type of communication, power control is an important aspect for higher data rate and low transmission power. In [57], the system sum rates are improved to allow power control in D2D communication, in which the optimization of the transmitting power of D2D users and cellular users (CUEs) takes place. To

TABLE 3.6 Summary of different interference management algorithms

SR. NO.	REFERENCES	ALGORITHM	DESCRIPTION
1	[51]	Clustering devices with bounded subset	It proposes a framework with new joint clustering interference management
2	[52]	1. Iterative distributed power allocation	It offers distributed power allocation for minimum power requirement
		2. Uniform pricing algorithm	In this algorithm, the BS sets and broadcasts a uniform price to all users
		3. Differentiated pricing	It defines the optimal differentiated pricing policy
3	[53]	CLPR	It offers link selection, mode selection, and power adaptation
4	[54]	Sub-band (SB) assignment and intracell resource block (RB) allocation	SB bandwidth is determined and based on interference-free environment and the minimum number of RBs is evaluated
5	[55]	Channel assignment	It allots a channel that has been engaged by a cellular device to a D2D pair in every round
6	[56]	Proposed matching algorithm	Distributed matching algorithm is proposed in which both MBS and UE D2D interact with maximum sum rate

increase system throughput by adjusting each D2D user transmitter power, the reinforcement learning (RL) algorithm is proposed in [58].

3.6 CONCLUSION

The next-generation cellular D2D communication is one of the emerging technologies that focuses on different prototypes, local data transmission, multiuser MIMO enhancement, cooperative relaying and virtual MIMO, IoT enhancement, cellular traffic offloading. Different integrated features associated with D2D support multiple interdisciplinary applications. D2D integrated with different next-generation trends like mmWave, AI, handover, ultra-dense network, HARQ, etc., results in the improvement of required cellular parameters like higher data rates, throughput, efficiency and more, for next-generation mobile network. According to recent research in D2D communication, resource allocation, basically divided into centralized and distributed allocation types, is the most promising challenges. For resource allocation, adoption of technologies like deep learning and machine learning is found to be more efficient and effective. With minimum power consumption, the discovery of nearby user is the most challenging aspect in peer discovery. For resource management, network overloading and single mode sharing problem, different types of modes have been proposed, such as partial, dedicated, pure cellular, and underlay modes.

REFERENCES

[1]. H. Hanan, H. A. Elsayed, and S. M. Abd El-kader, "Intensive benchmarking of d2d communication over 5G cellular networks: Prototype, integrated features, challenges, and main applications". Springer Science +Business Media, LLC, part of Springer Nature 2019, Sept. 2019.

[2]. M. H. Adnan and Z. A. Zukarnain, "Device-to-device communication in 5G environment: Issues, solutions, and challenges". *Symmetry*, vol. 12, p. 1762, 2020, doi: 10.3390/sym12111762

[3]. L. Song, X. Cheng, M. Chen, S. Zhang, and Y. Zhang, "Coordinated device-to-device local area networks". The approach of the China 973 Project D2D-LAN. *IEEE Netw.*, vol. 30, pp. 92–99, 2016.

[4]. M. Agarwal, A. Roy, and N. Saxena, "Next generation 5G wireless networks: A comprehensive survey (in English)". *IEEE Comm. Surv. Tutor.*, vol. 18, no. 3, pp. 1617–1655, 2016.

[5]. X. Wu, S. Tavildar, S. Shakkottai, T. Richardson, J. Li, R. Laroia, et al. "Flash LinQ: A synchronous distributed scheduler for peer-to-peer ad hoc networks". *IEEE/ACM Trans. Netw. (TON)*, vol. 21, pp. 1215–1228, 2013.

[6]. X., Lin, J. G., Andrews, A., Ghosh, and R., Ratasuk, "An overview of 3GPP device-to-device proximity services". *IEEE Commun. Mag.*, vol. 52, pp. 40–48, 2014.

[7]. Feng, D., Lu, L., Yuan-Wu, Y., Li, G. Y., Li, S., and Feng, G., "Device-to-device communications in cellular networks". *IEEE Commun. Mag.*, vol. 52, pp. 49–55, 2014.

[8]. K. Doppler, M. Rinne, C. Wijting, C. B. Ribeiro, and K. Hugl, "Device-to-device communication as an underlay to LTE-advanced networks". *IEEE Commun. Mag.*, vol. 47, pp. 42–49, 2009.

[9]. Zou, K. J., Wang, M., Yang, K. W., Zhang, J., Sheng, W., Chen, Q., et al., "Proximity discovery for device-to-device communications over a cellular network". *IEEE Commun. Mag.*, vol. 52, pp. 98–107, 2014.

[10]. S.-Y. Lien, C.-C. Chien, F.-M. Tseng, and T.-C. Ho, "3GPP device-to-device communication for beyond 4G cellular networks". *IEEE Commun. Mag.*, vol. 54, pp. 29–35, 2016.

[11]. U. N. Kar and D. K. Sanyal, "An overview of device-to-device communication in cellular networks". *Sci. Direct*, 2017. Available: www.sciencedirect.com

[12]. F. Jameel, Z. Hamid, F. Jabeen, S. Zeadally, and M. A. Javed, "A survey of device-to-device communications: Research issues and challenges". *IEEE Commun. Surv. Tut.*, vol. 20, no. 3, pp. 2133–2168, Third quarter 2018, doi: 10.1109/COMST.2018.2828120

[13]. Q. Wang, W. Wang, S. Jin, H. Zhu, and N. T. Zhang, "Mode selection for D2D communication underlaying a cellular network with shared relays". *2014 Sixth Int. Conf. Wirel. Commun. Signal Process.*, Hefei, China, pp. 1–6, 2014, doi: 10.1109/WCSP.2014.6992132

[14]. O. Hayat, R. Ngah, and Y. Zahedi, "In-band device to device (d2d) communication and device discovery: A survey". *Wirel. Pers. Commun*, vol. 106, pp. 451–472, 2019.

[15]. H. Tang, L. Wang, H. Wu, and G. L. Stüber, "Resource allocation for D2D communications underlay in Rayleigh Fading channels". *IEEE Trans*, Feb. 2017.

[16]. M. Elsherief, M. Elwekeil, and M. Abd-Elnaby, "Resource and power allocation for achieving rate fairness in D2D communications overlaying cellular networks". Springer Science+Business Media, LLC, part of Springer Nature, 2019.

[17]. P. Mach, Z. Becvar, and T. Vanek, "In-band device-to device communication in OFDMA cellular networks: A survey and challenges". *IEEE Commun. Surv. Tut.*, vol. 17, no. 4, pp. 1885–1922, 2015.

[18]. Yong, N, et al. "Exploiting device-to-device communications to enhance spatial reuse for popular content downloading in directional mmWave small cells". *IEEE Trans. Veh. Tech.*, no. 99, 2015, doi: 10.1109/TVT.2015.2466656

[19]. N. Bahadori, N. Namvar, B. Kelley, and A. Homaifar, "Device-to-device communications in the millimeter wave band: A novel distributed mechanism". *2018 Wirel. Telecommun. Symp (WTS)*, IEEE, pp. 1–6, 2018.

[20]. Y. Li, Y. Liang, Q. Liu, and H. Wang, "Resources allocation in multicell D2D communications for internet of things". *IEEE Internet Things J.*, vol. 5, no. 5, pp. 4100–4108, 2018.

[21]. M. Yao, M. Sohul, V. Marojevic, and J. H. Reed, "Artificial intelligence defined 5G radio access networks". *IEEE Comm. Mag.*, vol. 57, no. 3, pp. 14–20, 2019.

[22]. X. Wang, X. Li, and V. C. Leung, "Artificial intelligence based techniques for emerging heterogeneous network: State of the arts, opportunities, and challenges".*IEEE Access*, vol. 3, pp. 1379–1391, 2015.

[23]. M. Khan, M. Alam, Y. Moullec, and E. Yaacoub, "Throughput-aware cooperative reinforcement learning for adaptive resource allocation in device-to-device communication". *Future Internet*, vol. 9, no. 4, p. 722, 2017.

[24]. Kyocera Corporation. "Handover of device-to-device (D2d) user equipment (Ue) devices using D2d subframes with cell identifiers". U. S. Patent Documents-2015.

[25]. H. Y. Chen, M. J. Shih, and H. Y. Wei, "Method of performing handover procedure, making handover decision for D2D communications and control node". U.S. Patent, 2018.

[26]. C. G. Balaji, N. Varghese, and K. Murugan, "Optimal handover scheme for device-to-device communication in highly mobile LTE Het Nets". *Int. J. Commun.*, Aug. 9, 2019.

[27]. S. Mumtaz, et al. "Direct mobile-to-mobile communication: Paradigm for 5G". *Wirel. Commun. IEEE*, vol. 21, no. 5, p. 14, 2014.

[28]. J. Hwang, K. Jeong, Y. Chang, and H. Ryu, "Method and apparatus for (5 8) device-to-device HARQ process management". U. S. Patent Documents, Mar. 27, 2018.

[29]. M. F. Hashim, N. I. Abdul Razak, "Ultra-dense networks: Integration with device to device (D2D) communication". 28 February 2019 Springer Science+Business Media, LLC, part of Springer Nature, 2019.

[30]. Y., Cao, T. Jiang, and C. Wang, "Cooperative device-to-device communications in cellular networks". *Wirel. Commun., IEEE*, vol. 22.3, pp. 124–129, 2015.

[31]. D. Feng, L. Lu, Y. Yuan-Wu, G. Li, S. Li, and G. Feng, "Device-to-device communications in cellular networks". *IEEE Commun. Mag.*, vol. 52, no. 4, pp. 49–55, 2014.

[32]. O. Hayat, R. Ngah, S. Z. M. Hashim, M. H. Dahri, R. F.Malik, and Y. Rahayu, "Device discovery in d2d communication: A survey". *IEEE Access*, vol. 7, pp. 131114–131134.

[33]. O. Hayat, R. Ngah, Z. Kaleem, S. Z. M. Hashim, and J. J. P. C. Rodrigues, "A survey on security and privacy challenges in device discovery for next-generation systems". *IEEE Access*. doi: 10.1109/access.2020

[34]. H. Wu, X. Gao, S. Xu, D. O. Wu, and P. Gong, "Proximate device discovery for D2D communication in LTE advanced: Challenges and approaches". *IEEE Wirel. Commun.*, vol. 27, Aug. 2020.

[35]. O. Hayat, R. Ngah, and Y. Zahedi, "Cooperative device-to-device discovery model for multiuser and ofdma network base neighbour discovery in in-band 5G cellular networks "(in English). *Wirel. Pers. Commun.*, vol. 97, no. 3, pp. 4681–4695, 2017.

[36]. K. J. Zou, et al. "Proximity discovery for device-to-device communications over a cellular network". *Commun. Mag. IEEE*, vol. 52, no. 6, pp. 98–107, 2014.

[37]. F. Jameel, F. Jabeen, Z. Hamid, and M. Awais Javed, "A survey of device-to-device communications: Research issues and challenges". *IEEE Commun. Surv. Tut.*, April 2018.

[38]. G. L., Li, G. Manogaran, G. Mastorakis, and C. X. Mavromoustakis, "D2D communication mode selection and resource optimization algorithm with optimal throughput in 5G network". *IEEE Access*, vol. 7, pp. 25263–25273, 2019, doi: 10.1109/ACCESS.2019.2900422

[39]. S. Alemaishat, O. A. Saraereh, I. Khan, and B. J. Choi, "An efficient resource allocation algorithm for D2D communications based on NOMA". *IEEE Access*, vol. 7, pp. 120238–120247, 2019, doi: 10.1109/ACCESS.2019.2937401

[40]. J. Li, G. Lei, G. Manogaran, G. Mastorakis, and C. X. Mavromoustakis, "D2D communication mode selection and resource optimization algorithm with optimal throughput in 5G network". *IEEE Access*, vol. 7, pp. 25263–25273, 2019, doi: 10.1109/ACCESS.2019.2900422

[41]. Y. Chen, B. Ai, Y. Niu, K. Guan, and Z. Han, "Resource allocation for device-to-device communications underlaying heterogeneous cellular networks using coalitional games". *IEEE Trans. Wirel. Commun.*, vol. 17, no. 6, pp. 4163–4176, June 2018, doi: 10.1109/TWC.2018.2821151

[42]. B. Zhang, X. Mao, J. Yu, and Z. Han, "Resource allocation for 5g heterogeneous cloud radio access networks with D2D communication: A matching and coalition approach". *IEEE Trans. Veh. Technol.*, vol. 67, no. 7, pp. 5883–5894, July 2018, doi: 10.1109/TVT.2018.2802900

[43]. S. Guo, X. Zhou, S. Xiao, and M. Sun, "Fairness-aware energy-efficient resource allocation in D2D communication networks". *IEEE. Syst. J.*, vol. 13, no. 2, pp. 1273–1284, June 2019, doi: 10.1109/JSYST.2018.2838539

[44]. Y. Xu, "A mode selection scheme for D2D Communication in heterogeneous cellular networks". 2015 IEEE Global Communications Conference (GLOBECOM), San Diego, CA, USA, 2015, pp. 1–6, doi: 10.1109/GLOCOM.2015.7417701

[45]. P. S. Bithas, K. Maliatsos, and F. Foukalas, "An SINR-aware joint mode selection, scheduling, and resource allocation scheme for D2D communications". *IEEE Trans. Veh. Technol.*, vol. 68, no. 5, pp. 4949–4963, May 2019, doi: 10.1109/TVT.2019.2900176

[46]. C. Chen, C. Sung, and H. Chen, "Capacity maximization based on optimal mode selection in multi-mode and multi-pair D2D communications". *IEEE Trans. Veh. Technol.*, vol. 68, no. 7, pp. 6524–6534, July 2019, doi: 10.1109/TVT.2019.2913987

[47]. Y. Li, W. Song, Z. Su, L. Huang, and Z. Gao, "A distributed mode selection approach based on evolutionary game for device-to-device communications". *IEEE Access*, vol. 6, pp. 60045–60058, 2018, doi: 10.1109/ACCESS.2018.2874815

[48]. Y. Lianxin, D. Wu, H. Shi, Y. Long, and Y. Cai, "Social aware joint mode selection and link allocation for device to device communication underlaying cellular networks". *KSII TRANSACTION*, 2016.

[49]. K. Doppler, C. Yu, C. Ribeiro, and P. Janis, "Mode selection for device-to-device communication underlaying an LTE-advanced network". IEEE Wireless Communications and Networking Conference (WCNC), 2010.

[50]. L. Melki, S. Najeh, and H. Besbes, "Interference management scheme for network-assisted multi-hop D2D communications". 2016 IEEE 27th Annual International Symposium on Personal, Indoor, and Mobile Radio Communications (PIMRC), Valencia, Spain, 2016, pp. 1–5, doi: 10.1109/PIMRC.2016.7794834

[51]. S. Doumiati, M. Assaad, and H. A. Artail, "A framework of topological interference management and clustering for D2D networks". *IEEE Trans. Commun.*, vol. 67, no. 11, pp. 7856–7871, Nov. 2019, doi: 10.1109/TCOMM.2019.2931319

[52]. Y. Liu, R. Wang, and Z. Han, "Interference-constrained pricing for D2D networks". *IEEE Trans. Wirel. Commun.*, vol. 16, no. 1, pp. 475–486, Jan. 2017, doi: 10.1109/TWC.2016.2625255

[53]. T. Yang, X. Cheng, X. Shen, S. Chen, and L. Yang, "QoS-aware interference management for vehic-ular D2D relay networks". *J. Commun. Net.*, vol. 2, no. 2, Jun. 2017.

[54]. A. Celik, R. M. Radaydeh, F. S. Al-Qahtani, and M. Alouini, "Resource Allocation and Interference Management for D2D-Enabled DL/UL Decoupled Het-Nets". *IEEE Access*, vol. 5, pp. 22735–22749, 2017, doi: 10.1109/ACCESS.2017.2760350

[55]. L. Zhao, H. Wang, and X. Zhong, "Interference graph based channel assignment algorithm for D2D cellular networks". *IEEE Access*, vol. 6, pp. 3270–3279, 2018, doi: 10.1109/ACCESS.2018.2789423

[56]. S. Shamaei, S. Bayat, and A. M. A. Hemmatyar, "Interference management in D2D-enabled heterogeneous cellular networks using matching theory". *IEEE Trans. Mob. Comput.*, vol. 18, no. 9, pp. 2091–2102, Sept. 1, 2019, doi: 10.1109/TMC.2018.2871073

[57]. M. Zhao, Y. Wei, M. Song, and G. Da, "Power control for D2D communication using multi-agent reinforcement learning". *2018 IEEE/CIC International Conference on Communications in China (ICCC)*, 2018, pp. 563–567, doi: 10.1109/ICCChina.2018.8641165

[58]. H. H. Esmat, Mahmoud M. Elmesalawy, and I. I. Ibrahim, "Uplink resource allocation and power control for D2D communications underlaying multi-cell mobile networks". *AEU-Int. J. Electron. Commun.*, vol. 93, 2018.

Challenges, Opportunities, and Applications of 5G Network

4

Kalyani N Pampattiwar[1,2] and Pallavi Chavan[3]

[1] Research Scholar, Department of Computer Engineering, Ramrao Adik Institute of Technology, Chapter 4/ Challenges, Opportunities and, Applications of 5G Network, Full Mailing Address – 204, Anandam CHS, Plot No. 13, Sector 2, Koperkhairane, Navi Mumbai, Pincode-400709, MH, India

[2] Assistant Professor, SIES Graduate School of Technology, Nerul, Navi Mumbai, MH, India. Chapter 4/ Challenges, Opportunities and, Applications of 5G Network, Full Mailing Address – 204, Anandam CHS, Plot No. 13, Sector 2, Koperkhairane, Navi Mumbai, Pincode-400709, MH, India

[3] Department of Information Technology, Ramrao Adik Institute of Technology, Nerul, Navi Mumbai, MH, India. Chapter 4/ Challenges, Opportunities and, Applications of 5G Network, Full Mailing Address – C1004 Exotica, Casa Rio Gold, Palava City, Dombivli East, Dist: Thane, Pincode: 421204, MH, India

4.1 INTRODUCTION

Security plays a vital role in any organization. Security protects all the important and confidential data of the organization. It also protects the users' data and technologies being used by the organization. Thus, it enhances the ability of an organization to function as well as safeguards the various applications used.

4G technology is a huge success in the telecommunication sector. Whenever one speaks about any technology, it is always associated with the pros and cons. Obtaining information from people through data theft or other illegal ways is easier in 4G. There is a possibility of some data interference in 4G. It is vulnerable to attacks and hence the need for security and privacy is increased.

However, 4G has increased the connectivity. At the same time, connectivity has also evolved. This technology is not only providing internet connectivity, but also has moved ahead and has started connecting various equipment, vehicles, etc. Thus, creating and connecting various infrastructures enables remote operations. This is what happens in smart cities where not only our phones are connected through internet but also all equipment at homes, traffic signals within a city, and other devices and application based on IoT.

Due to numerous devices and availability of data, the need of the hour definitely is for a network which will be able to manage fast data speeds and also be reliable in terms of connectivity. Moreover, it should also have low latency as well have access to international mobile telecommunications. All this is required to send a vast amount of data via ultrafast broadband. 5G technology is the solution to such a requirement.

The world is connected and communication technology plays a significant role in making this happen. 5G is expected to change the way we communicate as it will not only provide data and voice communication but will also enhance connectivity. As the world transcends to embrace this new technology, it will give rise to newer Challenges, Opportunities, and Applications. The purpose of this chapter is to understand these aspects, beginning with the 5G architecture, as this will provide clarity on various aspects the chapter aims to address.

4.2 5G TECHNOLOGY: OVERALL ARCHITECTURE

The 5G system, which is shown in Figure 4.1, will have radio access nodes (RAN), distributed data centers, and central data centers. RAN's are flexible and overall, the architecture provides a lot of flexibility. This will enable flexible allocation of workload. The nodes and data centers will be connected by transport networks. The information flow from access nodes to data centers will be accomplished by transport networks. Most of the data will be stored in data centers. Network applications and all other applications will run over cloud; however, functions which are dedicated to access nodes will not be on the cloud. Applications, depending on the need, will be distributed or centralized.

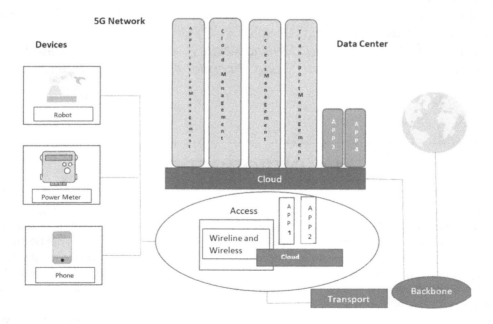

FIGURE 4.1 Overall 5G architecture [1].

5G architecture will provide a lot of flexibility and depending on the requirement, management applications, cloud, transport, and access nodes will be centralized or distributed. [1]

In 5G, all types of services will be flexible in terms of allocation, i.e., can be allocated anywhere in the network or can be allocated to any network or can be allocated to the end user device or external node. Such flexibility will be possible only with regard to the concept of network programmability node. At the same time, services will also have similar kinds of flexibility, which may not be confined to network operators. The service request may even originate from outside of the network.

Knowing that 5G will find many applications, services, devices, etc., end-to-end (E2E) orchestration is required to keep maintain the efficiency of the network with respect to service offerings. In other words, to deliver a certain service, orchestration will place the virtual network function (VNF) physically close to the user. Network slicing helps achieve this.

4.2.1 Network Slicing

In network slicing technique, various logical networks or slices allow the basic function to be on the first layer of the same physical infrastructure. According to the need of the service, infrastructure, and software resources can be allocated, i.e., it can be dedicated or it can be shared between different slices. Computing resources, storage resources, access equipment, transport network infrastructure, VNFs, etc., are the various resources.

A network slice is built to have the network behave as desired. The various aspects which need to be considered while building the slice are security, data-flow isolation, quality of service, reliability, independent charging, and so on. Thus, one or many services can be supported by a slice. This means that a virtual network operator is created to deliver a certain service. Thus, it can be understood that network slicing has several purposes and use in the 5G network [2].

In a scenario, when a private network is being set up in the form of a network slice, that is the slice is virtually completely isolated part of the public network, then the capabilities of the network in terms of its bandwidth, latency, availability, and so on will be tested fully. Once a new slice is created, then this network can be managed locally. For the owner of this slice, it will be like a complete network with storage, processing, and transport nodes. Thereafter, a slice owner can locally control this newly created slice. The slice owner will perceive the network slice as their own network as it is complete with all infrastructure and resources—transport nodes, processing, and storage. This slice can again have a mix of centrally located or distributed resources. Thus, a slice itself will also have a similar flexibility of the network and the slice owner will be able to create applications, store data centrally, or distributed or a combination of both. [3]

4.3 5G ENHANCED OVERALL SYSTEM ARCHITECTURE

After a brief overview of the 5G network, an in-depth understanding of the same enhances the understanding of the 5G network. This emphasizes on the requirement that the 5G network will have to support very diverse functions and performances to enable the coexistence of machine-driven applications and human-centric applications. The fundamental pillars of 5G network will hence be

1. E2E network slicing
2. Architecture that is service-driven
3. Software-Defined Networking (SDN)
4. Network Function Virtualizations (NFV) [4]

4.3.1 Overall Network Slicing

Network slicing, as understood from above, will need to execute numerous logical instances in a mobile network. This will happen on a shared infrastructure. Because of this, various service level agreements with customer and network performance capabilities will require a continuous reconciliation. In the previous technologies, service requirements were limited and hence the slice types were limited and network operators would fulfil this need manually. In 5G, there will be various and an increased number of such customer requests. Therefore, creation and operation will have to be automated through the lifecycle of the network slicing.

Lifecycle management would mean automation in a complete closed loop, i.e., from preparation, instant action, configuration and activation, run time, and decommissioning so that services assured are fulfilled through orchestration functions. This will be achieved with the help of software, namely network functions (NFs) need to be virtualized and all NFs and other resources need to be software-defined and programmed. Thus, all functions and resources will interact and operate in an E2E manner and deliver the service operation functions. Domain resources include a) Radio Access Network (RAN), b) Core and Transport Network, c) NFV, and d) Mobile Edge Network (MEN). While all of these have to be performed synchronously, orchestration is required so that the service operation is fulfilled, where each domain is an important part of the network and has to be automated. This will ensure that closed-loop procedures are deployed and intelligently functioning for fulfilment and assurance. This means that domain-specific controllers are programmed so that policies and rules are executed efficiently at resource and functional level.

Finally, from a common platform, all entities of the system and at all levels can access the data. This platform uses access control mechanisms with data exposure governance mechanism and provides details where data acquisition, processing, abstraction and distribution is involved. This, in turn, will provide subscriber data to the network resources, network slice and wherever else it is required [5,6].

Such architecture should be of a recursive nature, i.e., this structure—design, rule and procedure—should be so designed that it can be applied again and again wherever required. In a network when the service has to be delivered, this recursive goes on to mean that the structure can be applied only to a limited part of the network or only to the limited part of the deployment platform. In brief, it means that such a structure is capable enough to create and deliver a new service requirement out of the existing service and also to do it similarly for the next request of the same service and repeatedly. This will enable numerous service requests to be delivered simultaneously, thereby increasing the scale of operation.

Similarly, the virtualized infrastructure will allow a new slice instance service to utilize a slice instance operating on top of resources so that each instance utilizes the already existing system. It means that every service provider will become the owner and in turn is able to deploy its own Management and Orchestration (MANO) system. APIs which are homogeneous are needed to support the recursion. These APIs will provide a layer to the abstract and will manage each slice and also control the respective resources in a transparent manner. So, a standard template/blueprint/SLA will serve the purpose. This template/blueprint/SLA should elaborate the slice characteristics, namely topology, QoS, etc. and other attributes for the management, control and resiliency level and also for proper orchestration as desired.

4.3.2 End-to-End (E2E) Service Operations—Lifecycle Management

In 5G, infrastructure, platform, software, etc., there will be a service provider for all aspects and the platform will be cloud software platform. Network slicing will satisfy the customized need of the various services, combination of services and its components, NFs, and segments. Thus, an E2E service operation will be achieved.

4.3.3 Overall RAN Architecture

The architecture of the detailed RAN shown in Figure 4.2 is as per the latest 3GPP release specifications on NO-RAN, layer of Service Data Adaptation Protocol (SDAP), and interconnectivity with Centralized Unit–Distributed Unit (CU-DU) split. In this manner the coverage is provided in small cells as a service to many operators. The architecture further has two tiers—for low latency is the first tier and the second is the centralized tier. This centralized tier provides high processing power so that network applications that need high computation are executed. The architecture further becomes versatile by using superior performing visualization techniques. Thereby, isolation of data is attained, latency is reduced and efficiency of resource is increased, orchestration of virtual resources is achieved and VNFs operate more efficiently. Thus, this architecture separates small cell capacity and enhances network infrastructure along with edge cloud.

The Centralized Unit is again divided into 1) Control Plane (CP), also known as CU-C or CU-CP, and 2) the User Plane (UP), also known as CU-U or CU-UP. Thus CU-C and CU-U implementation at different locations is enabled. Additional split in the lower layer split can be applied to a DU. This can also work as a cell but is small. The functional models in the RAN architecture increases the base architecture as per the need. Controller layer is one such model, and the programmability in RAN is enabled. One such use can be in vehicular movement and safety by using various functional models to do the needful.

Deployment of multiple small cells is facilitated by NFV technology [7]. As a form of Multi-access Edge Computing (MEC) one can visualize various small cells as one small cell by using VNF in the cloud. It is also feasible to connect all small cells directly to the 5G core network (CN) and manage them through the core. However, the use of VNFs is better than the alternate possibility as this significantly reduces the signaling [8].

4.3.4 Core and Transport Architecture

The 5G System architecture has a CN and one or more access networks, e.g., an RAN. The CN consists of NFs, its services and the interaction between them, which helps provide mobile and converged networks. Use of NFV and SDN facilitates deployment of the CN services.

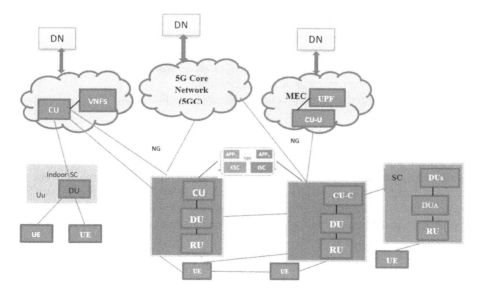

FIGURE 4.2 Overall RAN architecture [12].

To establish connectivity from the Access Points (APs) to the CN we need infrastructure. This infrastructure is called transport network connectivity. Transport networks are the basic building blocks of 5G systems as they lay out the network fabric inter-connecting NFs, CN and RAN, and the units of RAN. The 5G system is developed keeping in mind the new emerging services. This means that there will be new transport network requirements. The RAN architectures discussed in detail earlier can acquire Cloud-RAN (C-RAN) fundamentals that require infrastructure connectivity within the RAN. So, the connectivity infrastructure between the CUs and DUs can be called as front haul (FH). To support new operational network services over the transport network there is a need for new emerging services that will throw open newer challenges with respect to the Cloud-RAN. The transport connectivity will also need to support different RAN split options.

4.3.4.1 Core Network Architecture

Multicasting, Broadcasting, and Integrated data analytics in the 5G can be understood properly once we understand the interface between the CN and RAN. Then we can understand the requirements of the transport network and the protocols on these interfaces. Before that, let us look at the user and control plane separations. CN functions that use services from control network protocols rely on transport networks. CP and UP communicate using Packet Forwarding Control Protocol and General Packet Radio System Tunneling Protocol User Plane (GTP-U) for the respective portions of the interface. These protocols are transported on top of Internet Protocol/User Defined Protocol.

4.3.4.2 Transport Network Infrastructure

Transport Infrastructure—Wireline Technologies

Transport networks should be able to address the main challenge arising in the 5G communication, i.e., it should be able to provide connectivity between DUs and CUs in the common digitized format. Standardization is the solution which is already adopted and a flexible option like RAN allows to adapt the functional split between CU and DU.

As the service requirements will be different, RANs in 5G demand the development of a novel solution at the transport network stage for interconnectivity between CU, DUs, and remote users, etc., for the functions to be executed. The solutions, thus, provided need to be flexible, energy-efficient, and resourceful. Such solutions will meet the future requirements as well and hence are most desired.

Together, network software and advanced programmable hardware solutions will allow for the varied tasks to be smoothly executed by dynamic allocation between CUs and DUs. Thus, virtual and physical network will be properly employed on top of other network elements, e.g., programmable NF can be deployed at remote server or even at network nodes.

Some of the most advanced transport technologies are as below:

Transport Infrastructure—Programmable Elastic Frame-based Optical Transport

Considering the flexibility that 5G intends to offer, there will be a frequent need to reconfigure the network. Advanced optical networks are very flexible and adapt to the dynamically changing network requirements with respect to data needs, traffic flow and end-to-end connectivity. Optical networks also need to have programmability features to take into account the high bandwidth requirement, which itself is dynamic and diverse. Time-shared optical network is the solution which can address the various requirements.

Transport Infrastructure—Wireless Technologies

The network integration requirement that will be required in 5G, i.e., to connect diverse, heterogeneous, and various different end devices, is a challenge. Hence, various technologies like Sub-6, mmWave, massive MIMO, etc. will coexist with Long-Term Evolution, Wi-Fi, etc. Overall, all these will provide a larger reach, more density and mobility.

The transport network from the viewpoint of adapting a wireless technology will be possible with small cells operating in a dense layer. Satellite communication can also be considered as a transport network.

4.4 THREATS AND CHALLENGES

While availing features provided by 4G network, everyone is looking at 5G network in terms of faster speeds and more security. In this digital era, the number of devices is increasing so fast that establishing communication between these devices is a challenge. The major improvements anticipated by 5G networks is to ease the connection between these devices and to deliver high-quality services for all devices simultaneously. When we are talking about the devices it is not restricted only to the smartphones; but other IOT devices can also communicate through the network.

Security and privacy concerns in 5G networks are divided into three-tier architecture, namely the access networks, the backhaul networks, and the CN. Since there are a variety of nodes involved in communication; various access mechanisms are used to access them. This results in increased risk of attack, which gives rise to new security challenges. In case of backhaul network, wireless channels, microwaves, satellite links, as well as wired lines can be used to establish backhaul communication between the base station and the CN.

If the devices are not connected then the network is less vulnerable and threats are lesser than access networks. Mobile Management Entity in the CN is vulnerable as the elements of the access network are adjusted, even though additional security is guaranteed by the GPRS Tunnel Protocol (GTP). Security threats are transmitted to the CN because the backhaul network with the use of NFV and SDN techniques moves it into the data plane. Network is becoming more dynamic because of the use of NFV, SDN, and cloud techniques that results in a large number and variety of vulnerabilities [9].

For example, DoS or resource attack can take place as a large number of devices and services can engulf the signaling load. There are two methods which can discourse signaling overloads. The first method supports the communication between large number of devices using lightweight authentication and key agreement protocols. The second method supports group-based protocols that enable us to group devices together. Performance of the 5G network in terms of security can be improved using new techniques but it gives rise to security loopholes as well. For example, massive MIMOs assist in disguising passive and active eavesdropping. There is rise in threats from malicious applications or activities in OpenFlow implementation of SDN. Additionally, when a service is migrated from one resource to another, NFV raises security issues. Due to miscellany in business types and applications, new privacy issues exist in 5G networks. User's privacy easily and recurrently changes the state from being closed to being open due to openness of the network. Consequently, the risk of leaks increases as the contact state changes from offline to online. Therefore, we will certainly face privacy issues with 5G, which must be solved in the coming years. Fortunately, privacy preservation can be achieved to great extent using progressions in data mining and machine learning technologies [10].

4.4.1 5G Threat Landscape

5G Threat Landscape: 5G is a much developed technology and provides many improved network services. Thousands of millions of devices will be connected and will work with reliability, speed, capacity, etc., as compared with 4G devices. 5G will also allow these devices to function with better facilities,

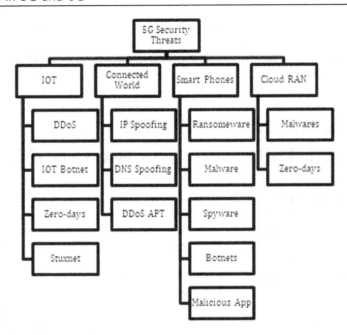

FIGURE 4.3 5G security threats.

band utilization, lesser tolerance towards fault, and latency as compared to 4G technology. The ecosystem thus formed because of various devices, IOT, connected world, and critical infrastructure facilities will make 5G a target for attackers.

Such attackers will be professionals who would have in-depth knowledge of technology and would have ample resources to exploit. Mostly such attacks are likely to be politically and financially motivated.

Criminals and professionals with ample number of resources and technical knowledge can undertake likely attacks to obtain political and financial motivated gains. Complex and sophisticated threats and malwares will be used to carry out such criminal activities. The outline of 5G security threats is shown in Figure 4.3. The rest of the chapter discusses the same in detail.

5G Evolved Security Model—Wireless communication systems support many other devices other than phones (audio and video calls) and numerous applications pertaining to gaming, shopping, bringing your own device, home appliances, cloud technology, etc. These have also given rise to many new challenges for researchers and developers. Thus, IOT has several vulnerabilities and is a great challenge in 5G. Hence, adequate and best security infrastructure and practices need to be put in place in this ever-evolving threat environment [11,12].

4.4.1.1 Threat Landscape of Internet of Things (IoT)

IOT has led to the creation of a smart world that is built on technologically smart computing devices and a combination of them will create a world like Social IoT, Industrial IoT, Healthcare IoT, smart water and power infrastructures, smart cities, etc. Such an environment will have connected devices, machines, household appliances, critical healthcare systems, power and water systems, and more. All of this will be on a connected infrastructure like a web of various technologies. It is noticed from the past and present that with increasing number of smart and intelligent systems, threats also have evolved and have become more complex. So, 5G will have an increased security threat, which will be the greatest challenge. 5G security faces significant challenges because of the high density and low latency nature of IoT.

4.4.1.2 Security Issues Related to SDN/SDMN

SDN technology came to the forefront with the requirement of the system to present the novelties quickly. It also streamlines and automates the huge network management. It logically operates the centralized control plane monitors and also controls the packet forwarding inside the network. SDN is one of the upcoming 5G technologies. SDN indeed offers exceptional reliability and speed for the network data and nodes. By segregating the network control function (SDN controller) and traffic delivery function (SDN switch) SDN technology manages the delivery function of the network. The control protocol between controller and switch is vulnerable to traffic bypass attacks. Similarly, an attacker will have an unauthorized access between switch and controller and overall depletion of the SDN system by DoS attacks will crumble the whole system. Numerous other attacks are also possible which could hamper the network operations [12]. Traffic bypass attacks make use of control protocol vulnerabilities between SDN controllers and switches. DoS attacks can immobilize services by unauthorized access between switches and controllers, and resource depletion of SDN systems [13]. For example, by attacking SDN controllers, SDN switch flow table can wear out if it is subjected to a saturation attack [14]. Network information from SDN controllers can easily be obtained by malicious third-party applications if proper safeguarding mechanism is not in place for the authentication and authorization of applications. [15]

4.4.1.3 Network Function Virtualization (NFV)-Related Security Issues

NFV is one of the basic and important technologies in 5G and will play a critical role in operation of 5G. The security issues here will impact the overall quality and resiliency of the network. Attacks can range from physical hardware level to the architecture level. If the software of the virtual infrastructure manager is exploited then it can have a cascading effect. So again, this is a very vulnerable infrastructure.

4.4.1.4 MEC and Cloud-Related Security Issues

MEC is a part of the network architecture, but is at the end of the network, i.e., it is at the edge of the network and operates close to the user end. It allows cloud computing processes at the edge of the network. So, physical attacks have a high chance in edge devices as compared to cloud devices. Billing data and changing data route take place through edge components and hence security of these components is also critical.

A huge amount of data is constantly generated in 5G technology by a number of applications. To handle this enormous data, one will surely need a cloud-based system. Due to vulnerabilities, storage limitation, energy, and reliability issues it is not prudent to save the data in end devices. So, an integration of mobile computing and cloud computing will be required. This leads to various risks with respect to privacy, data integrity, and security.

4.4.1.5 Network Slicing-Related Security Issues

In detail, we have understood the concept of Network Slicing and have realized that this is the most efficient and optimal way the 5G network will operate. It completely isolates the various virtual networks concerning access, transport, devices, and CN. Their deployment is with respect to various service mechanisms and situations. Network slicing logically divides the single physical network into various virtual, complete E2E networks. This makes it a very important concept of 5G network. The security challenges here are many as it forms the most critical part of the network structure.

4.4.2 5G Security Aspects

We have seen above the security threats and challenges in the key components of the 5G network. In a nutshell and if we have to speak about these challenges, they can be broadly classified into aspects related to Confidentiality, Integrity, and Availability (CIA). These security challenges pose difficulty in strengthening the security and protection of users' data. They form the pillars of security requirements in the 5G landscape.

A. Confidentiality: Data confidentiality is the key aspect and the primal requirement of security when it comes to 5G technology. Data transmission to unauthorized entities/persons and to protect it from attackers is what we mean by maintaining data confidentiality. Key encryption algorithms are used to realize the desired confidentiality in critical services and infrastructure.
B. Integrity: Integrity means that there is no tampering of data and there is no loss of information when it is transmitted from one point to another. With respect to 5G, it is very important as there are various devices, various critical infrastructures, etc. Hence in 5G integrity has to be protected at the Packet Data Convergence Protocol (PDCP) layer. Integrity protection signaling is used by 5G authentication scheme. This makes sure of the integrity and no unauthorized person can access or modify the data transmitted.
C. Availability: In 5G domain, availability again is a very important aspect of the network. Availability means that legitimate users can always use the network resources whenever they are required. It also exemplifies the reputation of the network operator. In other words, network should be available almost 100% times or very close to that figure. Also, it goes on to show that network is robust and is effectively secured against attacks. The performance of the network is degraded by DoS attack. The extreme mobile broadband provides 95% network availability whereas ultra-reliable machine-type-communication provides 99.99% network availability.

In addition to the above, centralized security policy and visibility are important macro-level aspects irrespective of the network operator.

D. Centralized security policy: The prevailing 4G architecture policies cannot be directly applied in 5G network. Hence, there needs to be a centralized security policy for users' interest to be protected and it also provides convenience to users to access applications and services.
E. Visibility: 5G operators must be able to manage visibility, inspection, and controls for all network layers so as to achieve a universal security mechanism. So, to manage the security policies, open APIs should be integrated with 5G technologies. This will ensure that policies are consistent with software and hardware in the network. Thus, visibility will help implement security mechanism suitable to 5G services. High visibility provides more security and threat prevention. It also helps find and quarantine the contaminated devices before any attacks take place.

We look further at the most demanding security issues with respect to the key security areas in 5G, i.e., access control, authentication, communication and encryption.

4.4.3 Security Challenges and Opportunities Related to 5G Communication

As shown in Figure 4.4, following are the security challenges related to 5G communication

Authentication

Authentication plays a very important role in security of any communication system. It is one of the primary requirements to verify the user's identity. There are several ways that have been used in the earlier technologies to provide this security aspect. In 5G, as the network has numerous devices, IoT, smart control systems, etc., and the technology to connect all of them is vast and complex, it is imperative that authentication is performed so that vulnerabilities are reduced. In 5G, authentication will be two stage—primary and secondary authentication. Primary authentication will provide device and network authentication. Secondary authentication will be outside the domain of the network operator's domain. Authentication at primary level can be done using Authentication and Key Agreement (AKA) and Extensible Authentication Protocol (EAP). Credentials and authentication methods related to EAP

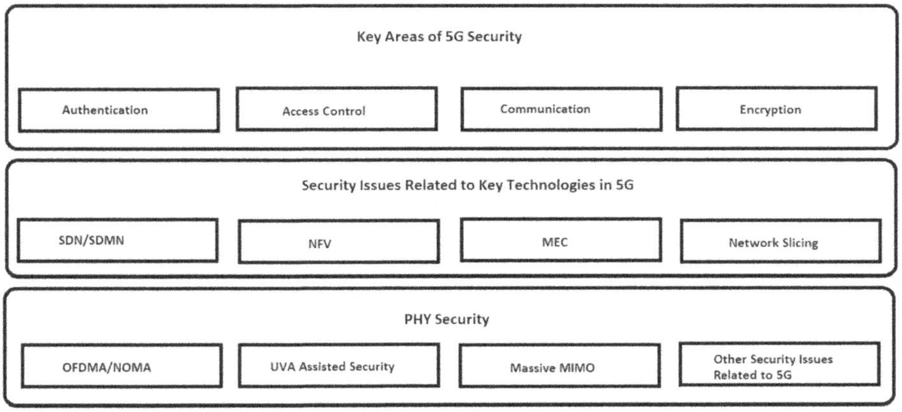

FIGURE 4.4 Key areas of 5G security [12].

are applicable in this two-stage authentication method. The facilitation of mutual authentication will be achieved by key management and primary authentication procedures. These are the opportunity areas with respect to 5G authentication.

Access Control

Access control plays a very important role in any type of network security system. Selective restriction of the access to the network is the main thing that access control has to perform. Network providers need to administer the access control networks to provide a secure and safe network environment. By means of providing access control one can guarantee that only the authorized users can access the system.

Communication Security

5G ecosystem comes with built-in vulnerabilities with respect to communication because of the way the signal is transmitted in this technology. There are various points from where the attacks can originate. Attacks can even get generated at access networks, operators end and even at the user end. Broadly, the two types of network traffic are: 1) Control traffic and 2) user data traffic. These are vulnerable to various security threats. Lack of IP level security is the most important security issue at the control traffic end. To manage channel communication securely, TLS and SSL are application layer security protocols that need to be used, this could be an opportunity in channel communication. Some of the IP level threats known are eavesdropping attacks, IP spoofing, TCP reset attacks, message modification attacks, TCP SYN DoS. Therefore, IP level security and higher-level security is necessary to make the network secure. Inter-Controller Communication (ICC) between the different SDN controllers helps in sharing the control information and carry out "n" number of NFs. The various NFs monitor the network, synchronization of the security policy, and manage mobility as well as traffic.

Encryption

Encryption of data is very important to ensure the confidentiality. E2E encryption is significantly important in 5G network because of the ecosystem of the network. This will protect the data in the mobile from getting accessed by unauthorized users. Encryption of radio transmission in 5G is done at the PDCP layer. Non-Access Stratum, Access Stratum, and the UP is secured by 128-bit encryption keys. Some of the encryption algorithms from current and previous technologies can continue to be used;

however, in 5G some new algorithms, namely Ultra-Reliable Low Latency Communication (URLLC) and RAN are far more secure and safe from various threats and attacks. Encryption will also play a vital role in privacy protection in 5G.

4.5 PRIVACY PERSPECTIVE

Another very important aspect which needs to be addressed in 5G is the privacy perspective, i.e., privacy of data, privacy of location, and privacy of identity.

Privacy of Data

Data privacy in 5G is of utmost importance as in this network users can consume high data and many smart on-demand services (e.g., streaming of high-resolution content, various healthcare services to monitor and keep track of health parameters, smart way of metering electricity, and water) using various high-tech devices. Network operators will need to provide these services to the customers and hence the individual data will be stored by them and thus personal data will be used only with the individual's permission. This personal data which is available with network operators may get distributed to other parties interested in this data. They may further analyze this to their advantage by using various tools/techniques. This compromises users' privacy. To maintain individual user's privacy service providers must share details with the users of where the individual's data have been stored. Also, users must be aware of how their data will be used and the purpose of it.

Privacy of Location

Location will be a very critical information and the location-based services and most devices will rely on this particular piece of information. Location-based services use location data and is nowadays used in several applications/services like government, entertainment, transportation, healthcare, food delivery, and others. Undoubtedly, life has become easier and more enjoyable, but this brings a lot of privacy issues—of being continuously tracked. At times, individuals may not even be aware of what are the implications of how their location is being used. This information is a very critical privacy issue.

Identity Privacy

Safeguarding the identity of a device or system or user against active attacks is known as identity privacy. There is a steady rise in the number of devices that are getting connected over internet and hence there is an increasing threat to the identity getting compromised. The theft of identity is one of the big threats in the 5G technology and also in the Internet of Things. Thus, it is the need of the hour to design a secure mechanism that is efficient in securing user identity.

4.6 APPLICATIONS OF 5G

The applicability of 5G technology is embellished with many discrete features, which serve wide range of people irrespective of their purposes. The various 5G applications in healthcare, smart mobility, social networking, and digital commerce aim for the betterment of end users' life. Internet services like mobile applications have shown a very rapid development over the past decades. Internet applications enable machine-to-machine communication by means of integrating digital domain and our physical environments. Many of these applications are context-based applications which need triggering actions [16], e.g., if the driver is driving a car, then the intelligent smartphone would not disturb the driver.

In spite of the various application domains of 5G, the IoT is the center of attraction; as development of smart city is a concern of 5G. A unified along with easy access for the administration of the city and services to residents can be provided by integrating traditional and modern Information and Communication Technology (ICT). This results in reduced operational expenses of public management. Also, the burden of administration on businesses is reduced and on citizens. For instance, reduced congestion or pollution in a city with the help of smart transportation [17].

Though there are many application domains of 5G, we cover the following five domains only.

Smart mobility: There exist a wide variety of applications that are portable in this technology, which range from the regular route planning to the newer driving services. To name a few, efficient routing, energy saving, traffic balancing, cost, emission reduction, and accident prevention are some of the advantages provided by smart mobility [18].

Smart energy: Smart energy applications help in the monitoring and management of power plants, detection of power failure and response to it, smart grid network, energy-saving services, power utilization ways for homes and offices, and commercial buildings and smart charging stations for electric vehicles. The ultimate objective of 5G is increase in efficiency of power systems and the reliability of the same with increase in use of renewable energy.

Smart health: Nowadays, for budding awareness of fitness and well-being, health mobile applications are becoming very much popular among mobile users. For example, smart wearable devices help in monitoring the health status of a person. Another emerging application in this domain is the use of augmented reality or virtual reality, e.g., in surgery. This needs low latency and high bandwidth along with the generic requirements of low power and security and privacy of data from 5G.

Applications in industry: Manufacturing, communication between the machines, 3D printing, and construction supported by AI are represented by industry IoT 4.0. The impact of these industrial applications benefits the entire society.

Consumer applications: The capability of 5G mobile business and technology innovations reflect a huge amount of consumer applications. The developing applications include pervasive gaming, ultra-HD mobile streaming, financial technology based on blockchain, gaming, mobile augmented reality or unmanned aerial devices supporting virtual reality, and holographic technology, such as HoloLens. All these advanced applications that are supported by 5G need a low energy footprint, comprehensive network, low energy, high bandwidth, link reliability, low latency, and security.

4.7 CONCLUSION

5G technology can help address issues with regard to large number of devices, continuously generating data and fast data speed. We have, thus, described in detail the overall and enhanced 5G architecture which is flexible and depending on the requirement, management applications, cloud, transport, and access nodes whether it will be centralized or distributed. Also, the chapter focuses on overall network slicing, end-to-end (E2E) service operations, overall RAN architecture, core, and transport architecture.

The landscape of the 5G network is continuously growing and so are the threats. This chapter further highlights the security attacks related to IOT, Cloud-RAN, connected world and smartphones. We have also emphasized the security issues in the 5G key technologies, i.e., Multi-access Edge Computing (MEC), Software-Defined Networking (SDN), Network Slicing (NS), Cloud Computing, and Network Function Virtualization (NFV) concepts. All these aspects lay the foundation towards the development of the 6G network. We have also discussed 5G privacy perspectives from the viewpoint of data privacy, location privacy and identity privacy. Although 5G can have multiple application domains, we majorly concentrated on Smart Mobility, Smart Energy, Smart Health, Industrial Applications and Consumer Applications. All these advanced applications, which are supported by 5G, need reliable links, low latency, extensive connectivity, high bandwidth, low energy footprint, and security.

REFERENCES

[1]. 5G architecture next mobile technology. Available: https://www.ericsson.com/en/reports-and-papers/white-papers, 2016.

[2]. N. Nikaein, E. Schiller, R. Favraud, K. Katsalis, D. Stavropoulos, I. Alyafawi, Z. Zhao, T. Braun, and T. Korakis, "Network store: Exploring slicing in future 5G networks". In *Proc. 10th Int. Workshop Mobility Evolving Internet Architecture*, pp. 8–13, 2015.

[3]. A. Osseiran, J. F. Monserrat, and P. Marsch, eds. *5g Mobile and Wireless Communications Technology*. Cambridge: Cambridge University Press, 2016.

[4]. J. Cosmas, N. Jawad, M. Salih, S. Redana, and O. Bulakci, "5G PPP architecture working group view on 5G architecture". In *Dept of Electronic and Computer Engineering Research Papers*, 2019. Available: http://bura.brunel.ac.uk/handle/2438/18546

[5]. N. G. M. N. Alliance, "5G security recommendations Package# 2: Network Slicing". *Ngmn*, pp. 1–12, 2016.

[6]. P. Rost, C. Mannweiler, D. S. Michalopoulos, C. Sartori, V. Sciancalepore, N. Sastry, O. Holland et al., "Network slicing to enable scalability and flexibility in 5G mobile networks". *IEEE Commun. Mag.*, 55, no. 5, pp. 72–79, 2017.

[7]. B., Chatras, U. Steve Tsang Kwong, and N. Bihannic, "NFV enabling network slicing for 5G". In *2017 20th Conf. Innov. Clouds, Internet Netw. (ICIN)*, pp. 219–225. IEEE, 2017.

[8]. ETSI, GRMEC, "Mobile edge computing (mec); deployment of mobile edge computing in an NFV environment". ETSI, DGS MEC 17, 2016.

[9]. M., Jaber, M. A. Imran, R. Tafazolli, and A. Tukmanov, "5G backhaul challenges and emerging research directions: A survey". *IEEE Access*, vol. 4, pp. 1743–1766, 2016.

[10]. M. Wang, T. Zhu, T. Zhang, J. Zhang, S. Yu, and W. Zhou, "Security and privacy in 6G networks: New areas and new challenges". *Digit. Commun. Netw.*, vol. 6, no. 3, pp. 281–291, 2020.

[11]. P. Bisson, and J. Waryet. "5G PPP Phase1 Security Landscape". In *5G PPP Security Group White Paper*, 2017.

[12]. R. Khan, P. Kumar, D. N. K. Jayakody, and M. Liyanage, "A survey on security and privacy of 5G technologies: Potential solutions, recent advancements, and future directions". *IEEE Commun. Surv. Tut.*, vol. 22, no. 1 (2019): pp. 196–248, 2019.

[13]. S. R. Hussain, M. Echeverria, A. Singla, O. Chowdhury, and E. Bertino, "Insecure connection bootstrapping in cellular networks: The root of all evil". In *Proc. 12th Conf. Secur. Privacy Wirel. Mobile Netw.*, pp. 1–11, 2019, 2019.

[14]. C. Cremers and M. Dehnel-Wild, "Component-based formal analysis of 5G-AKA: Channel assumptions and session confusion", 2019.

[15]. H. Kim, "5G core network security issues and attack classification from network protocol perspective". *J. Internet Services Inf. Secur.*, vol. 10, no. 2, pp. 1–15, 2020.

[16]. A. Y., Ding, and M. Janssen, "Opportunities for applications using 5G networks: Requirements, challenges, and outlook". In *Proc. Seventh Int. Conf. Telecommun. Remote Sens.*, pp. 27–34, 2018.

[17]. A., Zanella, N. Bui, A. Castellani, L. Vangelista, and M. Zorzi, "Internet of things for smart cities". *IIEEE Internet Things J.*, vol. 1, no. 1 (2014), pp. 22–32, 2014.

[18]. M. Pattaranantakul, R. He, Q. Song, Z. Zhang, and A. Meddahi, "NFV security survey: From use case driven threat analysis to state-of-the-art countermeasures". *IEEE Communications Surveys & Tutorials*, vol. 20, no. 4, pp. 3330–3368, 2018.

Machine Learning and Deep Learning for Intelligent and Smart Applications

5

Reena Thakur[1] and Dheeraj Rane[2]
[1]*Jhulelal Institute of Technology, Nagpur*
[2]*Medicaps University, Indore*

5.1 INTRODUCTION

Technology has become an embedded part of apps and the default catalyst for business development. Every day, new milestones are being reached with the advent of machine learning (ML). We are moving into an age of more and more convergence, making it an integral mediator between humans and machines. The rapid growth in the mobile industry seems like an unavoidable fusion of different technologies. New frontiers have been opened by the inherent capacity of such structures to enhance themselves, based on data analytics, IoT, and artificial intelligence (AI).

ML is a phenomenon that, while it has persisted for several years, has recently become a common subject, mostly due to advances in artificial neural network (ANN) and deep learning (DL). These are fundamental techniques of ML that have enabled us to move significantly forward in conceptual application of image processing, text analysis, and NLP. As seen in Figure 5.1, ML is also an area of AI.

Hand-crafted or rule-based methods in ML for dimensionality reduction are complex, time-consuming, and do not perform well and may yield inconsistent results for large and complex data [1]. Due to their great capacity to produce, store, and process vast volumes of data, the IoT (and big data) has enabled the development of ML and smart applications in the information age. For optimum functionality and benefits, it is the fuel required by the ML techniques.

Initially, the chapter gives an introduction and background/motivation. Second, different ML and DL techniques are discussed. They can be implemented in solving problems associated with the integration and management of smart techniques. In the next part, some applications regarding the security and efficiency of various applications based on ML and DL are provided. Further, a comparative

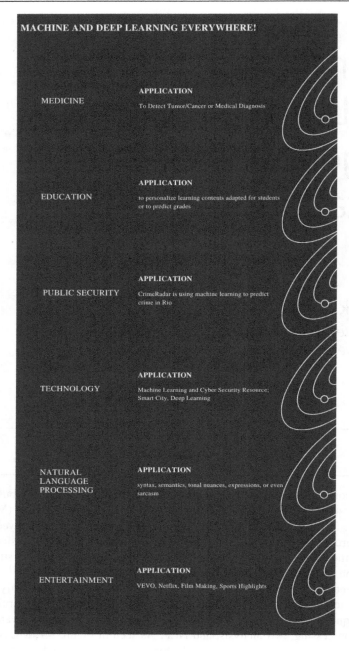

FIGURE 5.1 Machine and Deep Learning Everywhere

evaluation of characteristics is established to choose an algorithm. Finally, a series of conclusions is included.

ML as well as DL are the subfields of AI.

The list of subfields containing AI is given in Figure 5.2:

While ML is built on the basis that machines should be able to adapt from experience, AI refers to a broader idea where machines can "intelligently" learn to achieve things.

ML, DL, and other such methods are utilized by AI to solve practical problems.

ARTIFICIAL INTELLIGENCE

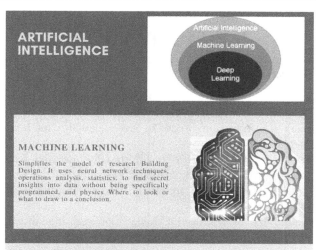

MACHINE LEARNING

Simplifies the model of research Building Design. It uses neural network techniques, operations analysis, statistics, to find secret insights into data without being specifically programmed, and physics Where to look or what to draw to a conclusion.

NEURAL NETWORK

It's a kind of machine learning influenced by the human brain's operations. The computer system consists of interconnected units that are Processes data by reacting to external inputs, relaying data Data between each device. Multiple passes are required for the process at the Data for finding links and deriving meaning from undefined knowledge.

DEEP LEARNING

Utilizes enormous Neural networks with multiple processing unit layers, which take advantage of Computing power advances and improved training methods to learn Complex patterns in vast quantities of knowledge. Popular applications include applications for Recognition of images and expression.

COMPUTER VISION

Relies on pattern recognition and deep learning to recognize what's in a picture or video. When machines can process, analyze and understand images, they can capture images or videos in real time and interpret their surroundings.

NATURAL LANGUAGE PROCESSING

Processing is the skill of Computers for human language processing, comprehension and generation, including expression.
Natural language interaction is the next stage of NLP, which enables humans to communicate with Communicate with computers to perform tasks using standard, everyday language.

FIGURE 5.2 Artificial Intelligence Subfields

5.2 MOTIVATION

DL is a recent ML field which has been launched with the aim of bringing ML and AI nearer.

It refers to the idea of "deep neural networks" (DNN) in the human brain and, from this viewpoint, DL aims to mimic the functions of the inner layers of human brain, such as extracting intelligence from multiple layers of processing of information. Since DL is modeled after the human brain, its capabilities are improved each time new data is available.

Recent developments in ML technologies over the past few decades have allowed an enormous amount of sensory information to be processed, analyzed, and interpreted. There is a new age emerging for "smart applications", which changes the technique used in traditional systems that are used to comprehend the environment.

Over the past few years, ML as well as DL have improved dramatically, and in many critical applications, machine intelligence has moved from laboratory interest to functional machinery. The ability to smartly track IoT devices offers a substantial response to new attacks. These are really useful methods for data exploration and for learning about "usual" and "abnormal" activities based on the performance of IoT components and devices in the IoT context. Additionally, the security of IoT systems, ML or DL methods is also a very crucial aspect to be considered.

Moreover, smart manufacturing involves the use of big data analytics to supplement physical science to boost the efficiency of the system and to make decisions. With the expanding use of the Internet of things (IoT) and sensors, the requirement for massive production facts data processing characterized by high-variety, velocity, and volume is becoming increasingly necessary. Deep learning is also becoming more popular. DL offers advanced analytics tools that help process and analyze large-scale data generated from various outputs.

Smart sensor systems make use of classic and evolving ML algorithms, as well as modern computer hardware, to build sophisticated "smart" models that are specially designed for sensing applications and fusing complex communication channels to achieve a more comprehensive appreciation of the device being monitored. An analysis of the recent sensing applications, which harness smart sensor systems using ML, is presented here.

In the information age, advancements in big data and IoT have facilitated the introduction of smart, intelligent, and ML applications. This is because of their huge capacities to store, process, and produce vast volumes of data. These technologies can be considered as fuel for optimum functioning of ML applications. ML is all around us, such as technology, public safety, schooling, medicine, public safety, trailer building, etc.

Cloud laundry is an e-commerce business model based on IoT for large-scale cloud laundry services. This model combines intelligent logistics management, big data analytics, and ML techniques. The model makes use of real-time big data and GPS data changes to assess the suitable transport path, update and, quickly and conveniently, access the logistic terminals redirectly.

With the advent of the IoT, systems are becoming smarter and have thereby increase the volume of linked devices in all facets of a modern city. ML approaches are used to further develop an application with expertise and capabilities as the volume of data collected increases. Many researchers have been drawn to the area of smart transportation, and both IoT and ML techniques have been used.

High-cost volatility may have a direct effect on the stability of the smart grid electricity market. Thus, to prevent severe repercussions of market dynamics, efficient and precise price forecasts must be enforced. This chapter discusses two smart approaches using ML to tackle the issue of electricity price forecasting. First, to predict the hourly price, a Support Vector Regression model is used. Second, it will explore and compare the DL model with the Support Vector Regression model.

Other applications such as smart cities, are aimed at managing urbanization, living conditions of their residents, preserving a green environment, using energy efficiently, enhancing the economy and the capacity of people to use and implement new information and communication technologies efficiently.

ICT plays a critical role in the idea of smart cities in policy formulation, decision-making implementation, and ultimate production services. The major aim of the chapter is to explore the role of DL, ML, and AI in the development of a smart city.

To make the network effective and autonomous, ML techniques are integrated with IoT. One of the forms of ML is DL, and it is computationally difficult and a costly technique. One of the challenges is to integrate DL approaches with IoT to increase the efficacy of IoT applications overall.

Fighting Webspam will address other smart and intelligent applications, where DL uses user data and applies NLP to infer about the emails it encounters. The second is imitation learning, something a person does as a child, which is somewhat close to observational learning. This is widely used in robotics in various fields and sectors such as agriculture, search, building, rescue, military, and others. The third is assistive robots, where robots are capable of processing sensory input in times of need and performing actions. There will be a discussion about the Smart Tissue Autonomous Robot, automatic translation, or recognition where ML is used to translate text into another language and smart applications that playing video games automatically.

5.3 NEED FOR MACHINE LEARNING

We have been generating immense data since the introduction of major innovations. According to a report, people produce 2.5 quadrillion bytes of data per day. It was predicted that by 2020 every person on the planet will produce 1.7 MB of data every second.

Finally, predictive models can be developed with so much data available for research and processing to find important information and provide more accurate outcomes.

Big firms such as Netflix and Amazon use considerable amount of data for development of ML models to find growth opportunities and eliminate unnecessary hazards.

Here, we list all the reasons why ML is so appropriate:

- *Data generation improvement*: Humans need a tool that can interpret, structure, and draw useful lessons from available data as a result of excessive data output. Here is where ML is involved.
- *Boost decision-making*: Using multiple algorithms, ML can be used to achieve a company's business goals. ML is being used to predict revenue, stock market downturns, risks, and phenomena, etc. (Figure 5.3).
- *Discover patterns & data trends*: The most critical aspect in ML is to identify the underlying knowledge and derive key insights from the available data. ML includes searching under the surface and evaluating data on a microscale by creating predictive models and using statistical methods. This will take several days for data to be understood and patterns to be extracted physically, whereas ML approaches can achieve similar calculations within a small amount of time.
- *Complex problem solving*: From discovering the genes linked to fatal amyotrophic lateral sclerosis (ALS) infection to developing self-driving cars, ML can be used to solve some of the most challenging problems.

5.4 MACHINE LEARNING (ML)—DEFINITIONS

- *Algorithm (step-by-step procedure):* A set of rules and analysis tools are termed as ML algorithms as they help learn data patterns and derive important inputs from them. That is the concept behind the model of ML. The Linear Regression algorithm is a ML approach.

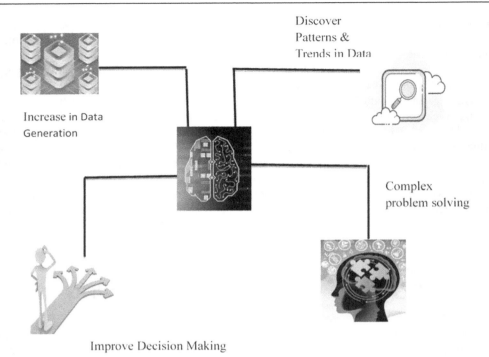

FIGURE 5.3 Importance of Machine Learning

- *Model (prototype):* The most significant element in ML is its design. An ML algorithm is used to train models. To find the effective outcome, an algorithm maps all of the decisions that a model is supposed to make based on the given input.
- *Predictor variable:* It is a data characteristic(s) to estimate the outcome.
- *Response variable:* The predictor variable must be used to estimate the function or the output variable(s).
- *Training data:* Training data is used to develop a ML (prototype) model. It allows the model to recognize significant patterns and trends required for the expected performance.
- *Testing data:* It must be put to the test to see how well it can predict an outcome after the model has been conditioned. This is achieved by the Analysis Data Collection (Figure 5.4).

An ML process starts by inputting a large amount of data in the machine for training. These findings are then used to create a ML model, which is then applied to a problem and solved using an algorithm.

FIGURE 5.4 Basic structure - Machine Learning

5.5 MACHINE LEARNING—PROCESS

- The ML approach necessarily involves the creation of a predictive model that can be used to find a proposed solution. Assume you have been given a problem to deal using ML to better understand the process (Figure 5.5).

Challenge—To predict the rain likelihood in the resident zone using ML.
The steps in an ML process are as follows:

Step 1. Objective definition

At this point, everyone wants to know what needs to be inferred precisely. The goal in our case is to use weather conditions to forecast the likelihood of rain. At this point, detailed notes on what type of information can be used to resolve this challenge or the type of process you need to use to find the solution must also be taken into consideration.

Step 2. Data collection (Gathering)

The following is the list of questions while surveying:

- What type of data is necessary for problem solving?
- What is the feasibility of study?

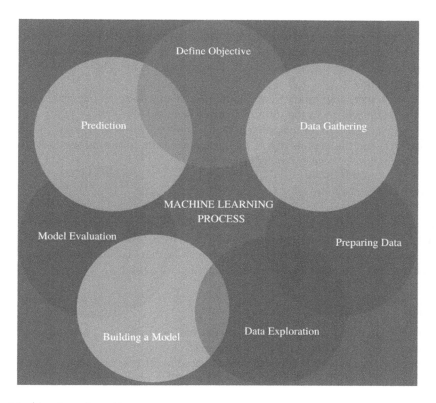

FIGURE 5.5 Machine Learning - Process

- Is data easily accessible?
- How to obtain data?

Surveyor need to understand how this knowledge can be extracted so as to know the kind of data needed. Data processing can be achieved automatically or by data extraction.

The data required for weather forecasting includes factors like pressure, humidity level, location, temperature, etc. This type of information should be compiled and then processed to interpret the data.

Step 3. Data arranging (Preparation)

You never get the details you need in the correct setting. You will encounter several anomalies in the collection of data, such as redundant variables, duplicate values, missing values, etc. It is very important to avoid such discrepancies because they may lead to unfair predictions as well as forecasts. Furthermore, search the dataset for any anomalies at this point and repair them immediately.

Step 4. Analyze exploratory data

This phase is about digging into data and discovering all the hidden secrets. ML's brainstorming stage is Exploratory Data Analysis and Exploratory Data Review. The analysis of data trends and patterns is needed for data exploration. At this point, all of the useful conclusions have been drawn, and the relationships between the variables have been identified.

For beginners, in view of forecasting precipitation, we know that if the temperature has dropped down, there is a strong likelihood of rain. At this stage, such similarities must be mapped and understood.

Step 5. Model building (ML)

Both the observations and patterns produced during data analysis are used to create ML models. Usually, this phase begins by dividing the dataset into two sections: Training data and testing data. The training data will be used to build and evaluate the model. The logic of the model is based on the ML algorithm that has been implemented.

In the case of forecasting rainfall, we might use a classification algorithm such as Logistic Regression, so the result will be in the form of Correct or Incorrect.

Choosing the right algorithm for the problem being attempted to solve is based on the location of the problem, the data set, and the complexity of the problem. In the following sections, we highlight the various types of problems that can be solved using ML.

Step 6. ML model evaluation

Following the creation of a model, it is able to place it to the test by collecting training data. The research data set is used to assess the model's efficacy and ability in predicting the outcome accurately. After the accuracy has been measured, the model can be improved further. To improve the model's efficiency, techniques such as cross validation and feature selection can be used.

Step 7. ML predictions

The model is finally used to make predictions after it has been validated and strengthened. The final result may be a categorical variable or a fixed quantity.

In our case, performance will be the categorical variable for estimating the probability of rainfall. But that was the whole method of ML. Now it is time to reflect on the various ways in which machines can learn.

5.6 ML TECHNIQUES

The algorithms in ML can be supervised or unsupervised, based on their training mode. If we indicate the expected performance, also called label, target, or answer, for each input while training our model, the algorithm is regulated. In this case, if the output has a numeric value among an infinite number of possible values, we talk about a regression problem. In comparison, if the value is discrete or finite, each output is called a class, and the problem to be solved is that of classification. The problem is called binary classification (yes/no, true/false, etc.), if there were only two outcomes.

The algorithm, on the other hand, is unsupervised if the predicted outputs for the various inputs are not taught while training our model. In this case, the algorithms that are not monitored are helpful in understanding the data organization and data structure. They are usually used to organize the inputs into groups or clusters or to reduce dimensionality. It is difficult to test unsupervised algorithms as there are no predicted outcomes to compare a prediction with.

The definition of ML is old. AI is thoroughly linked to ML. Through ML, AI becomes viable. Computer structures learn via ML to accomplish tasks such as grouping, forecasting, clustering, identification of patterns, etc. Usually, the sample data is specified by measurable features, and an ML algorithm takes effort to link the features to unique output values known as labels [2]. The knowledge collected during the learning process is then used to identify trends or make decisions based on new data. Clustering, sorting, regression and deciding association laws are all problems that can be solved using ML. It is possible to divide ML algorithms into four groups, depending on the learning style:

- *Supervised learning:* Supervised learning, using algorithms such as Linear Regression or Random Forest, solves regression problems such as population growth prediction and predicting real experience, weather forecasting. Additionally, with the help of algorithms like Random Forest, Nearest Neighbor, Support Vector Machines, and others, supervised learning solves classification problems such as identifying fraud detection, digital recognition, diagnostics, and speech recognition. In supervised learning, there are two levels. The preparation process and the stage of research. The data sets used in the training process must contain known labels. The algorithms attempt to determine the test data output values by studying the relationship between the parameter functions and the labels [3].
- *Unsupervised learning:* This type of learning deals with issues including the decrease in dimensionality used for big data representation and for evocation of identification of unseen structures or features. Furthermore, this method is being used to cluster issues like targeted marketing, recommendation systems and customer segmentation. There are no labels available in this form, unlike supervised learning. Unsupervised learning algorithms seek to classify data by predicting future values, testing patterns or clustering the data [3].
- *Semi-supervised learning:* Semi-supervised learning is a blend of supervised meaning labeled as well as unsupervised meaning unlabeled classifications. It primarily deals with the consequences that a less quantity of labeled data may have, such as unsupervised learning [2].
- *Reinforcement learning:* The algorithms in this learning style attempt to predict the performance of a problem using a series of tuning parameters. After that, the expected output is used as an input variable, and a new output is calculated before evaluating the best output. DL and ANNs use this style of learning, which will be discussed later. This type is mostly used for applications such as AI gaming, reinforcement learning, real-time judgment, robot navigation, and skill acquisition [3]. There are two key parameters to remember when using ML techniques: How fast a given technique and how computer-intensive it is. Based on the application form, the best ML algorithm is selected. For example, if real-time analysis is needed, the algorithm chosen must be able to track changes in the input data and produce the desired output in a reasonable time (Figure 5.6).

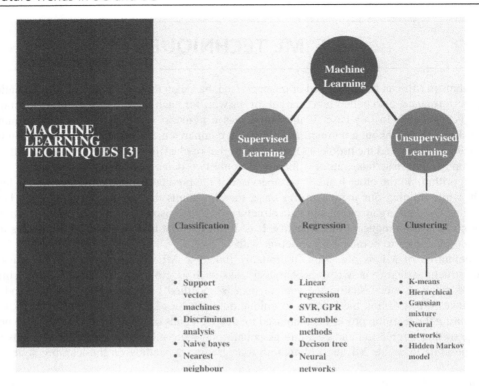

FIGURE 5.6 Machine Learning Techniques

5.6.1 Supervised and Unsupervised Techniques

The following section describes some supervised and unsupervised techniques.

Support vector machine (SVM): SVM is a widespread supervised learning approach which is extensively used in pattern recognition and classification. This was developed by Vapnik and others [4,5]. The basic idea behind SVM is to transform the input into a n-dimensional space where features can be found. Various types of kernel functions, like radial, polynomial, and linear, are used to achieve this mapping. The classification accuracy is influenced by the kernel function selection which is an significant SVM job. The training dataset affects the kernel function selection. If the dataset is separated linearly then the linear kernel function works; radial basis function (RBF) and polynomial kernel are the two common kernel functions. In most instances, the RBF-based SVM classifier outperforms the other two kernel functions [6,7]. To improve the margin between different groups, SVM intended to identify a different hyperplane in the feature space. For each class, the margin is the distance between the data points and the hyperplane nearest to it. Support vectors are described as the neighboring closest data points. Figure 5.7 shows an SVM Classifier in action. There are several possible hyperplane separations between two groups, as seen in the figure, but only one perfect separating hyperplane will increase the margin. See [8–10] for a more detailed discussion on SVM.

Descriptive discriminant analysis (DDA): The DDA is a multivariable statistical technique that seeks to analyze whether there are substantial differences between the groups of objects with respect to a collection of variables evaluated on those objects. If they exist, their existence must be clarified and, thus, the procedures for the systematic classification of new findings of unknown origin in one of the groups analyzed [9] should be facilitated. When it is important to describe a qualitative variable based on a certain number of quantitative variables called explained variables or predictors, a discrimination analysis problem is posed. DDA can be interpreted as a set of two goals-oriented methods and statistical

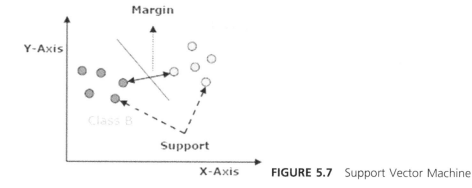
FIGURE 5.7 Support Vector Machine

procedures that may be complementary [10]: 1) Decide whether the variables observed may a priori differentiate (discriminate) the r classes. This goal has a descriptive component and is related to the analysis of the main elements. At this point, it is evident that the construction of two-dimensional representations of individuals, variables, and groups a priori is essential. 2) Construct classification rules to assign one of the classes a priori to each new object. This goal is definitive and is connected to probabilistic techniques. Both the construction of decision rules and the protocols for the subsequent review are essential for this approach.

Bayes' Theory: The conditional probability is used in Bayes' theory to assess the likelihood of an occurrence taking place given prior knowledge of conditions that may be important to the event. The Bayes' theory is defined mathematically by the following equation:

$$P(H|E) = \frac{P(E|H)P(H)}{P(E)} \quad (5.1)$$

Where $P(H|E)$ represents the likelihood of the hypothesis H carrying considering the current proof E, $P(E|H)$ represents the probability of evidence E depending on the hypothesis H, $P(H)$ represents the prior probability of hypothesis H individual of evidence E, and $P(E)$ represents the probability of evidence E. The probability model can be used to measure multiple posterior probabilities for various categories when classifying a new input data set. The data sample will be sorted into the $P(H|E)$ class with the greatest posterior probability. The advantage of Bayes' theory is that learning the probability model only requires a small number of training datasets [11]. Using the Bayes principle, however, there is an implicit presumption of freedom. The features of data samples in the training dataset are considered to be independent of each other in order to measure $P(E|H)$ [12]. See [13,14] and [15] for more details on Bayes' theory.

Neural Network (NN): The human brain, which uses core elements known as neurons to conduct parallel, highly complex, and nonlinear computations, is the inspiration for NNs. The points of a NN refer to the characteristics of biological NNs. Nonlinear computations are performed by these nodes using activation functions. The hyperbolic tangent and sigmoid functions seem to be the most commonly used activation functions [16,17]. Variable link weights bind the nodes in a NN, simulating the connections between neurons in the human brain. ANN is made up of n-layers. The input layer is the first layer and the output layer is the second. The layers between the input and output layers are the ones that are hidden. The output of each layer is the input of the next layer, and the effect is the output of the last layer. By changing the number of hidden layers and nodes in each layer, complex models can be trained to improve the performance of NNs.

NNs are widely used in many applications, such as pattern recognition. The simplest NN has three layers, an input layer, a hidden layer, and an output layer, as shown in Figure 5.8.

K-Nearest Neighbor: The classification of a data set is calculated when using k closest neighbor of the unclassified sample in K-NN, a supervised learning technique. The process of the K-NN algorithm is

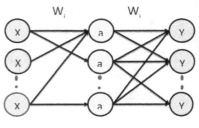

FIGURE 5.8 An input, a hidden, and an output–layer showing a basic neural network

very easy: It refers an unclassified sample to a class if the majority of its K neighbors are members of that class. The higher the K, the less the classification affects the noise. Since spacing is a useful factor in the K-NN algorithm, a number of functions can also be used to calculate the distance between the unlabeled sample and its neighbors, including Euclidean and Euclidean square, City-block, and Chebyshev [18].

Decision tree (DT): The DT is a supervised ML method which is used for classification. Each leaf node represents a class code, while each branch describes the set of features that leads to classifications. By comparing the key parameters of an unlabeled sample to the DT nodes [19,20], an unlabeled sample can be defined. The DT offers a number of benefits, which include intuitive knowledge expression, ease of use, and high classification precision. The widely used DT algorithms for automatically classifying training datasets are C,5.4, ID3, CART by the authors [21–23], respectively. The dividing criteria used to create DTs are the most important differences between them. The splitting parameters used by C4.5, CART, and ID3 are information benefit, impurity, and gain ratio, respectively [24] gives a complete evaluation of the three DT algorithms.

5.7 APPLICATIONS OF MACHINE LEARNING

- **Customized Endorsement Engine of Netflix**: Its infamous recommendation engine is the cornerstone of Netflix. Netflix recommends over 75% of what you watch and these suggestions are based on data interpretation by ML.
- **Auto tagging showcase of Facebook:** Face testing by the DeepMind system of Facebook is focusing on NNs and ML. DeepMind evaluates a photo's facial characteristics to classify your relatives and family.
- **Alexa of Amazon**: Alexa, a virtual assistant that is based on ML and NLP, is a high-level virtual assistant that does more than playing songs from a favorite playlist.
- **The Spam Blocker on Google**: ML is used by Gmail to filter spam messages. It evaluates emails in real time and classifies them as spam or not using algorithms based on NLP and ML.

5.8 APPLICATIONS OF DEEP LEARNING

We would never have imagined a DL software a few years ago that would be used in self-driving cars and virtual assistants such as Alexa, Google Assistant, and Siri. All these inventions are part of our daily life now. With its infinite possibilities, such as fraud detection and pixel recovery, DL continues to fascinate us. Let us further understand the applications through industries with the DL usage.

5.8.1 Self-Driving Cars

Autonomous driving is now possible due to DL. A billion data sets are placed into a computer to construct a model, then the machines are trained to learn and the outcomes are checked in a secure environment. Pittsburg and the Uber Artificial Intelligence Labs are focused on more than just making mundane driverless vehicles; they are integrating many clever features with the usage of cars without drivers, such as food delivery choices. Engineers working on self-driving cars face the most challenging task of all: Coping with extraordinary circumstances. A regular test and implementation cycle, which is the characteristic of DL algorithms, guarantees safe driving as the number of scenarios is increased. Sensor, camera, and geo-mapping data are used to construct easy and sophisticated models for traffic navigation, route identification, pedestrian-only roads, signage, and real-time components like road blockages and traffic volume.

5.8.2 Fraud News and Aggregation News Detection

This method scans all the horrible and hideous news in a newsfeed. Enormous use of DL in aggregation of news supports attempts to alter news according to a person's preference. While it may not seem new, filtering news based on social, economic, and geographical criteria, as well as the individual preferences of a reader to identify a reader person, introduces new levels of complexity. Moreover, in today's world, everywhere, the internet has become the key source of both true and/(or) false facts; the ability to detect fraud news is a valuable asset. Identifying fake news becomes increasingly difficult as bots instantly replicate it through networks.

5.8.3 Natural Language Processing (NLP)

Understanding the dynamics of a language, whether grammar, semantics, sarcasm, tonal nuances, or gestures, is one of the most difficult tasks for humans. As a result of continuous training since birth and exposure to different social environments, individuals may develop acceptable responses and a personalized mode of speech to each situation. NLP, a branch of DL, aims to accomplish the same purpose by teaching machines to identify linguistic nuances and frame unique responses. Document summarization is widely used and checked in the legal profession, making paralegals obsolete. Answering questions, emotion analysis, text classification, twitter analysis, and language modeling at an extensive level are all subsets of NLP where DL is getting more attraction. Distributed representations, reinforcement learning, memory augmentation, convolutional neural networks, recurrent, and recursive neural networks strategies have all been used to construct composite, time-taking models.

5.8.4 Virtual (Tacit) Assistants

Virtual assistant is one of the popular applications of DL, including digital assistants such as Alexa, Google Assistant, and Siri. Each contact with these supporters helps them to know about the voice and vocabulary, giving you a second chance to communicate with people. DL is used by virtual assistants to learn more about their topics, from your eating habits to your favorite locations and songs. They need to understand the directives to carry them out by determining natural human language. Another capability that virtual assistants have is the ability to convert your voice to text, write notes, and arrange appointments for you. Since they can do everything from run errands to automatically responding to your specific calls to arranging activities between you and your team members, digital assistants are basically your Alex Jones. Digital assistants will help create or send accurate email copies using DL applications like text creation and document summaries.

5.8.5 Entertainment

IBM Watson is used by Wimbledon 2018 to evaluate expressions and movements of the players to create telecast moments based on hundreds of hours of footage. This has saved a lot of time and money as a result. DL, to come up with a more detailed outline, can differentiate the audience reaction and player or match performance. Amazon and Netflix are developing their DL competences to create a more modified practice for their customers by developing personals that take into account show preferences, access time, history, and other variables to suggest shows that a particular viewer enjoys.

5.8.6 Visual—Recognition

Consider working through a slew of childhood photos that drive you an unhappy nostalgic rabbit hole. You want to frame some of them, but first you need to figure them out. The only way to do this back in the days when we lacked metadata was to do it manually. The best you could do was group them by date, but downloaded photos lack metadata as well. DL is available now. Images can be organized with the help of locations in the pictures that are known, names, a mix of individuals, events, dates, and so on. The quest for a specific image from a library uses cutting-edge visual recognition systems with multiple layers to classify items from basic to advanced. Deep neural network visual recognition boosts development in the digital media measurement types, according to a large-scale image.

5.8.7 Fraud Detection

Another sector that can benefit from DL is finance and banking, as money transactions grow increasingly digital, it is struggling with the issue of fraud detection. Credit card fraud is being tracked using Keras and Tensorflow auto-encoders, which would save financial institutions billions of dollars in recovery and insurance premiums. The prevention and detection of fraud is focused on the recognition of consumer purchase trends and credit ratings, anomalous conduct, and outliers. ML methods for NNs, classification as well as regression are used to detect fraud. Although ML has been most frequently seemed to classify cases of fraud concerning human reevaluation, DL seeks to reduce such attempts by scaling.

5.8.8 Healthcare

NVIDIA says that the entire healthcare sector is undergoing a revolution, from medical imaging to genome research to drug discovery, and General Processing Unit computing is the core of it. GPU-enhanced software and services offer new competences and resources to doctors, healthcare practitioners, and researchers who are passionate about bettering the lives of others. DL-based applications, which are gradually increasing advancements in the field of healthcare—helping in initial, precise, and rapid diagnosis of chronic diseases, are supporting to address the healthcare industry's shortage of reliable doctors and healthcare providers, to standardize pathology results and treatment courses and to understand genetics in a better way to predict the future risk of disease and adverse health effects, which also, ultimately, save millions in expenses. However, DL and NNs in public healthcare services are lowering costs while reducing health risks associated with readmissions. Regulatory authorities are gradually using AI in clinical research to find solutions for incurable disorders, but doctors' scepticism and a lack of broad data sets are both stumbling blocks to the application of DL in the field of medicine.

5.8.9 Personalization

To provide personalized experiences to their customers, many industries are now using chatbots. DL enables e-commerce behemoths like Alibaba, E-Bay and Amazon, among others, to offer seamless, personalized services such as product reviews, customizable bundles, and discounts, as well as to spot big revenue opportunities during the holiday season. In addition, recovery in new markets is accomplished by introducing goods, deals, or outlines, which lead to micro market growth. E-learning self-systems are on the rise, and successful workflows have been made available all due to internet's resources that were previously only visible and usable at a single point in time. Robots, which specialize in different activities, help customize experiences in real time by providing you with the best resources available, whether it is insurance plans or any customized food.

5.8.10 Detecting Developmental Delay in Children

Children with oration delays, schizophrenia, and behavioral insanity can struggle to maintain a better living standard. Early identification and care may have a positive impact on the emotional, mental, or physical well-being of children of varying abilities. As a consequence, identification and treatment of such problems in babies and young children is one of the most noble applications of DL.

5.8.11 Black and White Image Colorization

Picture-colorized version, which takes grayscale images as input and generates colored images as output, represents the semantic of input colors and tones. This method, assessing the complexity of the situation, was historically discovered by humans with considerable human efforts. However, today's DL technology, like the manual method, is now capable of coloring the picture by applying it to objects and their context within the photograph. The image is recreated with the inclusion of color using increased convolutional NNs in supervised layers (Figure 5.9).

5.8.12 Adding Sounds to Silent Movies

Synthesizing sounds to fit silent images is a function of both LSTM recurrent neural networks and CNNs. A DL model has been used to link video frames with a repository of pre-recorded signals to select appropriate signals for the situation. This is achieved by training 1,000 videos with drum sticks that strike and emit different sounds on various surfaces. DL models can then use these videos to predict the best sound for the game. Later, a test will be built to get the best output in deciding whether the sound is fake or real.

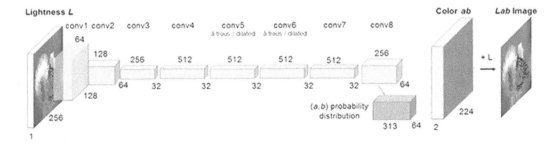

FIGURE 5.9 Classification (Black images and White images)

5.8.13 Machine Translation that Is Automatic

CNNs are effective in labeling photos of letters which can be named. They can then be converted into text, thus interpreted and replicated along with an image by the help of translated text after they have been discovered. Instant visual translation is the name of this technique. Automated conversions from one language with a given word, expression, or a sentence in another language is also a feature of this application. Regardless of the fact that automated machine translation has existed for quite some time, DL outperforms it in two areas:

1. Text conversion that is completed automatic
2. Image translation by computer

Text translations are typically performed without a pre-processing stage. This aids the model's learning of word connections and mapping into a different language.

5.8.14 Handwriting Generation by Machine

The introduction of innovative handwriting sets for a given word or sentence corpus is part of this DL framework. When taking samples, handwriting is generally represented as a sequence of data points generated by a pen. New examples are established as a result of the discovery of the relationship between the movement of the pen and the letters.

5.8.15 Playing Games Automatically

Here, a text database is decoded, and new text is generated character-by-character or word-by-word. DL technique can learn to read, spell and even interpret the style of a text in corpus phrases. To learn text generation from objects in sequences of input strings, broad recurrent NNs are commonly used. However, using a character-based model that produces one character at a time, LSTM recurrent NNs have recently showed excellent achievement in this issue. Here are a few examples of the application, according to Andrej Karpathy:

1. Essays about Paul Graham
2. Shakespeare Business
3. Papers from Wikipedia
4. Geometry of Algebraic
5. Source Code for Linux
6. Names for Newborn

5.8.16 Image–Language Translations

The image–language translations provide a fascinating DL application. With the Google Translate app, photographic images with text can now be automatically converted into the language of your choice in real time. Computer uses a DL network to read the image, then recognize the image, and then translate it into text in the desired language. The person has to put the object's top under the camera. This is a very useful application since languages would gradually become extinct, allowing universal human communication (Figure 5.10).

FIGURE 5.10 Example image-language translation.

5.8.17 Restoration of Pixel

Before DL, the thought of swooping into videos over their resolution seemed unimaginable. Researchers at Google Brain trained a DL network to identify a person's face from very low-resolution photos of faces in 2017. The method is now known as Super Resolution of Pixel Recursive method. It improves image resolution dramatically, identifying prominent features to the point that only the identification of personalities is required.

5.8.18 Photo Descriptions

Computers prefer to identify images automatically. Facebook produces galleries of tagged photos, smartphone uploads, and timeline images, for example. Similarly, for simplified searches, Google Images automatically marks all uploaded photos. These are, however, merely stickers. DL takes several steps forward and onto another stage. It has the capacity to identify in an image all the current elements. DL is used by Li Fei-Fei and Andrej Karpathy to train a network to identify hundreds of incredible places in a picture and compose a phrase to describe each one. This implies that the machine learns not only how to identify the elements in the image, as well as how to describe them using English grammar.

5.8.19 Deep Dreaming

Researchers from Google discovered an approach that enhanced computer image features using DL networks in 2015. Although the Deep Dreaming idea is basically used in one of the DL applications today, it is used in a number of ways. As the name implies, this technique allows the machine to hallucinate over an actual image, resulting in a reconvened dream.

Deep Dreaming method was used by a researcher's group at the Sussex University to construct a virtual reality vision device that permits users to encounter psychopathological symptoms or psychoactive drugs. Furthermore, by employing deep NN algorithms, this fantastic effort opens up the possibility of more induced dreaming experiences (Figure 5.11).

5.8.20 Demographic and Election Predictions

The author used a DL network and observed what a DL network could do with 1.5 billion Google maps images. As expected, the results were outstanding. It was possible for the machine to understand and identify the cars with their specifications and provisions. It was able to recognize over millions of vehicles, as well as their make, model, body style. and year of manufacture. The discoveries there were not halted as a result of the progress of this DL capability.

FIGURE 5.11 Deep learning thinking (before and after).

SUMMARY

Initially, introduction and background/motivation are discussed. The various machine learning (ML) and deep learning (DL) techniques are highlighted. They can be implemented in solving problems associated with the integration and management of smart techniques. In the next part, some applications regarding the security and efficiency of various applications based on machine and deep learning are shown. Further, a comparative evaluation of characteristics is established to choose an algorithm. Finally, a series of conclusions is included.

The chapter also explained how these techniques can be implemented in solving problems associated with the integration and management of smart techniques. Some applications regarding the security and efficiency of various applications based on machine and deep learning are described.

REFERENCES

[1]. N. H. Khan and A. Adnan, "Urdu optical character recognition systems: Present contributions and future directions." *IEEE Access*, vol. 6, pp. 46019–46046, 2018. doi:10.1109/ACCESS.2018.2865532.

[2]. M. Mohammed, M. B. Khan, and E. B. M. Bashier, *Machine Learning: Algorithms and Applications*. Boca Raton, FL: CRC Press, 2016.

[3]. M. Kubat, *An Introduction to Machine Learning*. Cham, Switzerland: Springer, 2017.

[4]. MathWorks, "Aprendizajeautomático: Tres cosas que es necesario saber - MATLAB & Simulink." [Online]. Available: https://la.mathworks.com/discovery/machinelearning.html. [Accessed: 23-Jun-2018].

[5]. V. N. Vapnik and V. Vapnik, *Statistical Learning Theory*. New York: Wiley, vol. 1, 1998.

[6]. B. Yekkehkhany, A. Safari, S. Homayouni, and M. Hasanlou, "A comparison study of different kernel functions for SVM-based classification of multi-temporal polarimetry SAR data," *Int. Arch. Photogramm. Remote Sens. Spat. Inf. Sci.*, vol. 40, no. 2, p. 281, 2014.

[7]. A. Patle and D. S. Chouhan, "SVM kernel functions for classification," in *Proc. IEEE ICATE'13*, Mumbai, India, pp. 1–9, Jan 2013.

[8]. I. Steinwart and A. Christmann, *Support Vector Machines*. New York: Springer Science & Business Media, 2008.

[9]. M. Martınez-Ramon and C. Christodoulou, "Support vector machines for antenna array processing and electromagnetics," *Synthesis Lectures on Computational Electromagnetics*, vol. 1, no. 1, pp. 1–120, 2005.

[10]. H. Hu, Y. Wang, and J. Song, "Signal classification based on spectral correlation analysis and SVM in cognitive radio," in *Proc. IEEE AINA'08*, Okinawa, Japan, pp. 883–887, March. 2008.

[11]. G. E. Box and G. C. Tiao, *Bayesian Inference in Statistical Analysis*. New York: John Wiley & Sons, vol. 40, 2011.

[12]. M. A. Alsheikh, S. Lin, D. Niyato, and H. P. Tan, "Machine learning in wireless sensor networks: Algorithms, strategies, and applications," *IEEE Commun. Surveys Tutorials*, vol. 16, no. 4, pp. 1996–2018, Fourth Quarter 2014.

[13]. X. Wang, X. Li, and V. C. M. Leung, "Artificial intelligence-based techniques for emerging heterogeneous network: State of the arts, opportunities, and challenges," *IEEE Access*, vol. 3, pp. 1379–1391, 2015.

[14]. A. L. Buczak and E. Guven, "A survey of data mining and machine learning methods for cyber security intrusion detection," *IEEE Commun. Surv. Tut.*, vol. 18, no. 2, pp. 1153–1176, Second Quarter 2016.

[15]. E. Hodo, X. Bellekens, A. Hamilton, C. Tachtatzis, and R. Atkinson, "Shallow and deep networks intrusion detection system: A taxonomy and survey," arXiv preprint arXiv:1701.02145, 2017.

[16]. S. Haykin and N. Network, "A comprehensive foundation," *Neural Netw.*, vol. 2, no. 2004, p. 41, 2004.

[17]. K. Lee, D. Booth, and P. Alam, "A comparison of supervised and unsupervised neural networks in predicting bankruptcy of Korean firms," *Expert Syst. Appl.*, vol. 29, no. 1, pp. 1–16, 2005.

[18]. T. Cover and P. Hart, "Nearest neighbor pattern classification," *IEEE Trans. Inf. Theory*, vol. 13, no. 1, pp. 21–27, Jan. 1967.

[19]. L. Breiman, J. Friedman, C. J. Stone, and R. A. Olshen, *Classification and Regression Trees*. CRC Press, 1984.

[20]. J. Han, J. Pei, and M. Kamber, *Data Mining: Concepts and Techniques*. Amsterdam: Elsevier, 2011.

[21]. J. R.Quinlan, *"Induction of decision trees,"* Mach. Learn., vol. 1, no. 1, pp. 81–106, 1986.

[22]. S. Karatsiolis and C. N. Schizas, "Region based support vector machine algorithm for medical diagnosis on Pima Indian Diabetes dataset," in *Proc. IEEE BIBE'12*, Larnaca, Cyprus, Nov. 2012, pp. 139–144.

[23]. W. R. Burrows, M. Benjamin, S. Beauchamp, E. R. Lord, D. McCollor, and B. Thomson, "CART decision-tree statistical analysis and prediction of summer season maximum surface ozone for the Vancouver, Montreal, and Atlantic regions of Canada," *J. Appl. Meteorol.*, vol. 34, no. 8, pp. 1848–1862, 1995.

[24]. A. Kumar, P. Bhatia, A. Goel, and S. Kole, "Implementation and comparison of decision tree based algorithms," *Int. J. Innov. Advancement Comput. Sci.*, vol. 4, pp. 190–196, 2015.

Key Parameters in 5G for Optimized Performance

6

Dhanashree A. Kulkarni[1] and Anju V. Kulkarni[2]

[1]Department of Electronics and Telecommunication, Assistant Professor, Dr. D. Y. Patil Institute of Technology, Pimpri, Pune -18.
[2]Department of Electronics and Telecommunication, Professor and Dean (Research), Dr. D. Y. Patil Institute of Technology, Pimpri, Pune -18.

6.1 INTRODUCTION

During the past four decades, mobile network era had a tremendous change with a gap of approximately every 10 years in wireless communication. The various emerging technology in mobile networks are now a part of our lives, society, and economy too. The First Generation (1G) of mobile cellular technology was commercialized in 1980 with analog technique catered to voice communication with FDMA. The Second Generation (2G) came up in 1990 with digital voice using TDMA/CDMA with a data rate of 64 kBps. With this 2G technology next commercialization were e-mails, web browsing named as (2.5G) technology. Even today there are millions of subscribers using 2G technology. It is one of the most important evolutions in the mobile communication system. In 2002 with Third Generation (3G) with digital voice, data was predesigned into the system and was into the circuit switch networks with WCDMA technology which brought applications like mobile television (TV), video conference, fast downloading with a data rate of up to 2 Mbps. Currently, Fourth Generation (4G) systems, LTE and WiMAX, use OFDMA technology with packet switched network and provides a data rate of up to 200 Mbps. This brought the applications like cloud computing, wireless communication is shown in Figure 6.1. Now all the researchers are working towards the next and the biggest transformation in the cellular networks, i.e., Fifth Generation (5G) mobile networks.

Why Do We Need 5G?

The 5G wireless network will address the evolution from mobile internet to Internet of Things (IoT)-based devices. 5G is a new generation of wireless communication which will improve the existing

DOI: 10.1201/9781003175155-6

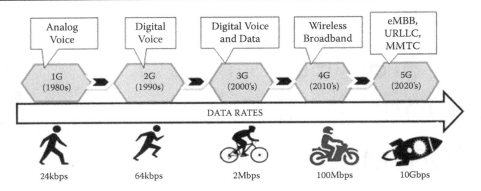

FIGURE 6.1 Evolution of wireless communication.

broadband technology with the increase in the data rates but at the same time with massive number of connections. The increase in subscriptions, data consumption demanded a system which can sustain with more number of customers and the data which will be initiated with this technology. Following are the reasons we need 5G:

Subscription Trends: According to the case study by Global Analyst of Mobile Networks [1] in the year 2017–2018 addition of 170 million subscribers were noted. Figure 6.1 gives an insight of the subscription trends in the global market. In developing country like India the subscribers from 439.9 million is expected to increase to 916.6 million in 2027 with a CAGR of 6.3% from 2019 to 2027 [2].

Data Consumption Trends: The statistics shows a rise of data consumption in India from 2017 to 2022 will be growing at a CAGR of 72.6% [3]. We need a system which can handle a huge amount of increase in data volume. With the data consumption it is already IoT trends which are also contributing to the 5G system. This explains a great need of an era which has capability to handle all such communication. With this increase in demand Mobile and wireless communications Enablers for Twenty-twenty (2020) Information Society (METIS) 2020 in 2013 laid of the following requirements:

- 1000 times higher mobile data volume per area.
- 10–100 times higher number of connected devices.
- 10–100 times higher typical user data rate
- 10 times longer battery life for low power MMC
- 5 times reduced End-to-End (E2E) latency

These all should be achieved at the similar cost and energy consumption, i.e., 2013 networks.

Chapter Organization

Section 2 provides the overview of the 5G technology. Section 3 provides the network optimization parameters in the 5G system. Section 4 provides a survey on the latency parameter of the 5G technology and we have also discussed the various ways to reduce the latency in 5G cellular networks. Section 5 focuses on the energy efficiency (EE) aspect and different areas in which we can work on this aspect. Section 6 focuses on the capacity issue of the 5G system and discusses the work related to the solution for capacity. Section 7 is the state-of-the-art of the chapter which focuses on network softwarization. Section 8 categorizes network slicing with different technologies and focuses on the recent trends and also tries to relate the effect on these key parameters with network slicing. In the last section, we discuss about the future challenges with regard to all the parameters discussed in the chapter. The diagrammatic representation of all the sections is given in Figure 6.2.

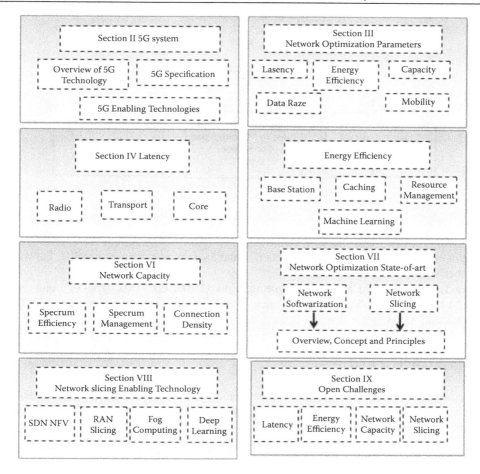

FIGURE 6.2 Diagrammatic view of the origination of the survey.

6.2 5G SYSTEMS

6.2.1 Overview of 5G System

5G network is considered the evolution of the next-generation mobile networks. If compared with the LTE or LTE advanced, 5G will provide the key performance parameters according to the requirement. Thus, 5G network will not only work for mobile applications, but also is expected to help connect all the cities through one network with not only the mobile users but also other devices and machines. In this regard, the International Telecommunication Union (ITU) broadly categorized the 5G network into three horizons as shown in Figure 6.3.

- **Enhanced Mobile Broadband (EmBB):** High data rate and greater coverage for mobile cellular devices. This also covers watching 4K or 8K videos, watching virtual reality, streaming video games. This can also provide fixed wireless access.
- **Massive Machine Type Connections (MMTC):** This includes machine-to-machine communication, i.e., IoT devices at a very large scale. It is designed to provide wide area coverage.

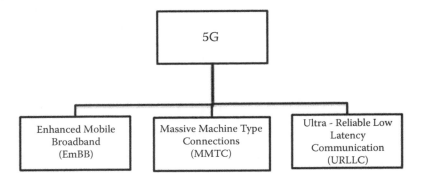

FIGURE 6.3 Use cases of 5G system.

This will bring very new business opportunities in Industry 4.0, agriculture, logistics, etc. This type of communication requires low data rate and data volume, long battery life, but with large number of devices.
- **Ultra-Reliable Low Latency Communication (URLLC):** This type of communication should be highly reliable and low latency. This includes remote level surgeries, V2V communication, autonomous driving, etc.

6.2.2 5G Specifications

Table 6.1 shows the International Mobile Telecommunications-2020 (IMT 2020) requirements and also compares them with LTE advanced requirements which were very high [M-2083]. Following are the 5G requirements:

- **User Experienced Data Rate (100 Mbps):** The number of bits transferred per second.
- **Spectrum Efficiency (Downlink-30 bps/Hz, Uplink-15 bps/Hz):** Average data throughput per unit of spectrum resource and per cell (bit/s/Hz).
- **Mobility (500 km/hr):** Maximum speed at which a defined QoS is achieved and seamless transfer between radio nodes.
- **Latency (1 ms):** The time from when the source sends a packet to when the destination receives it.
- **Connection Density (10^6 devices/km^2):** The number of devices connected per km^2.

TABLE 6.1 Comparative table of LTE-advanced and 5G requirements

PARAMETERS	LTE-ADVANCED	METIS REQUIREMENTS FOR 5G	5G REQUIREMENTS
User experienced data rate	10 Mbps	10 times higher	100 Mbps
Peak data rate	1 Gbps	20 times higher	20 Gbps
Connected devices	2000 devices/ km^2	500 times higher	10^6 devices/km^2
Latency	10 ms	10 times less	URLLC –1 msEmBB –4 ms
Mobility	350 km/hr		500 km/hr
Area capacity	1 Mb/s/m^2	10 times higher	10 Mb/s/m^2
Spectral efficiency	UL –15 bps/HzDL –6.75 bps/Hz	20 times better	UL –30 bps/HzDL –15 bps/Hz

- **Network Energy Efficiency (100 times):** The performance per unit energy consumption.
- **Area Traffic Capacity (10 Mbit/s/m^2):** Total traffic throughput served per geographic area.

Table 6.2 depicts the key capabilities important for the different use cases, where some key capabilities are significant [M-2083]. Table 6.3 shows the 5G applications and their requirements.

6.2.3 5G Enabling Technologies

The 5G mobile network is a global wireless standard which comprises many enabling technologies to fulfill the user requirements. These technologies are the solution for the transformation of the wireless connectivity [4]. Figure 6.4 depicts the technologies of 5G system and Table 6.4 depicts the 5G requirements and their enabling technologies. Below are the enabling technologies:

- Wireless Software Defined Networks(WSDN)
- Network Function Virtualization (NFV)
- Ultra-Dense Networks (UDN)
- Device-to-Device Communication(D2D)
- Cloud Computing
- Massive Multiple Input Multiple Output (MIMO)

TABLE 6.2 5G Requirements according to the use cases

PARAMETERS CATEGORY	LATENCY	MOBILITY	SPECTRUM EFFICIENCY	USER EXPERIENCED DATA RATE	PEAK DATA RATE	AREA TRAFFIC CAPACITY	ENERGY EFFICIENCY	CONNECTION DENSITY
eMBB	Med	High	High	High	High	High	High	High
URLLC	High	High	Low	Low	Low	Low	Low	Low
mMTC	Low	Low	Low	Low	Low	Low	Med	High

TABLE 6.3 5G applications

5G REQUIREMENTS	VERTICALS
• Large bandwidth • Low latency	Virtual reality
• High data rate • Greater coverage	Smartphones
• Large bandwidth • Low latency	Education
• Broad coverage • Low latency	Automotive
• Low power • Low latency • High data rate	Healthcare
• High reliability • Broad coverage • Low latency	Smart energy

120 Future Trends in 5G and 6G

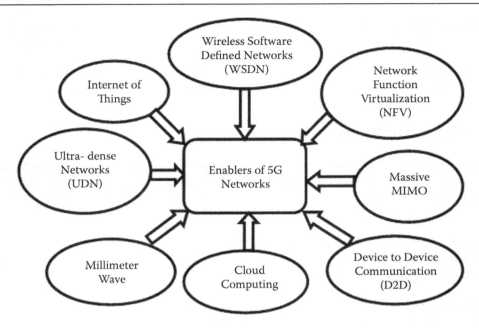

FIGURE 6.4 5G enabling technologies.

TABLE 6.4 5G enabling technologies

5G REQUIREMENTS	ENABLING TECHNOLOGIES
• High data rate • low latency • Less network bandwidth	Wireless Software Defined Networks (WSDN)
• This will virtualize physical infrastructure allowing various virtual networks to enable network slicing in 5G	Network Function Virtualization (NFV)
• Better user experience • High network capacity	Ultra-Dense Networks (UDN)
• Increase in the capability • Energy efficiency	Device-to-Device Communication (D2D)
• Cloud computing can enable hotspots, mobile internet connectivity in the area where there is lack of connectivity	Cloud Computing
• Increase the capacity • High data rate	Massive MIMO
• High data rate • Large capacity	Millimeter wave (mmWave)

- Millimeter Wave
- Internet of Things (IoT)

Wireless Software Defined Networks (WSDN): SDN is an approach to design, build, and manage networks. This approach is obtained through control and data plane, where control plane is programmable. In 5G, SDN provides an intelligent architecture for network programmability through which the network would gain better data flows, low latency, and minimize network bandwidth.

Network Function Virtualization (NFV): NFV is designed so as to separate hardware and software. This will enable the service providers to add firewall, virtual machines in the network without a new setup of hardware. In 5G this will virtualize the physical infrastructure allowing various virtual networks to enable network slicing in 5G.

Ultra-dense Networks (UDN): UDN is a network consisting of small cells which are densely populated, i.e., the network with more radio resource with more number of base stations. This technology is one of the most important enabler for better user experience and increase the capacity of the network.

Device-to-Device Communication (D2D): D2D communication enables data exchange between the two devices without network connection. This technology will enable 5G mobile networks to coordinate between each other, which may offload the network and ultimately increase the capability and EE.

Cloud Computing: The digital world demands to store data for each business where Cloud Computing will help to store the data needs with very high data rates. This data needs to be processed and transfer it to the business holders. Cloud computing can enable hotspots, mobile internet connectivity in the area where there is lack of connectivity.

Massive MIMO: Massive MIMO is a technology which uses a group of antennas to transmit and receive data to increase the capacity and data rates in the 5G system.

Millimeter wave (mmWave): mmWave band ranges between 30 and 300 GHz, which is much higher bandwidth when compared with today's cellular networks. This will allow a higher data rate with large capacity which is expected to reach in Gigabits per second.

Internet of Things: The most part of the 5G paradigm are these IoT devices with a greater speed of communication these IoT devices can communicate between each other for required actions.

Table 6.4 gives the overview of the technologies which are used for gaining the 5G requirements. Even though these techniques have their own benefits and detriments, proper standardization of these technologies will finally lead to the 5G network. In the below paper, the work related to these technologies is studied in detail.

6.3 NETWORK OPTIMIZATION PARAMETERS

In all the mobile networks planning the three most important factors are capacity, coverage and quality of service (QoS). Aforementioned 5G specifications are responsible for overall quality of experience (QoE) of the network by the users through balancing of these parameters.

6.3.1 High Data Rates

Along with the increase in number of subscribers in the cellular world, the needs of internet of services on mobile networks have also increased. The applications like high streaming videos, online gaming requires high data rates. The required data rate has increased by 100 times in 5G than the IMT-advanced.

6.3.2 Latency

The latency requirement of 5G system varies with the type of service to be delivered. The latency requirement for eMBB can be of 4 ms where the mobiles are connected. But the applications like mMTC and URLLC the latency expected is 1 ms with more reliability. Latency plays a very vital role in the applications which require real-time interactions like autonomous driving, healthcare. This technical aspect requires basic changes with the frame structure and coding techniques.

6.3.3 Capacity

In 5G system along with the increase in mobile customers there are machines and critical communications which will take place through the same network. The networks traffic capacity needs to be increased by ten times compared with 4G for the traffic diversifications.

6.3.4 Energy Efficiency

EE aspect came into the consideration as one of the important parameters in 5G system. With number of sensors being connected through this system it has to sustain the better battery life as these devices will not always be connected to the base station. EE should be 100 times better. The small cells are the wise options but the interference of signals is also a point of concern.

6.3.5 Mobility

Mobility aspect is very important parameter of 5G system. Mobility management traces the mobile devices and also tracks the mobile system. The mobility can be of two types, on the Spectrum side and on the User Equipment (UE) side. The mobility parameter uses a handover technique where the call should not drop during the changes in the locations of the mobile device.

In this section we have introduced all the parameters of 5G network. Further we give a detailed study of latency, EE, and capacity.

6.4 LATENCY

Advancements in wireless communication have brought about a lot of challenges with 5G system. One of the important parameters is latency. According to the IMT 2020 the latency requirement for URLLC is 1 ms and eMBB is 4 ms. For applications like remote level surgeries, autonomous driving latency is the key parameter which should not be more than the standards. Latency is divided into two parts: 1) User Plane Latency and 2) Control Plane Latency. User plane latency is the area of interest because the delay in the transmission is due to the Radio Access Network (RAN), backhaul, transport, and the core. Total T required for one time transmission in the LTE system is referred as

$$T = T_R + T_B + T_T + T_C;$$

where, T_{Radio} = packet Transmission time,
$T_{Backhaul}$ = Connection time between eNB and Core Network
$T_{Transport}$ = Processing time by the core network
T_{Core} = Time required for data communication between core and cloud.

In 5G System the total T should not be more than 1ms. For this, researchers have prescribed various solutions on different layers of the system. For TRadio enhancements are done on the Physical Layer, TBackhaul, and TTransport in Transport Layer and for TCore in Network Layer. In the following section we have described the relative solutions according to the layers of the system.

6.4.1 Physical Layer Design

Achieving the latency for URLLC applications is very challenging in the Physical Layer. The latency in radio processing is due to the long decoding time. The ways of handling latency in Physical Layer are 1) Frame structure and 2) Coding and Multiple Access Schemes. The present frame structure can be used for the EmBB applications as it requires 4 ms. But the latency of 1 ms cannot be satisfied with the present frame structure that of IMT-advanced. A new frame structure has been discussed in detail through which the packet success probability is improved to improve the latency [5]. The frame structure which has fixed size slots can be used for eMBB, but the concept of minislot [6] and wider sub-carrier spacing in [7] are designed for applications with URLLC. Minislot is the smaller version of the slot. The difference is that there is no need to wait for the slot boundary and it can carry 2, 4 or 7 symbols. This allows the devices to communicate without slot boundaries and thus helps to achieve low latency. The sub-carrier spacing is the technique used for flexible numerology in 5GNR where the smaller slots with larger sub carrier spacing varying from 15 KHz to 240 KHz the lower the latency. The Self Contained Slot in 5G NR is a technique through which data and the corresponding ACK/NACK can all be fit in 500 μsec. In UDN, a sub-carrier spacing of 312.5 kHz is used [7]. The above specified works are all computer simulations study. There are also field test trials of some of the architectures in particular scenarios using 60 kHz sub carrier spacing [8]. The authors in [9] have proposed short Transmission Time Interval (TTI) and more switching points for low latency in Time Division Duplexing (TDD) systems. This flexibility allows for different deployments according to the most suitable properties like for 1) outdoor and small cell deployments with TDD > 3 GHz, 30 kHz of sub-carrier spacing can be used. 2) Indoor wideband deployments with TDD = 5 GHz, 60 kHz of sub-carrier spacing can be used. With this randomness of packet transmission it is very difficult to deal with the collision of packets, the proposed scheme of contention based transmission is the one in which the collision probability is decreased with multi-channel and repetitive transmission with reliable latency [10].

Coding schemes and multiple access schemes also play a very important role in latency related to Physical Layer. There are many orthogonal and non-orthogonal techniques discussed in [11]. Further described technologies are some of the important technologies in 5G system. Sparse Code Multiple Access (SCMA) is a technology which performs modulation and spreading jointly [12]. A joint node binary low-density parity-check codes (LDPC) and SCMA is proposed to improve the processing latency [13]. The polar code technique has advantages over LDPC codes. A method is proposed in [14] for a prior information based on message passing algorithm (MPA) for the next outer loop iteration. The other technique is universal filtered multi-carrier (UFMC) which is known for its robustness and low latency and filter bank based multi-carrier (FBMC) is known for its frequency selectivity. Authors in [15] have proposed a binomial filter for FBMC and Fractional powered binomial filter for UFMC for reducing PAPR to 0.81 dB with 256-QAM level. Generalized frequency division multiplexing (GFDM) is used in 5G for inter carrier interference and inter symbol interference. Authors in [16] used lifting wavelet transform (LWT) with GFDM for latency and peak-to-average power ratio (PAPR). Different methods to reduce latency of interleave division multiple access (IDMA) are in [17,18]. IDMA is proposed in [19] where the proposed scheme works with the multi-user detection directly without deinterleaving the frame which reduces the latency by 50%. The circuit area has reduced by 53% and power consumption by 58%. Latency-sensitive applications suffer from the latency due to its long decoding time. The successive cancellation polar decoder suffers from long decoding time due to its serialized processing. This successive cancellation (SC) polar decoder has also been proven for short-length codes [10,20]. But due to its latency requirement in processing the codes authors in [21] worked on SC parallel polar decoder to reduce the processing time. The authors in [22] have also proposed SC parallel decoder with the change in parallel factor for the merging times fates than that of in [21]. A detailed survey of these schemes is done in Table 6.5 in which application scenarios are also described.

TABLE 6.5 Detailed survey of multiple access techniques w.r.t. complexity and latency

REF. NO.	TECHNIQUE	APPROACH	RELATED FEATURES/ OBJECTIVES		SUITABLE 5G SCENARIO RECOMMENDED
			LESS COMPUTATIONAL COMPLEXITY	LOW LATENCY	
[14]	SCMA	Message passing algorithm is implemented with logarithms	√	√	URLLc
[13]		Joint sparse graph is designed with LDPC and SCMA	√	√	mMTC
[15]	FBMC+ UFMC	Combination of two techniques is used to reduce PAPR	-	√	URLLc
[16]	GFDM	GFDM is utilized with LWT	-	√	mMTC
[23]		GFDM is combined with IM for IM numerology	-	√	mMTC
[19]	IDMA	Proposed a scheme of multi-user detection without deinterleaving the frame	√	√	eMBB, URLLc, mMTC
[24]		The proposed architecture needs 1 cycle to store and process the LLRs	√	√	eMBB, URLLc, mMTC

6.4.2 Transport Layer Design

The present scenario of 4G LTE comprises radio, transport, and core. Transport layer is responsible for significant amount of latency. 5G transport network consist of fronthaul, middle haul, and backhaul system. This transport network plays a very important role in satisfying the data rate with high capacity and maintaining the QoS. The fronthaul system is the link between the baseband unit (BBU) and the remote radio head. The middlehaul is the linkage between the distribution unit (DU) and the centralized unit (CU). The backhaul is the link between the centralized unit and the RAN. Densification of small cell deployments had changed the scenario from wired backhaul system as in 4G LTE to wireless backhaul system in 5G. The proposed scheme in [25] is the joint traffic offloading scheme for eMBB and URLLC. A puncturing technique is used to reduce latency for URLLC. The eMBB request is allotted to satellite backhaul and URLLc to terrestrial networks, whereas wavelength assignment problem is illustrated as a mathematics model to reduce the wavelength tuning overhead for heterogeneous networks [26]. Tailored dynamic bandwidth allocation PON-based architecture in [27] is used to manage the traffic offloading. In the mobile backhaul system there are many protocols working for the synchronization to move from the TDM to packet switch operation which increases the delay in processing, the authors in [28] have proposed a digital signal processing clock distribution and recovery scheme to reduce the latency.

Caching is also one of the efficient approaches to reduce latency by decreasing the backhaul load so as to meet the QoS. Proposed scheme is implemented using Q-learning for D2D caching problem [29]. The authors in [30] have proposed a content-aware cache technique on SDN platform aiming for latency-sensitive applications. The authors in [31] have improved the flexibility of cache placement strategies through which 27% of the total latency for mobile users was obtained. In [32] to reduce latency authors have discussed about the content based networking through the distributed algorithm applied at the backhaul switches.

6.4.2.1 Mobile Edge Computing

Mobile Edge Computing (MEC) is also a technique which is used to reduce latency in the required processing time [33]. MEC has gained a significant interest by all the researchers since it provides a platform to load the applications at the very edge of the mobile network [34]. This also helps in reducing the energy consumption of the mobile devices and also reduces congestion. In [35–37] and [38] single-users and multi-user MEC systems are studied. In [35] and [36] the resource utilization schemes are derived with single-users to minimize the latency. Aforementioned papers have worked on the computational offloading either at the mobile device or at the edge cloud. The authors of [39], [40] have focused on the algorithm for partial offloading but no such system has been formed with all the results. In [41] partial offloading with piecewise convex problem is focused, which is further solved by sub-gradient method to reduce the scheduling time and significantly the weighted-sum latency. In [42] the author has focused on partial offloading and has also contributed in three types of compression: Local compression, edge compression, cloud compression, and partial compression. In local compression data is compressed at mobile devices for which a convex optimization problem is designed to minimize the weighted sum delay. At the edge cloud compression joint resource allocation is formulated. In partial compression offload model piecewise optimization problem has worked on reducing latency at all the parts on MEC.

6.4.3 Network Layer Design

The current 4G system cellular techniques support only the conventional mobile broadband services and cannot support the heterogeneous requirements of 5G system. These SDN, NFV and the backhaul techniques are the three entities which will provide a platform to reduce latency in the core network. These enhancements in the network layer leads to reduce the process time and also bypass many protocol layers.

In today's scenario with so much of devices in the network it becomes a crucial issue about the routing protocols in the backhaul system since they are responsible for the data packet flow and creation of paths. A mesh based backhaul is specified in [43] to calculate the best path through the improvisation in harnessing the control messages so as to find the available neighbor nodes for the performance of network when the traffic overloads. However static mesh backhaul system comes with delay in scheduling and routing algorithms. In [44] a network link scheduling and routing schemes are designed for small-scale backhaul system which is capable to cope up with the frequent link degradation. The SDN/NFV boosts the efficiency of network service deployment. With this technology several isolated subnets are designed for ultra-low latency [45]. In [46] different forwarding routes are selected to satisfy the requirements. This dynamic routing provides significant improvement in latency. LOADng routing protocol is used for generating a path between nodes whenever necessary. In [47] the mobility feature of the protocol is enhanced through control messages and the understanding of short paths to reduce the latency and also increase the packet delivery ratio. In these scenarios the authors have observed the increase in memory consumption.

This SDN technology has also enabled satellite communication through 5G; still there is a research which has started for satellite terrestrial systems. But in satellite communication since the end to end latency is too high due to the limited capacity and uneven distribution of traffic of feeder. For this a load balancing algorithm for fluctuation and a resource allocation algorithm based on dynamic queue is proposed to reduce the latency [48].

6.5 ENERGY EFFICIENCY

In 5G and 5G beyond networks will contain millions of devices connected to it like the mobiles, machines, IoT devices. To support all these devices the 5G system should have 10 times better battery life

[49]. In 4G system, parameters like data rate, latency, and throughput are considered as the prime for network optimization. But in 5G EE is also introduced as the important parameter. Saving the energy only by considering the transmit power will not be able to handle the networks. In 5G, deployments of small cells are the key enabler to achieve the speed required by the user. With this ultra-dense deployment more number of BS is required which corresponds to more energy consumption [50].

In recent years, cell switch-off techniques are the ones used at the base station level so as to save energy at the sites or sector level accessing less number of active devices.

Caching is one of the techniques where many researchers are working where the popular files can be stored to reduce the traffic. Resource allocation or resource sharing has also gained a great importance for EE in the network as these algorithms help manage the resources and reduce the consumption at the network level. Machine Learning (ML) algorithms or deep learning algorithms with predictive analysis are the major demands expected from the network. There are researchers working on the energy consumption through traffic analysis. The study presents the highlights of the papers focusing on EE or SE. The survey can be divided as follows:

5.1. Energy Efficiency at Base Station Level

5.2. Energy Efficiency with Cached Technique

5.3. Energy Efficiency with Resource Allocation

5.4. Energy Efficiency with Machine Learning Techniques

6.5.1 Energy Efficiency at Base Station Level

Deployment of small cells UDN is the most promising way to achieve the high data rates [51]. These networks introduce the challenges of more energy consumption and interference. The most effective way to approach these challenges are cell switch-off techniques online CSO and offline CSO on these small cells. These sleep mode techniques are also used with femtocells to increase the capacity of the network. The below section introduces the recent work in EE at the base station level.

The authors have worked with offline CSO patterns, considering the regular CSO patterns. In general till now researchers have worked on regular CSO but with site level, i.e., only at base station. In [52] the researchers have focused on site level as well as on the sector level. By using a breakpoint they have plotted the SINR to CDF graph. Authors have also stated that EE is dependent on power consumption of the sector per user. In [53] focus of work is on sleep mode concept of femtocells to save the power consumption. In this authors have introduced a concept where femtocells with active user can also sleep if the relocation of active users is possible to other active femtocells considering three cases: 1) Relocation of users, 2) Increase in total data rates offered by these femtocells, 3) Increase in the total capacity of femtocells. Results show that they are able to achieve 5% to 9% more power savings than the conventional methods.

The focus is on the performance of the SE and EE depending upon the intra-tier interference and high energy consumption [54]. Using game theoretic framework, the EE maximization problem is formulated to find the relationship between EE and SE. The proposed algorithm CE2MG outperforms with the decrease in tradeoff and fairness for large number of small cells.

There are many researchers which focus on spectrum utilization for 5G cellular networks [55–57]. Full duplex relaying schemes are best utilized to obtain better outage performance rather than the half duplex schemes through dividing the access link and backhaul link considering the different scenarios like I and Y relaying channels. The other way is wireless transfer of energy from the base station (BS) to radio remote unit (RRU). The challenge for optimizing the transmission power at RRU level for improving QoS with EE can also be done through energy harvesting small cells which can work as personalized active cells for priority users [58–60]. Authors in [61–63] have focused on energy harvesting with full duplex transmission at the base level. The mean field game theory model is used for strategic decision-making in wireless networks. This model is used in [64–66] to energy efficient strategies. In [66] the author has used a hybrid of these techniques to get the better EE of the network. The author has assumed the wireless transfer of energy with full duplex transmission at the RRU, the field game equation has been formulated to find the RRU transmission power strategy (EMFG). This algorithm proves to be more stable for average network coverage probability at different receive signal power, i.e., -100 and 10 dB compared with fixed RRU transmission power strategy.

6.5.2 Energy Efficiency with Cached Technique

5G has many enabling technologies like MIMO, UDN, D2D communication, but all these technologies require a support from backhaul system which increases the overhead on the network. This introduces the concept of caching which priori saves the popular files to reduce the backhaul traffic. The basic idea of content storage at the user end reduces the cost and increases the SE by resisting duplicate transmission of popular files. This strategy subsequently decreases the EE and latency in the network. To improve the EE there are two aspects in the cache technique: Improvement in cache hit rate and optimization of the cache size.

In [67] the authors have elaborated how caching at the base station can prove to be energy efficient. Successful content delivery probability depends on access delivery probability, backhaul delivery probability, and cache hit ratio. Content delivery and request arrival density is considered to derive EE [68]. Small contents like catalogue size and cached contents at small cells base station can improve the EE [69].

In the above study authors have assumed the content popularity is fixed and known. The content popularity is always dynamic, i.e., the frequency of request appears during the certain period of time or at specific location. In [70] the investigation is done on the EE problem by jointly considering the cache hit rate and optimal cache. An optimization problem formulation and a close-form solution of optimal cache capacity is determined for maximizing the EE. The numerical results shows increase in throughput with decrease in power consumption.

6.5.3 Energy Efficiency by Resource Allocation

In the past decade, in the wireless communication system Orthogonal Multiple Access (OMA) schemes were widespread. Later on, Non-Orthogonal Multiple Access (NOMA) with its new perspective allowed multiple users to share the same frequency bandwidth with different power levels. The number of applications of NOMA is the cache-based technique [71]. In [72], the authors suggested the use of index coding (IC) than superposition coding (SC) at the transmitters. The application of XOR operation to the packets, prove that IC requires minimum power to satisfy the data rate than SC. The focus is on the clustering algorithm with optimum power allocation method for clusters. By using Monte Carlo simulation comparison of conventional cache-based NOMA is formed and could successfully gain 79.87% power saving technique which reduces the signaling overhead.

In [73] NOMA has been introduced with grant-free transmission scheduling technique where the user becomes active only when it has data in the buffer. This technique reduces the signaling overhead for low power consumption. There are many researchers working on energy consumption through resource allocation. In [74] for maximization of weighted sum-rate authors have used user selection, channel assignment, and power allocation problem. In [75] the authors have addressed resource allocation problem for maximization of EE by optimal sub-channel and power allocation with perfect CSI; whereas in [76] with imperfect CSI. In all this work authors have addressed data rate model from information theoretic point of view. In [77] focus on energy consumption in power domain NOMA system is applied through linear programming model for optimal transmission scheduling and resource allocation. This has shown an increase in EE by 40% and reduced energy consumption by 96%. While considering the performance evaluation of all the research above focuses on resource allocation problem for EE considering NOMA transmission which is not energy efficient for the users away from the base station.

6.5.4 Energy Efficiency by Machine Learning Techniques

Traditional techniques such as game theory, linear programming model are the models which are used by the researchers for minimizing the power consumption. But, these models always work with some assumptions to achieve the parameters and on the scenario been created [78]. With the growth in more number of connected devices will make the network more complex. This complexity of the networks might not be solved by the traditional techniques.

This increases the need for ML techniques which are able to analyse huge amount of data as well as learn from the data to take corrective actions [79]. Authors in [80] have approached the problem of balancing between the EE and high bandwidth utilization. Proposal of game frame work iterative algorithm is used with learning ability for user equipment association and Orthogonal Frequency-Division Multiple Access (OFDMA) scheduling. This model also take into account the inter cell interference and QoS. Since switch ON/OFF of the small cell is a traditional technique for EE.

Self-Organized Networks (SON) reduce human intervention for management of small cells by integrating intelligence and autonomous adaptability for improvement in overall system efficiency. In [81] SCs independently learn a radio resource management scheme through multi-agent reinforcement learning. By extending this work of proposing an offline trained algorithm by switching ON/OFF policies for the solution on capacity and EE [82]. Two algorithms are proposed: 1) Reckless Algorithm and 2) Neighbor Aware Algorithm. The first algorithm switches OFF the base station and the second predicts the condition of the neighboring BS. These algorithms jointly work for the reduction of energy consumption. The authors have simulated the algorithms on two scenarios, one with 4G and the other with the 5G system [83].

6.6 NETWORK CAPACITY

Cellular communication networks are evolving to keep up with the recent requirements of high data rates. The one way to achieve such throughput is the Massive MIMO concept [84,85] which is the key factor to increase the SE. With this MIMO as spectrum sharing is also important to increase the capacity of the networks. The recent enabling technologies like D2D and NOMA are important for network density. Figure 6.5 shows the trends in 5G network capacity.

6.6.1 Spectrum Efficiency

MIMO is a wireless technology that increases the capacity with multiple transmissions and multiple receptions. The data rate depends on the number of transmit and receive antennas. Many research papers have proposed the highest SE of 145.6 bits/s/Hz based on the channel state information, Shannon capacity calculations, and large number of antennas at the base station. The highest SE can be gained by required number of antennas and decrease the number of users before it reaches the peak value of SE. This is achieved by Error Vector Magnitude (EVM). In MIMO the SE also depends on the number of users decided in one time slot, but practically due to delay violation due to the transmission fluctuations. EC framework is adopted in [86] to determine the relationship between delay violation and transmission rate fluctuation.

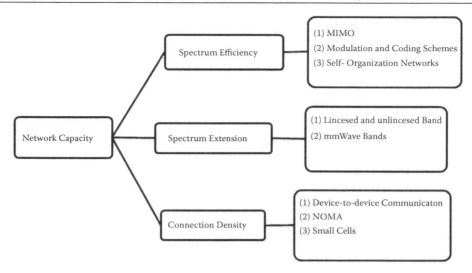

FIGURE 6.5 Trends in 5G network capacity.

In spectrum efficiency with MIMO systems, channel estimation is important for sequent signal processing. For channel estimation in an MIMO system, the solution of extrapolating the channel frequency and frequency bands of uplink and downlink respectively is explored and also compared the results with time division duplex system [87]. Modulation techniques and coding schemes also play very important role in SE. OFDM subcarrier and spacing duration supports high data rates as described in [88,89]. The key feature of 5G system is the scalability, where OFDM supports the various use cases. Spectrally Efficient FDM (SEFDM) where the proposed system generates continuous and special pilot signals to improve the efficiency by 20% than with the OFDM frame structure [90]. According to the new release on 5G NR Polar codes have been accepted for error detection. A joint detection and decoding receivers for reducing interference is proposed and have introduced metric to improve computational efficiency [91]. The authors in [92] have concentrated on the scheduling of wireless constraints. Spectrum sharing is a very well-known technique of efficient use of spectrum [93,94] like spectrum sharing protocol, generally primary transmitter performs above the target and hence tolerates secondary system interference. Spectrum sharing without the primary system assistance is used for increasing the spectrum efficiency [95].

6.6.2 Spectrum Extension

In 3GPP, the outline specified for 5G requirements demanded a lot of spectrum. This need introduces the mmWave technology and use of unlicensed frequency band. Below is the spectrum allocation which can be used for unlicensed band in 5G and also the licensed band which is being used in 4G. Spectrum sharing can be done on the two horizons vertical and horizontal basis. Vertical sharing works with the priority to the availability of resources according to the use cases. The horizontal sharing where the system will have the same priority. The typical use cases can be home Wi-Fi system and hotspot network.

6.6.3 Connection Density

There are many solutions to increase the SE, but while working with the frequency band variations there are limitations on the deployed infrastructure. In this context device-to-device communication, small cells, NOMA technology are some of the device-related solutions for SE. D2D communication is an

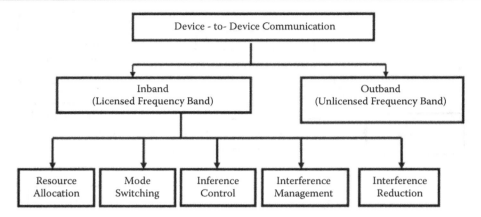

FIGURE 6.6 Device-to-device communication operating bands.

effective way of traffic offloading the base station through cache technique. These D2D links can also be used for coverage extensions, where the UEs will act as a relay to connect the other UEs to the base station [96]. D2D can also be used for sharing files with high data rates and low energy. Below research paper focuses on increasing the connection density of the network.

As in the above Figure 6.6, the inband D2D uses the licensed band for communication. The interference control is feasible while using the inband communication but with outband, i.e., unlicensed free band of 2.4 GHz ISM or 38 GHz mm wave band can also be used for communication. In outband there are many interference from the devices like Bluetooth and Wi-Fi. The SE can be gained in Inband with proper resource and power allocation schemes. The usage of inband also brings the interference due to the reuse of the channels. Using shared scheme will bring more spectrum efficiency rather than using D2D in dedicated mode or cellular mode.

Table 6.6 shows that currently researchers are working towards the hybrid solution of EE and SE. In [102–104], the authors have worked only on SE with resource allocation algorithms. Currently, the system design issue is dealing with various requirements like increasing the throughput with optimum SE and EE. Multiple Objective Maximization Problem (MOOP) which focuses on tradeoff between the ergodic EE and SE maximization through CU's interference power received and transmit power of D2D users. Both sparsely and densely deployed network scenarios are used to prove the analysis [97]. Small cells are promising solutions to enhance the capacity of spectrum, but these also introduce inter and intra cells solutions to enhance the capacity of spectrum.

To this Fractional Frequency Reuse (FFR) was introduced where the frequency band is allotted to particular call and it reduces interferences [98]. But FFR is not the efficient way for spectrum efficiency since the frequency band is not reused. Directional antennas for interference reduction and FFR framework is used where the different frequency sub bands are allotted to each tier leading to EE and mathematical analysis of SE [99]. With 5G technology V2V communication is also one of the leading applications. D2D communication benefits are also utilized to gain end-to-end throughput [100]. Clustering algorithm is proposed where the cluster heads are selected such that it reduces the intra-cluster interference. Different time slots are used for communication with D2D pairs to increase throughput. A cross-layer model is proposed for the interference neutralization in multihop VANETs. In [101] destination blind relay selection algorithm is used for selection of set of relays for error free transmitter and receiver. On second level destination aware relay selection algorithm is proposed to maximize the SE. Comparison of the simulation results of outage probability and SE is widely done.

To increase the network capacity in 5G the major task is to increase the Machine Type Communication (MTC) [105]. The promising technique is to use the Ultra Dense Network, i.e., the deployment of small cells [106]. In recent years researchers are working with the joint deployment of

TABLE 6.6 Survey of algorithms for D2D communication for different scenario

REFERENCE NO.	PROPOSED ALGORITHM	OBJECTIVE	NETWORK SCENARIO
[97]	Multi-objective optimization problem (MOOP)	Maximization of ergodic sum rate of D2D and minimization of transmit power	Limited interference with sparse and dense deployment
[98]	Non-convex mixed algorithm multi-objective optimization	Resource allocation and energy harvesting time slot allocation in D2D communication network	–
[99]	Fractional frequency release based spectrum partitioning	Different optimization problem based on the EE and SE at D2D UE (macro cell and femtocell), respectively	–
[100]	Clustering mechanism—Cross layer optimization frame for multi hop VANETs	Intra cluster resource efficiency and nullify the intra cluster spectral efficiency	Multi hops VANETs
[101]	Relay Selection	Algorithm selects the relay for transmission for finding the relay and spectral efficiency	Deployment of frequency base station in small geographical area

NOMA and UDN for incrementing the network capacity. NOMA allows more number of users by serving the available resources in Power domain and Code domain [107]. The authors in [108], [109,110] have compared the NOMA with OMA to support multiple users on the same frequency time. The results proved that NOMA outperformed with massive number of connections having the same resources. In [111] NOMA is combined with beam forming which allows reusing the available resources orthogonally and thus increasing the number of available resources. The other way is to use the advanced receiver with iterative coding and detection so as to gain the high connection density. Interleave grid multiple access (IGMA) in combination with DFT-OFDM can reduce the PAPR [112]. IGMA is been proposed in [113] used for multi-user gain and low complexity but with high PAPR. To further support the IoT devices narrow-band IoT systems are used where it can support an uplink data rate up to 250 Kbps [114] and subcarrier (3.75 kHz or 15 kHz). For NB-IoT system the authors in [115] have considered the joint subcarrier and power allocation algorithm while transmission using uplink power domain NOMA. This algorithm achieves 87% more connection density than OMA.

6.7 NETWORK SOFTWARIZATION

In current network the topology of network depends on the static configuration. These techniques are able to complete the different customer requirements. With the 5G and beyond 5G systems these pre-defined techniques are more expensive, inflexible, and are inefficient for resource allocation strategies. This makes them incapable to adopt the changes in the network according to the customer requirement [116].

SDN and NFV are the promising enabling technologies that fulfill the customer demands. Virtualization is not a new concept in networking. This concept in 1960 was used to divide the system resources in the mainframe between the applications. Deployment of these virtual applications inside virtual switches made it faster and scalable.

6.7.1 Software Defined Networking (SDN)

With increase in size of networks individually setting the virtual switches became difficult. This brought SDN which enables these virtual switches to be programmed. SDN architecture decouples the data plane and control plane. The network infrastructure can be abstracted from applications and network services. The use of SDN makes the network more dynamic, cost-effective, and manageable. Table 6.7 provides the benefits to the network.

With SDN much functionality is provided, adoption into networking enables many features as shown in Figure 6.7. Some of the key concepts are stated below:

1. **Forwarding and Control Function Separated:** The control plane and data plane are separated. The control plane is responsible for administration. Data plane encompasses the actual packet processing for the applications.
2. **Programmability:** SDN enables networks through programmability. The network behavior is controlled through the software. Network operators can update the software to support new services.
3. **Central Management:** SDN is based on logically centralized networks, which supports central management of all the schemes. This makes the network provides optimization in bandwidth management policies.

TABLE 6.7 SDN benefits

SR. NO.	NEED OF SDN	FUNCTIONS
1.	Virtualization	Virtually utilizing network resources
2.	Orchestration	Controls and manages many devices with one command
3.	Programmable	Variable according to the on demand requirement
4.	Dynamic scaling	Changes the size and quantity
5.	Automation	To minimize OPEX
6.	Visibility	Resources monitoring and connectivity
7.	Performance	Network optimization
8.	Multi-tenancy	Tenants need control over topology and routing
9.	Openness	Openness to the plug-ins and self-organized tasks to the networks by APIs

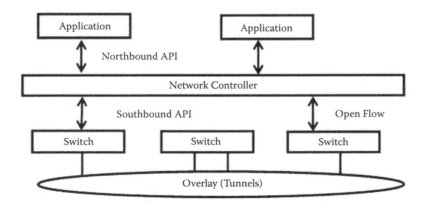

FIGURE 6.7 Framework of SDN.

4. **Open Standard Usage:** SDN came up with a concept of openness. SDN originated from OpenFlow. This concept can be used by number of vendors when a single user wants the probability. This was possible only when the open standards were used. This makes the system more easy and flexible.

- Southbound API: XMPP(Juniper), OnePK (Cisco)
- Northbound API: I2RS, I2AEX, ALTO
- Overlay: VxLAN, TRILL, LISP, SIT, NVO3, PWE3, L2VPN, L3VPN
- Configuration API: NET, CONF
- Controller: PCE, ForCES

6.7.2 Network Function Virtualization (NFV)

NFV is an approach in network where the network entities which used dedicated hardware could be now replaced by the software. By downloading this software the service providers could manage then own platforms over the same existing hardware. ETSI ISG first introduced this NFV in 2012 to provide hardware reductions in Capital Expenditure (CAPEX) and Operation Expenditure (OPEX) [117]. A network with NFV includes a number of elements. These elements provide better stability and interoperability. Figure 6.8 gives an overview of high level NFV framework.

Virtualized Network Functions (VNF): This is the software used to create various network functions such as directory services, file sharing, and IP configuration.

Network Function Virtualization Infrastructure NFVI: NFVI consists of the infrastructure components like compute, storage, and networking components that host the VNFs. NFVI can be located across several physical locations. VMware is an example of VIM.

Network Function Virtualization Management and Orchestration Architectural Framework: NFV-MANO provides the framework that enables information, manipulation needed to manage NFVI and VNFs. 5G's vision is to support the vertical applications with different requirements. The

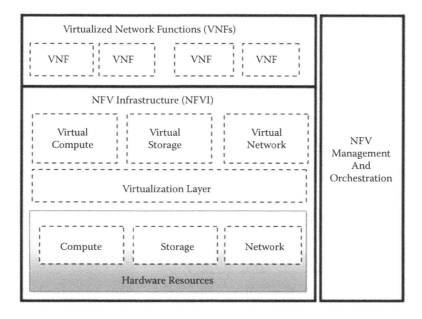

FIGURE 6.8 NFV framework.

system demands the flexibility, interoperability, and programmability. Network slicing is one of the use cases of network virtualization. Network slicing is the best solution to meet the requirements from wide range of enterprises. The network slicing leverages the principles of SDN and NFV to create the multiple virtual networks on top of the physical infrastructure. In the further section we provide the concept of network slicing in 5G and other enablers like SDN, NFV, Fog Radio Access, and RAN slicing.

6.7.3 Network Slicing

In today's world mobile devices are the essential part of our daily routine. The era of mobile from 1G to 4G has brought lot many changes in terms of technologies and customer demands. As with the new era, the customer requirements went on increasing, e.g., throughput, latency, etc. With 5G system the demands of throughput, latency, capacity increased by around 100 times better than 4G. In 4G it was one-size-fits-all system, but the same is not possible with 5G and beyond 5G requirements.

With 5G and beyond 5G, the network has to satisfy a wide range of services with variations in requirements. These networks not only support mobile devices but also the machine-to-machine communications and ultra-low latency communication [118]. To support all these demands on the same infrastructure with a flexible framework brings us to the concept of "Network Slicing". This service-oriented network requires flexible and programmable technologies like SDN and NFV, which will help the network to make a slice on demand from the customer by ensuring which of the requirements is fulfilled. Figure 6.9 shows the applications related to each type of slice.

6.7.3.1 Concept of Network Slicing

Figure 6.10 shows the complete cycle of the slicing technique starting from service provider to the termination of the slice.

- **Abstraction** of network function enables the creation of virtual network according to the customer by the service provider.
- During **service creation** the slice templates are formed and are also mapped with the SLA between the customer and the service provider.

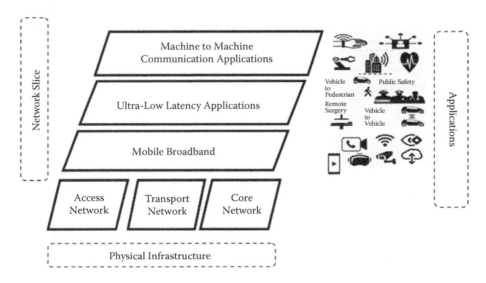

FIGURE 6.9 Network slicing scenario.

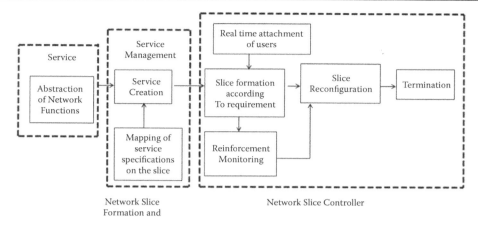

FIGURE 6.10 Lifecycle of network slicing.

- In **network slice controller** in run time the slice is formed resources and network environment are configured and the slice is allotted to the customer according to the requirements.
- In run time the traffic handling management performance like QoS is been monitored.
- If any reconfiguration or up gradation of the slice is required, is also done in this phase.
- Termination phase includes deactivation or decomposition of the network slice been allocated are also reclamation.

6.8 NETWORK SLICING ENABLING TECHNOLOGIES

6.8.1 SDN/NFV

Since the emergence of network slicing in 5G system, it has become the topic of interest for various industries and academia. There are many papers which refer to the study of SDN/NFV technologies and also to the introduction of such architecture, [119,120] and [121]. When it comes to network slicing resource allocations and configuration of the slices according to the diverse use cases becomes the prime objective. In [122] and [123], sharing of radio resources various resource allocation schemes are used for isolation between the slices for dynamic environment. A resource pricing mechanism was solved by joint optimization on allocation of shared and dedicated resources. All these research does not take into consideration the reconfiguration overhead. In [124] column generation problem has been used to maximize the profit of service provider. The total cost is maximized related to energy consumption and reconfiguration by using Multicast Dissemination Protocol (MDP) in [125]. In [126] different from all these the authors have considered elastic services and have proposed Hybrid Slice Reconfiguration (HSR) framework in which Fast Slice Reconfiguration (FSR) takes care about the minimum time of individual slice of arrival/departure. Dimensioning Slice with Reconfiguration (DSR) is to adjust the allocated resources according to the traffic demand. This framework reduces the reconfiguration overhead and high profit for slices. Even though in above papers they have concentrated on overhead time of reconfiguration, this reconfiguration of slices can also be due to the failures of VNF and the links. These can be due to the uncertain traffic demands. In [127] authors have proposed an optimal slice recovery for deterministic traffic demands. With the deployment of heuristic algorithm and joint slice recovery and reconfiguration algorithm an adjustable tolerance of link uncertainties can be provided. While we talk about all these algorithms, in [128] the authors have proposed an optimizing resource allocation and extra bandwidth to the slice in each TTI scheduling time. The authors have used Optimized Network Engineering Tools (OPNET) modeler for the scenario operation and simulation.

The research in network slicing with SDN/NFV technologies there are papers which have focused on some use cases. In [129] resource algorithm which is used in use case of an electric fence in a farm and demonstrated the proper use of network slicing by the IoT devices. The other application of Autonomous Driving was executed with a framework which meets the low latency reconfiguration [130].

6.8.2 RAN Slicing

SDN/NFV brings flexibility in 5G network through which the network can adopt the concept of network slicing. A slice is formed with different resources according to the requirements. For end to end network slicing, slicing at RAN and Core network level play a vital role in acquiring the QoS. There are many survey papers [131,132] which focus on the network slicing at core network level. In end to end latency, capacity of the network the resource availability defines RAN level Slicing. As shown in Figure 6.11 a network slice is built on top of the virtual physical infrastructure where the resources, the infrastructure in it are shared. RAN slice decides how the radio resources are allocated to fulfill the requirement. Below are the various components of RAN slicing system.

Slice Context Manager: This is responsible for performing lifecycle management of each slice.

Virtualization Manager: This provides a virtual set of radio resources and data plane and is also responsible for isolation of control plane.

Radio Resource Manager: This allocates the physical radio resources among different slices.

UE Association Manager: This performs mapping of UEs with the slices.

In multi-domain infrastructure multi-network slices can be efficiently allocated through slicing [133,134]. In [135] RAN configuration parameters are also referred to as RAN slice descriptors which characterize the features of the slice. There are studies where depending on the bit rate requirements slices are created as in [135]. In [136] authors have considered guaranteed bit rate or the creation of slice.

Once the slices are created, commissioning schemes are used to allocate radio resources to the slices. Binary Integer Programming Problem for partitioning of slices in [137]. In [135] authors have designed an overall resource utilization schemes by reinforcement learning. In [138] the resource utilization is maximized through offline resource allocation algorithm reinforcement learning followed by a heuristic algorithm with low complexity. Heterogeneous Non-Orthogonal Multiple Access Technique (H-NOMA) involves homogeneous requirements and the results in [139] are compared with NOMA technique. The

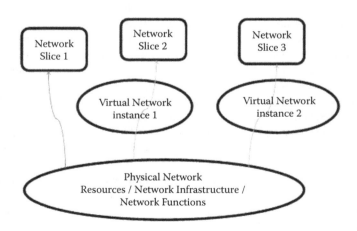

FIGURE 6.11 RAN slicing scenario.

proposed framework provides isolation between the slices with heterogeneous requirements. Orthogonal slicing approach used for better reliability for eMBB and URLLC. The authors have focused on power consumption and revenue maximization. In this the power consumption is done on priority basis on importance of slice admission or not [140]. A prototype is presented to create a RUN time slice according to the requirements with isolation in three different cases [141]. In [142] the focus is on admission control schemes using Markovian approach. All these approaches have not considered latency while creation of slices and partitioning of resources. In [143] the authors have proposed a rate and latency sensitive slicing solution for Industry 4.0. For simulation they have considered deterministic aperiodic, deterministic periodic, and non-deterministic traffic for the use case of 4.0. All these above survey shows the creation, isolation of network slice. But all these schemes with predictive features and SONs will make the network more precise. In 5G and beyond 5G system so many new technologies being introduced, such as Fog computing, ML, and Deep Learning, can help the networks to be more perfect.

6.8.3 Fog Radio Access Networks

Internet of vehicles is an emerging technique in which the self-driving service and there are other transportation related applications [144,145]. The amount of road traffic affects the amount of transmission delay. In such scenarios where due to static resources some of the resources are underutilized and some is overexhausted. This brings the need of Fog RANs to satisfy the needs of resources with minimum latency for safe vehicles driving. V-FRANs consist of vehicular fog and edge fog [146]. Network slicing technique can be utilized here for reducing the OPEX [147]. The resource utilized is kept on the Fog. The conventional slice scheduling cannot be used for such requirements. So the authors in [146] have proposed a joint scheduling scheme between the vehicles and edge resources. Perception reaction time [148] mechanism is used for the relationship between the road traffic and resources.

In Fog-RANs energy harvesting has also gained interest of many researchers. Since the communication infrastructure is responsible for the carbon dioxide emissions [149,150] is the first one to work with fog node cooperation for energy harvesting. This method has significantly increased the energy utilization of each fog node and ultimately the network.

The game theory is widely used for resource allocation problems in network slicing. The other method used in resource allocation is also widely known and that is Hierarchical Resources Allocation. In which Global Radio Resource Manager (GRRM) allocates the resources to the Local Radio Resource Manager (LRRM) is a follower which then allocates these resources within the slices. In [151,152] authors have used GRRM and LRRM to control the interference. BS and relay act as a leader in the [153] and [154], respectively, which only take care of their own resources. Whereas the authors in [155] have assigned each follower as a resource manager to allocate resources to the slice GRRM directly does not allocate resources. Authors have attempted to solve this issue of resource allocation by considering each LRRM as a Mixed Integer Non-Linear Programming (MINLP). The above study gives an overview of resource allocation strategies for Fog-RANs. Table 6.8 shows the overview of the above papers with regard to the different parameters achieved.

6.8.4 Deep Learning

To support the heterogeneous requirements of 5G system, the network slicing technique enables dynamic resource management in real time. But the challenge while allocating these resources in real time is to check the SLAs before allocation, demand depends upon the traffic spectrum efficiency [156]. With all these challenges "Deep Learning" can be used to investigate these dynamic demands and respond to either creation of the slice or reconfiguration of the available slices.

Reinforcement Learning interacts with the environment by trying alternative actions, learning from those actions and rewarding the network with liable actions. Deep Reinforcement Learning

TABLE 6.8 Overview of network slicing papers w.r.t. the parameters

REF. NO.	TECHNOLOGY	PARAMETERS					
		LATENCY	ENERGY EFFICIENCY	THROUGHPUT	CAPACITY	COST	ISOLATION
[130]	SDN	✓	✓	–	–	–	–
[129]		✓	✓	–	✓	–	–
[127]		–	✓	–	✓	–	–
[126]		✓	✓	✓	✓	–	–
[128]		–	–	✓	–	–	–
[150]	Fog computing	–	✓	–	–	–	–
[146]		✓	–	✓	–	–	–
[155]		✓	–	–	✓	–	–
[138]	RAN slicing	✓	–	✓	–	–	–
[141]		–	–	–	–	–	✓
[143]		✓	–	–	–	–	–
[140]		–	✓	–	–	✓	–

(DRL) has been applied to lot of work in power control, CRAN [157], edge computing [158,159] and researchers have observed better rewards through simulation or through use cases. Algorithms use genetic algorithm for inter-slice resource management, i.e., bandwidth, but it did not consider Service Level Agreement (SLA). In [160] the authors used Round Robin scheduling algorithm and bandwidth allocation algorithm to the slice, which is done per second. The authors compared the results with baseline algorithms and have also tested it by changing number of antennas with satisfactory QoE. [161] authors have proposed Markov Decision Process for RL problem, REINFORCE algorithm is used for learning and been evaluated on real data sets. Considering QoS parameter as the feedback from the environment the authors have proposed adaptive bandwidth allocation using RL in [162]. In [163] the authors have allotted radio resource slicing and core network slicing which require different waiting time and compute time.

6.9 OPEN ISSUES AND CHALLENGES

6.9.1 Latency

There are many existing solutions to reduce latency to 1 ms, as discussed in Section 6.4. But these technologies open several challenges for future work. In Physical Layer design transmission of small packet brings channel modeling as an open challenge. Researchers can work on field test for this issue. There are many described multiple access schemes but still there is a need of the scheme with less complexity with low latency. [164] caching, AI-enabled cache technique will help to get the proper content on the cache transmission schemes for proper coordination of the content should also be redesigned. Grant free access schemes should be reframed to meet the URLLC requirements [165]. With SDN/NFV technology used to reduce latency, orchestration, and management play the most important role in the network. Mentioned techniques for latency should be more practical oriented and investigated before standardization of 5G.

6.9.2 Energy Efficiency

Even though NOMA has proved itself over the OMA technique in various scenarios, further way to enhance the capacity of NOMA is to introduce full duplex technique [166]. To improve the EE in MMTC scenarios, backscattering communication technology communication technology can be used to increase the battery life of the IOT devices [167]. In today's scenario, with small cell deployments instead of focusing on the transmission side the researchers should concentrate on the computational side [168]. Self-learning mechanisms are still less explored in the aspect of EE. In 5G due to the random approach of the traffic, energy-efficient models should be such that they should be able to handle the random behavior of the network. Still there is no end-to-end energy efficient model which can successfully work with all the enabling technology on a unified platform. Scheduling and resource allocation algorithms need to be improved to meet the expectation from 5G of 100–1000 times better than 4G LTE.

6.9.3 Network Capacity

The potential to offer greater capacity over current communication systems, mmWave Communications is one of the promising candidates for the 5G mobile networks. The current solutions in mmWave for 5G communications has been discussed in Section 6.6. The current challenges demand the software defined architecture, better control mechanisms and efficient interference management [169]. D2D communication is also surveyed for greater connection density; in this the biggest challenge is the synchronization of the UE [170]. Interference management, resource allocation and mode selection are very closely related. The optimization schemes with considering all these parameters into one can be the solution. There are algorithms available but still the researchers are working to get the appropriate results [171].

6.9.4 Network Slicing

Aforementioned research shows a detailed survey on network slicing. For end-to-end network slicing the network can be divided into two parts RAN and core network. The researchers are focusing on either the core network slicing or the RAN slicing. Some authors have worked with end-to-end network slicing, but with this type of resource allocation SLA monitoring, SLA decomposition, and multiple subnet coordination are the main challenges. For proper reuse slices from the higher level to formation and termination slices is again an open issue with meeting all the SLA slices, orchestration and management of these slices from the higher level to formation and termination slices is again an open issue with meeting all the SLA between the users and the subscribers. With heterogeneous networks 5G requirements, since the slices also needs to be isolated. Isolation among the slices on the basis of traffic and radio electrical isolation authors in [172] have proposed four levels of RAN slicing. But the qualitative levels like flexibility and complexity may advocate to other solutions.

With the above discussions iSection 8.4 of Deep learning and Section 7.3 of network slicing there are number of algorithms which have proven Deep Learning as the solution for resource management schemes in network slicing. However, there still are a number of issues that need to be addressed: 1) Slice admission control for new slices. 2) Learning algorithms applied are only flexible to the particular scenario. There needs to be unified learning scheme for end-to-end network slicing. 3) Lack of cost-effective Data Retrieval Language (DRL) and Deep Q-Learning (DQL) time complexity leads to reduction in the latency of the whole system.

REFERENCES

[1]. Mobile Data Consumption Trends Market to 2027 - Global Analysis and Forecasts By Subscribers, Technology and Geography Dublin, May 14, 2019 (GLOBE NEWSWIRE).
[2]. The India Mobile Data Consumption Trends Market to 2027 - Regional Analysis and Forecasts By Subscribers and Connection Type India, May 23, 2019 (GLOBE NEWSWIRE).
[3]. Data Usage in India to Grow at 73% CAGR by 2022: Study, India, March 19, 2019 (Economic Times).
[4]. IEEE 5G and Beyond Technology Roadmap White Paper, IEEE 5g and Beyond Technology Roadmap White Paper (2017).
[5]. 3GPP, R1-168371, WF on URLLC Evaluation Parameter and LLS method, Aug. 2016.
[6]. 3GPP, R1-1609664, Comparison of Slot and Mini-slot Based Approaches for URLLC, Oct. 2016.
[7]. P. Kala, M. Casta, J. Salmi, K. Leppännen, T. Turkka, T. Hiltunen, and M. Hronec, "A novel radio frame structure for 5g dense outdoor radio frame structure for 5g dense outdoor radio access networks", *2015 IEEE 81st Veh.r Technol. Conf. (VTC Spring)*, May 2015.
[8]. M. Iwabuchi, A. Benjebbour, Y. Kishiyama, G. Ren, C. Tang, T. Tian, L. Gu, and T. Takada, T. Kashima, "5G field experimental trials on URLLC using new frame structure", 2017 IEEE Globecom Workshops (GC Wkshps), 2017.
[9]. S. Guo, X. Hou, H. Wang, and X. Wang, "Investigation of TDD Frame Structure Design for Latency Reduction of LTE Enhancement", *2017 23rd Asia-Pacific Conf. Commun. (APCC)*, 2017.
[10]. K. Chen, K. Niu, and J. Lin, "List successive cancellation decoding of polar codes", *Electron. Lett.*, vol. 48, no. 9, pp. 500–501, 2012.
[11]. G. Wunder, P. Jung, M. Kasparick, T. Wild, F. Schaich, Y. Chen, S. T. Brink, I. Gaspar, N. Michailow, A. Festag, L. Mendes, N. Cassiau, D. Ktenas, M. Dryjanski, S. Pietrzyk, B. Eged, P. Vago, and F. Wiedmann, "5GNOW: Non-orthogonal, asynchronous waveforms for future mobile applications", *IEEE Commun. Mag.*, vol. 52, no. 2, pp. 97–105, Feb. 2014.
[12]. H. Nikopour and H. Baligh, "Sparse code multiple access", *2013 IEEE 24th Annu. Int. Symp. Pers. Indoor Mobile Radio Commun.(PIMRC)*, London, pp. 332–336, 2013.
[13]. K. Lai, L. Wen, J. Lei, P. Xiao, A. Maaref and M. A. Imran, "Sub-graph based joint sparse graph for sparse code multiple access systems", *IEEE Access*, vol. 6, pp. 25066–25080, 2018, doi: 10.1109/ACCESS.2018.2828126.
[14]. Xi Wu, Zhanji Wu, "Performance analysis of SCMA system based on polar codes", *2018 IEEE Globecom Workshops (GC Wkshps)*, 2018.
[15]. A. K. M. Baki, R. A. Ahsan and A. Awsaf, "Novel methods of filtering for FBMC/UFMC based 5G communication systems", *2019 7th Int. C. Smart Comp. Commun. (ICSCC)*, Sarawak, Malaysia, Malaysia, pp. 1–4, 2019, doi: 10.1109/ICSCC.2019.8843617.
[16]. M. Maraş, E. N. Ayvaz, M. Gömeç and A. Özen, "Improving the performance of GFDM waveform employing lifting wavelet transform", *2019 27th Signal Process. Commun. Appl. Conf. (SIU)*, Sivas, Turkey, pp. 1–4, 2019, doi: 10.1109/SIU.2019.8806553.
[17]. S. Wu, X. Chen, and S. Zhou, "A parallel interleaver design for Idma systems", *Proc. Int. Conf. Wireless Commun. Signal Process. (WCSP)*, Nanjing, China, pp. 1–5, Nov. 2009.
[18]. S. Yoshizawa, M. Nozaki, and H. Tanimoto, "VLSI implementation of an interference canceller using dual-frame processing for OFDMIDMA systems", *IEICE Trans. Fundam. Electron. Commun. Comput. Sci.*, vol. E98-A, no. 3, pp. 811–819, Mar. 2015.
[19]. T. T. T. Nguyen, L. Lanante, S. Yoshizawa, and H. Ochi, "Low latency IDMA with interleaved domain architecture for 5G communications", *IEEE Trans. Emerg. Sel. Topics Circuits Sys*t., vol. 7, no. 4, pp. 582–593, Dec. 2017, doi: .1109/JETCAS.2017.2776979.
[20]. I. Tal and A. Vardy, "List decoding of polar codes", *IEEE Trans. Inf. Theory*, vol. 61, no. 5, pp. 2213–2226, 2015.
[21]. H. S. B. Li and D. Tse, "Parallel decoders of polar codes", Sep. 2013 [Online]. Available: http://arxiv.org/abs/1309.1026.
[22]. D. Kam and Y. Lee, "Ultra-low-latency parallel SC polar decoding architecture for 5G wireless communications", *2019 IEEE Int. Symp. Circuits Syst (ISCAS)*, Sapporo, Japan, 2019, pp. 1–5, doi: 10.1109/ISCAS.2019.8702786.
[23]. E. Öztürk, E. Basar and H. A. Çırpan, "Generalized frequency division multiplexing with flexible index modulation numerology", *IEEE Signal Process. Lett.*, vol. 25, no. 10, pp. 1480–1484, Oct. 2018, doi: 10.1109/LSP.2018.2864601.

[24]. B. Y. Kong and I. Park, "A memory-efficient IDMA architecture based on on-the-fly despreading", *IEEE J. Solid-State Circuits*, vol. 53, no. 11, pp. 3327–3337, Nov. 2018, doi: 10.1109/JSSC.2018.2863950.
[25]. B. Singh, O. Tirkkonen, Z. Li and M. A. Uusitalo, "Contention-based access for ultra-reliable low latency uplink transmissions", *IEEE Wirel. Commun. Lett.*, vol. 7, no. 2, pp. 182–185, April 2018, doi: 10.1109/LWC.2017.2763594.
[26]. W. Abderrahim, O. Amin, M Alouini, and B. Shihada, "Latency-aware offloading in integrated satellite terrestrial networks", *IEEE Open J. Commun. Soc.*, vol. 1, 2020, pp. 490–500, 2020.
[27]. H. Zhang, C. Huang, J. Zhou, and L. Chen, "QoS-aware virtualization resource management mechanism in 5G backhaul heterogeneous networks", *IEEE Access*, vol. 8, pp. 19479–19489, 2020, doi: 10.1109/ACCESS.2020.2967101.
[28]. M. Zhu, N. Cvijetic, M.-F. Huang, T. Wang, and G.-K. Chang, "Low-Latency synchronous clock distribution and recovery for DWDM-OFDMA-based optical mobile backhaul". *J. Light. Technol.*, vol. 32, no. 10, pp. 2012–2018, 2014.
[29]. H. Y. R. Im, K. Thar and C. S. Hong, "Q-learning based social community-aware energy efficient cooperative caching in 5G networks", *Int. Conf. Ubiquitous Future Netw. ICUFN. (ICUFN)*, Zagreb, Croatia, pp. 500–503, 2019, doi: 10.1109/ICUFN.2019.8806093.
[30]. G. Sun, H. Al-Ward, G. O. Boateng and W. Jiang, "Content-aware caching in SDN-enabled virtualized wireless D2D networks to reduce visiting latency", *2018 IEEE 15th Int. Conf. Mobile Ad Hoc Sensor Syst. (MASS)*, Chengdu, pp. 149–150, 2018, doi: 10.1109/MASS.2018.00032.
[31]. S. Rezvani, N. Mokari, M. R. Javan, and E. A. Jorswieck, "Fairness and transmission-aware caching and delivery policies in OFDMA-based hetnets",*IEEE Trans. Mob. Comput.*, vol. 19, no. 2, pp. 331–346, 1 Feb. 2020, doi: 10.1109/TMC.2019.2892978.
[32]. S. Vakilinia and H. Elbiaze, "Latency control of ICN enabled 5G networks", *J. Netw. Syst. Manag.*, vol. 28, no. 1, 2020.
[33]. D. Feng et al., "Toward ultrareliable low-latency communications: Typical scenarios, possible solutions, and open issues", *IEEE Veh. Technol. Mag.*, vol. 14, no. 2, pp. 94–102, June 2019, doi: 10.1109/MVT.2019.2903657. (latency survey)
[34]. European Telecommunications Standards Institute, "Mobile-edge computing- introductory technical white paper", Sep. 2014. Available:https://portal.etsi.org/portals/0/tbpages/mec/docs/mobile-edge computing introductory technical white paper v1%2018-09-14.pdf.
[35]. J. Liu, Y. Mao, J. Zhang, and K. B. Letaief, "Delay-optimal computationtask scheduling for mobile-edge computing systems", *Proc. IEEE Int. Symp. Inf. Theory (ISIT)*, Barcelona, Spain, Jul. 2016, pp. 1451–1455.
[36]. C. You, K. Huang, and H. Chae, "Energy efficient mobile cloud computing powered by wireless energy transfer", *IEEE J. Select. Areas Commun.*, vol. 34, no. 5, pp. 1757–1771, May 2016.
[37]. Y. Mao, J. Zhang, S. Song, and K. B. Letaief, "Stochastic joint radio and computational resource management for multi-user mobile-edge computing systems", *IEEE Trans. Wirel. Commun.*, vol. 16, no. 9, pp. 5994–6009, Sep. 2017.
[38]. X. Chen, L. Jiao, W. Li, and X. Fu, "Efficient multi-user computation offloading for mobile-edge cloud computing", *IEEE Trans. Netw.*, vol.24, no. 5, pp. 2795–2808, Oct. 2016.
[39]. C. You, K. Huang, H. Chae, and B.-H. Kim, "Energy-efficient resource allocation for mobile-edge computation offloading", *IEEE Trans. Wireless Commun.*, vol. 16, no. 3, pp. 1397–1411, Mar. 2017.
[40]. Y. H. Kao, B. Krishnamachari, M. R. Ra, and F. Bai, "Hermes: Latency optimal task assignment for resource-constrained mobile computing", *IEEE Trans. Mobile Comput.*, vol. 16, no. 11, pp. 3056–3069, Nov. 2017.
[41]. G. Chi, Y. Wang, X. Liu and Y. Qiu, "Latency-optimal task offloading for mobile-edge computing system in 5G heterogeneous networks", *2018 IEEE 87th Veh. Technol. Conf. (VTC Spring)*, Porto, 2018, pp. 1–5, doi: 10.1109/VTCSpring.2018.8417606.
[42]. J. Ren, G. Yu, Y. Cai and Y. He, "Latency optimization for resource allocation in mobile-edge computation offloading", *IEEE Trans. Wirel. Commun.*, vol. 17, no. 8, pp. 5506–5519, Aug. 2018, doi: 10.1109/TWC.2018.2845360.
[43]. F. Tang et al., "On Removing routing protocol from future wireless networks: A real-time deep learning approach for intelligent traffic control", *IEEE Wirel. Commun.*, vol. 25, no. 1, pp. 154–160, Feb. 2018.
[44]. Kari Seppanen, Jorma Kilpi, Jori Paananen, Tapio Suihko, Pekka Wainio, Jouko Kapanen, "Multipath routing for mmWave WMN backhaul, communications workshops (ICC)", *2016 IEEE Int. Conf.*, pp. 246–253, 2016.
[45]. S.-C. Lin, P. Wang, and M. Luo, "Jointly optimized QoSaware virtualization and routing in software defined networks", *Comput. Netw.*, vol. 96, pp. 69–78, Feb. 2016.

[46]. Q. Meng, M. Jiang, W. Yue, and Y. Meng, "A beyond 5G edge network for ultra-low latency services", *2018 15th Int. Symp. Pervasive Syst., Algorithms Netw. (I-SPAN)*, Yichang, China, 2018, pp. 121–126, doi: 10.1109/I-SPAN.2018.00028.

[47]. J. V. V. Sobral, J. J. P. C. Rodrigues, R. A. L. Rabêlo, K. Saleem, and S. A. Kozlov, "Improving the performance of LOADng routing protocol in mobile IoT scenarios", *IEEE Access*, vol. 7, pp. 107032–107046, 2019, doi: 10.1109/ACCESS.2019.2932718.

[48]. S. Zheng, Z. Gao, X. Shan, W. Zhou, Y. Wang, X. Zhang,"End-to-end latency optimization in software defined LEO satellite terrestrial systems". In: Yu Q. (eds) Space Information Networks. SINC 2018. Communications in Computer and Information Science, vol. 972. Springer, Singapore,2019.

[49]. A. A. R. Alsaeedy and E. K. P. Chong, "Mobility management for 5G IoT devices: Improving Power consumption with lightweight signaling overhead", *IEEE Internet Things J.*, vol. 6, no. 5, pp. 8237–8247, Oct. 2019, doi: 10.1109/JIOT.2019.2920628.

[50]. T. Beitelmal, S. S. Szyszkowicz, D. G. González and H. Yanikomeroglu, "Sector and site switch-off regular patterns for energy saving in cellular networks", *IEEE Trans. Wirel. Commun.*, vol. 17, no. 5, pp. 2932–2945, May 2018, doi: 10.1109/TWC.2018.2804397.

[51]. T. Beitelmal, S. S. Szyszkowicz, and H. Yanikomeroglu, "Regular and static sector-based cell switch-off patterns", *Proc. IEEE Veh. Technol. Conf. (VTC-Fall)*, pp. 1–5, Sep. 2016.

[52]. C. Bouras and G. Diles, "Energy efficiency in sleep mode for 5g femtocells", *2017 Wirel. Days*, Porto, pp. 143–145, 2017, doi: 10.1109/WD.2017.7918130.

[53]. C. Yang, J. Li, Q. Ni, A. Anpalagan and M. Guizani, "Interference-aware energy efficiency maximization in 5G ultra-dense networks", *IEEE Trans. Commun.*, vol. 65, no. 2, pp. 728–739, Feb. 2017, doi: 10.1109/TCOMM.2016.2638906.

[54]. J. Thompson, X. Ge, H.-C. Wu, et al., "5G wireless communication systems: Prospects and challenges", *IEEE Commun. Mag.*, vol. 52, no. 2, pp. 122–130, Feb. 2014.

[55]. Z. Zhang, Z. Ma, M. Xiao, et al., "Two-timeslot two-way full-duplex relaying for 5G wireless communication networks", *IEEE Trans. Wirel. Commun.*, vol. 64, no. 7, pp. 2873–2887, Jul. 2016.

[56]. A. Yadav, O. A. Dobre, N. Ansari, "Energy and traffic aware full-duplex communications for 5G systems", *IEEE Access*, vol. 5, pp. 11278–11290, 2017.

[57]. O. Ozel, K. Tutuncuoglu, J. Yang, et al., "Transmission with energy harvesting nodes in fading wireless channels: Optimal policies", *IEEE J. Sel. Areas iCommun.*, vol. 29, no. 8, pp. 1732–1743, Sep. 2011.

[58]. W. Cheng, X. Zhang, H. Zhang, "Statistical-QoS driven energy-efficiency optimization over green 5G mobile wireless networks", *IEEE J. Sel. Areas Commun.*, vol. 34, no. 12, pp. 3092–3107, Dec. 2016.

[59]. Y. Mao, Y. Luo, J. Zhang, et al., "Energy harvesting small cell networks: Feasibility, deployment, and operation", *IEEE Commun. Mag.*, vol. 53, no. 6, pp. 94–101, Jun. 2015.

[60]. L. Chen, F. R. Yu, H. Ji, et al., "Green full-duplex self-backhaul and energy harvesting small cell networks with massive MIMO", *IEEE J. Selected Areas. Commun.*, vol. 34, no. 12, pp. 3709–3724, Dec. 2016.

[61]. P. S. Yu, J. Lee, T. Q. S. Quek, et al., "Traffic offloading in heterogeneous networks with energy harvesting personal cells-network throughput and energy efficiency", *IEEE Trans. Wirel. Commun.*, vol. 15, no. 2, pp.1146–1161, Feb. 2016.

[62]. X. Ge, S. Tu, G. Mao, et al., "5G ultra-dense cellular networks", *IEEE Wirel. Commun.*, vol. 23, no. 1, pp. 72–79, Jun. 2016.

[63]. J. M. Lasry, P.-L. Lions, "Mean field games", *Japanese J. Math.*, vol. 2, pp. 229–260, 2007.

[64]. M. Huang, P. E. Caines, R. P. Malhame, "Large-population cost-coupled LQG problems with nonuniform agents: individualmass behavior and decentralized -Nash equilibria", *IEEE Trans. Autom. Control*, vol. 52, no. 9, pp. 1560–1570, Sep. 2007.

[65]. F. Meriaux, V. Varma, S. Lasaulce, "Mean field energy games in wireless networks", *Proc. 46th Asilomar Conf. Signals, Syst. Comput.*, pp. 671–675, 2012.

[66]. X. Ge, H. Jia, Y. Zhong, Y. Xiao, Y. Li and B. Vucetic, "Energy efficient optimization of wireless-powered 5G full duplex cellular networks: A mean field game approach", *IEEE Trans. Green Commun. Netw.*, vol. 3, no. 2, pp. 455–467, June 2019, doi: 10.1109/TGCN.2019.2904093.

[67]. X. Wang, M. Chen, T. Taleb, A. Ksentini, and V. C.M. Leung, "Cache in the air: Exploiting content caching and delivery techniques for 5G systems", *IEEE Commun. Mag.*, vol. 52, no. 2, pp. 131–139, 2014.

[68]. M. Ji, G. Caire, and A. F. Molisch, "Wireless device-to-device caching networks: Basic principles and system performance", *IEEE J. Sel. Areas Commun.*, vol. 34, no.1, pp. 176–189, 2016.

[69]. B. A. Ramanan, L. M. Drabeck, M. Haner, N. Nithi, T. E. Klein, and C. Sawkar, "Cacheability analysis of HTTP traffic in an operational LTE network", *Proc. 12th Annual Wirel. Telecommun. Symp.: Global Wirel. Commun.- Future Directions*, WTS 2013, USA, April 2013.

[70]. Jiequ Ji, Kun Zhu, Ran Wang, Bing Chen, and Chen Dai, "Energy efficient caching in backhaul-aware cellular networks with dynamic content popularity", *Hindawi Wirel. Commun. Comput.*, vol. 2018, 2018.

[71]. Y. Saito et al., "Non-orthogonal multiple access (NOMA) for cellular future radio access", *Proc. IEEE VTC*, pp. 1–5, Jun. 2013.

[72]. Y. Fu, Y. Liu, H. Wang, Z. Shi and Y. Liu, "Mode selection between index coding and superposition coding in cache-based NOMA networks", *IEEE Commun. Lett.*, vol. 23, no. 3, pp. 478–481, Mar. 2019, doi: 10.1109/LCOMM.2019.2892468.

[73]. K. Yang, N. Yang, N. Ye, M. Jia, Z. Gao and R. Fan, "Non-orthogonal multiple access: Achieving sustainable future radio access", *IEEE Commun. Mag.*, vol. 57, no. 2, pp. 116–121, Feb. 2019, doi: 10.1109/MCOM.2018.1800179.

[74]. B. Di, L. Song, Y. Li, "Sub-channel assignment, power allocation, and user scheduling for non-orthogonal multiple access networks", *IEEE Trans. Wireless Commun*, Mar. 2017, doi: 10.1109/ICIN.2017.7899244.

[75]. F. Fang, H. Zhang, J. Cheng, V. C. M. Leung, "Energy efficiency of resource scheduling for non-orthogonal multiple access (NOMA) wireless network", Proc. IEEE ICC, pp. 1–5, 2016.

[76]. F. Fang, H. Zhang, J. Cheng, V.C.M. Leung, "Joint user scheduling and power allocation optimization for energy-efficient NOMA systems with imperfect CSI", *IEEE J. Sel. Areas Commun.*, vol. 35, no. 12, pp. 2874–2785, 2017.

[77]. Md. Forkan Uddin, "Energy efficiency maximization by joint transmission scheduling and resource allocation in downlink NOMA cellular networks", *Comp. Net.*, vol. 159, pp. 37–50, 2019.

[78]. C. Jiang, H. Zhang, Y. Ren, Z. Han, K. Chen and L. Hanzo, "Machine learning paradigms for next-generation wireless networks", *IEEE Wirel. Commun.*, vol. 24, no. 2, pp. 98–105, April 2017, doi: 10.1109/MWC.2016.1500356WC.

[79]. M. Ali Imran, A. F. dos Reis, G. Brante, P. V. Klaine, and R. D. Souza, *Machine Learning in Energy Efficiency Optimization*, Wiley online Library, 2019.

[80]. Y. Wang, X. Dai, J. M. Wang, and B. Bensaou, "A reinforcement learning approach to energy efficiency and QoS in 5G wireless networks", *IEEE J. Sel. Areas Commun.*, vol. 37, no. 6, pp. 1413–1423, June 2019, doi: 10.1109/JSAC.2019.2904365.

[81]. M. Miozzo, L. Giupponi, M. Rossi, and P. Dini, "Distributed QLearning for energy harvesting heterogeneous networks", IEEE ICC 2015 workshop on Green Communications and Networks with Energy Harvesting, Smart Grids and Renewable Energies, London, United Kingdom, 2015.

[82]. M. Miozzo, L. Giupponi, M. Rossi and P. Dini, "Switch-on/off policies for energy harvesting small cells through distributed Q-learning", 2017 IEEE Wireless Communications and Networking Conference Workshops (WCNCW), San Francisco, CA, pp. 1–6, 2017, doi: 10.1109/WCNCW.2017.7919075.

[83]. D. Sesto-Castilla, E. Garcia-Villegas, G. Lyberopoulos and E. Theodoropoulou, "Use of Machine Learning for energy efficiency in present and future mobile networks", 2019 IEEE Wireless Communications and Networking Conference (WCNC), Marrakesh, Morocco, pp. 1–6, 2019, doi: 10.1109/WCNC.2019.8885478.

[84]. T. L. Marzetta, "Noncooperative cellular wireless with unlimited numbers of base station antennas", *IEEE Trans. Wireless Commun.*, vol. 9, no. 11, pp. 3590–3600, Nov. 2010.

[85]. W. Boukley Hasan, P. Harris, A. Doufexi and M. Beach, "Real-time maximum spectral efficiency for massive MIMO and its limits", *IEEE Access*, vol. 6, pp. 46122–46133, 2018.

[86]. L. Chen and L. Zhang, Spectral efficiency analysis for massive MIMO system under QoS constraint: An effective capacity perspective, *Mobile. Netw. Appl.*, 2020.

[87]. F. Rottenberg, T. Choi, P. Luo, C. J. Zhang and A. F. Molisch, "Performance analysis of channel extrapolation in FDD massive MIMO systems", *IEEE Trans. Wirel. Commun.*, vol. 19, no. 4, pp. 2728–2741, April 2020.

[88]. 3rd Generation Partnership Project; Technical Specification Group Radio Access Network; Study on New Radio (NR) access technology (Release 14), 3GPP, Tech. Rep., 06 2017.

[89]. Characteristics template for NR RIT of 5G (Release 15 and beyond), 3GPP, Tech. Rep., 2017.

[90]. H. Ghannam and I. Darwazeh, "A proposal for scalable 5G new radio frames with enhanced throughput", *2019 IEEE 89th Veh. Technol. Conf. (VTC2019-Spring)*, Kuala Lumpur, Malaysia, 2019, pp. 1–6.

[91]. A. Jalali and, Z. D. "A new algorithm for improved blind detection of polar coded PDCCH in 5G new radio", arXiv, 2019.

[92]. A. Sinha, M. Andrews, and P. Ananth, "Scheduling algorithms for 5G networks with mid-haul capacity constraints", arXiv, 2019.

[93]. Kour, H., Jha, R.K., Jain, S. et al. "Protocol design and resource allocation for power optimization using spectrum sharing for 5G networks". *Telecommun. Syst*, vol. 72, pp. 95–113, 2019.

[94]. Gao, L., et al., "Spectrum trading in cognitive radio networks: A contract-theoretic modeling approach", *IEEE JSAC.*, vol. 29, no. 4, pp. 843–855, 2011.

[95]. Simeone, O., et al, "Spectrum leasing to cooperating secondary ad hoc networks", *IEEE JSAC.*, vol. 26, no. 1, pp. 203–213, 2012.

[96]. A. Aijaz, H. Aghvami, M. Amani, "A survey on mobile data offloading: technical and business perspectives", *IEEE Wirel. Commun.* vol. 20, no. 2, pp. 104–112, 2013.

[97]. M. Robat Mili, A. Khalili, N. Mokari, S. Wittevrongel, D. W. K. Ng, and H. Steendam, "Tradeoff between ergodic energy efficiency and spectral efficiency in D2D communications under rician fading channel", *IEEE Trans. Veh. Technol*, vol. 69, no. 9. pp. 9750–9766, 2020

[98]. S. Sobhi-Givi, A. Khazali, H. Kalbkhani, M. G. Shayesteh, andV. Solouk, "Resource allocation and power control for underlay device-to-device communication in fractional frequency reuse cellular networks", *Telecommun. Syst.*, vol. 65, pp. 677–697, 2017.

[99]. A. Hazali, S. Sobhi-Givi, H. Kalbkhani et al. "Energy-spectral efficient resource allocation and power control in heterogeneous networks with D2D communication". *Wireless Netw.*, vol. 26, pp. 253–267, 2020.

[100]. J. Yang, X. Han, Z. Ding, and X. Wei, "Spectral efficiency optimization and interference management for multi-hop D2D communications in VANETs", *IEEE Trans. Veh.Technol.*, vol. 69, no. 6, pp. 6422–6436, June 2020.

[101]. H. Kalbkhani and M.G. Shayesteh, Relay selection for multi-source network-coded D2D multicast communications in heterogeneous networks, *Wireless Netw.*, 2020.

[102]. B. Kaufman and B. Aazhang, "Cellular networks with an overlaid device to device network", *Proc. 42nd Asilomar Conf. Signals, Syst. Comput.*, pp. 1537–1541, Oct. 2008.

[103]. C.-H. Yu and O. Tirkkonen, "Device-to-device underlay cellular network based on rate splitting", *Proc. IEEE Wireless Commun. Netw.Conf. (WCNC)*, Apr. 2012, pp. 262–266.

[104]. Z. Liu, T. Peng, H. Chen, and W. Wang, "Optimal D2D user allocation over multi-bands under heterogeneous networks", *Proc. IEEE Global Commun. Conf. (GLOBECOM)*, pp. 1339–1344, Dec. 2012.

[105]. S. K. Sharma and X. Wang, "Towards massive machine type communications in ultra-dense cellular iot networks: Current issues and machine learning-assisted solutions", *IEEE Commun. Surveys Tuts.*, vol. 22, no. 1, pp. 426– 471, 2019.

[106]. M. Kamel, W. Hamouda, and A. Youssef, "Ultra-dense networks: A survey", IEEE Commun. Surveys Tuts., vol. 18, no. 4, pp. 2522–2545, Fourthquarter, 2016.

[107]. Z. Zhang, G. Yang, Z. Ma, M. Xiao, Z. Ding, and P. Fan, "Heterogeneous ultradense networks with noma: System architecture, coordination framework, and performance evaluation", *IEEE Veh. Technol.Mag.*, vol. 13, no. 2, pp. 110–120, June 2018.

[108]. Z. Zhang, Y. Hou, Q. Wang, and X. Tao, "Joint sub-carrier and transmission power allocation for mtc under power-domain noma", *2018 IEEE Int. Conf. on Commun. Workshops (ICC Workshops)*, May 2018, pp. 1–6.

[109]. A. E. Mostafa, Y. Zhou, and V. W. S. Wong, "Connection density maximization of narrowband iot systems with noma", *IEEE Trans. Wirel. Commun.*, vol. 18, no. 10, pp. 4708–4722, 2019.

[110]. T. Lv, Y. Ma, J. Zeng, and P. T. Mathiopoulos, "Millimeter-wave noma transmission in cellular m2m communications for internet of things", IEEE Internet Things J., vol. 5, no. 3, pp. 1989–2000, June 2018.

[111]. J. Zeng, T. Lv, W. Ni, R. P. Liu, N. C. Beaulieu, and Y. J. Guo, "Ensuring max-min fairness of ul simo-noma: A rate splitting approach", *IEEE Trans. on Veh. Tech.*, vol. 68, no. 11, pp. 11080–11093, 2019.

[112]. C. Qian, Q. Xiong, B. Yu and C. Sun, "Low complexity detection algorithm for low PAPR interleaving based NOMA schemes", *2017 IEEE 86th Veh. Technol. Conf. (VTC-Fall)*, Toronto, ON, 2017, pp. 1–5.

[113]. Samsung, R1-163992, Non-orthogonal multiple access candidate for NR 3GPP RAN1 #85, Nanjing.

[114]. Y. P. E. Wang, X. Lin, A. Adhikary, A. Grovlen, Y. Sui, Y. Blankenship, J. Bergman, and H. S. Razaghi, "A primer on 3GPP narrowband Internet of Things", *IEEE Commun. Mag.*, vol. 55, no. 3, pp. 117–123, Mar. 2017.

[115]. A. E. Mostafa, Y. Zhou, and V. W. S. Wong, "Connection density maximization of narrowband IoT systems with NOMA", *IEEE Trans. Wirel. Commun.*, vol. 18, no. 10, pp. 4708–4722, Oct. 2019

[116]. J. Chen et al., "Realizing dynamic network slice resource management based on SDN networks", *2019 International Conference on Intelligent Computing and its Emerging Applications (ICEA)*, Tainan, Taiwan, pp. 120–125, 2019, doi: 10.1109/ICEA.2019.8858288.

[117]. M. Condoluci and T. Mahmoodi, "Softwarization and virtualization in 5G mobile networks: benefits, trends and challenges", *Computer Netw.*, vol. 146, pp. 65–84, 9 December 2018.

[118]. X. Foukas, G. Patounas, A. Elmokashfi and M. K. Marina, "Network slicing in 5G: Survey and challenges", *IEEE Commun. Mag.*, vol. 55, no. 5, pp. 94–100, May 2017, doi: 10.1109/MCOM.2017.1600951.

[119]. J. Ordonez-Lucena et al., "Network slicing for 5G with SDN/NFV: Concepts, architectures, and challenges", *IEEE Commun. Mag.*, vol. 55, no. 5, pp. 80–87, May 2017.
[120]. 5G network architecture a high-level perspective," Shenzhen, China, Huawei Technol., White Paper, 2016.
[121]. H. Zhang, "Future wireless network: MyNET platform and end-to-end network slicing", arXiv Preprint arXiv: 1611.07601, pp. 1–54, 2016.
[122]. P. Rost et al., "Network slicing to enable scalability and flexibility in 5G mobile networks", *IEEE Commun. Mag.*, vol. 55, no. 5, pp. 72–79, May 2017.
[123]. P. Caballero, A. Banchs, G. de Veciana, and X. Costa-Pérez, "Multitenant radio access network slicing: Statistical multiplexing of spatial loads", *IEEE/ACM Trans. Netw.*, vol. 25, no. 5, pp. 3044–3058, Oct. 2017.
[124]. J. Liu, W. Lu, F. Zhou, P. Lu, and Z. Zhu, "On dynamic service function chain deployment and readjustment", *IEEE Trans. Netw. Service Manag.*, vol. 14, no. 3, pp. 543–553, Sep. 2017.
[125]. V. Eramo, E. Miucci, M. Ammar, and F. G. Lavacca, "An approach for service function chain routing and virtual function network instance migration in network function virtualization architectures", *IEEE/ACM Trans. Netw.*, vol. 25, no. 4, pp. 2008–2025, Aug. 2017.
[126]. G. Wang, G. Feng, T. Q. S. Quek, S. Qin, R. Wen and W. Tan, "Reconfiguration in network slicing—optimizing the profit and performance", *IEEE Trans. Netw. Service Manage.*, vol. 16, no. 2, pp. 591–605, June 2019.
[127]. R. Wen et al., "On robustness of network slicing for next-generation mobile networks", *IEEE Trans. Commun.*, vol. 67, no. 1, pp. 430–444, Jan. 2019.
[128]. A. S. D. Alfoudi, S. H. S. Newaz, A. Otebolaku, G. M. Lee and R. Pereira, "An efficient resource management mechanism for network slicing in a LTE network", *IEEE Access*, vol. 7, pp. 89441–89457, 2019.
[129]. J. Chen et al., "Realizing dynamic network slice resource management based on SDN networks", 2019 International Conference on Intelligent Computing and its Emerging Applications (ICEA), Tainan, Taiwan, 2019.
[130]. D. A. Chekired, M. A. Togou, L. Khoukhi and A. Ksentini, "5G-slicing-enabled scalable SDN core network: Toward an ultra-low latency of autonomous driving service", *IEEE J. Sel. Areas Commun*, vol. 37, no. 8, pp. 1769–1782, Aug. 2019
[131]. D. Sattar and A. Matrawy, "Optimal slice allocation in 5G core networks", *IEEE Netw. Lett.*, vol. 1, no. 2, pp. 48–51, Jun. 2019.
[132]. D. A. Chekired, M. A. Togou, L. Khoukhi, and A. Ksentini, "5G-slicingenabled scalable SDN core network: Toward an ultra-low latency of autonomous driving service", *IEEE J. Sel. Areas Commun.*, vol. 37, no. 8, pp. 1769–1782, Aug. 2019.
[133]. R. Ferrus, O. Sallent, J. P. Romero, and R. Agusti, "On 5G radio access network slicing: Radio interface protocol features and con_guration", *IEEE Commun. Mag.*, vol. 56, no. 5, pp. 184–192, May 2018.
[134]. I. Vilà, O. Sallent, A. Umbert, and J. Pérez-Romero, "An analytical model for multi-tenant radio access networks supporting guaranteed bit rate services", *IEEE Access*, vol. 7, pp. 57651–57662, 2019.
[135]. A. Aijaz, "Hap-SliceR: A radio resource slicing framework for 5G networks with haptic communications", *IEEE Syst. J.*, vol. 12, no. 3, pp. 22852296, Sep. 2018.
[136]. System Architecture for the 5G System; Stage 2 (Release 15), document 3GPP TS 23.501 v1.0.0, Sep. 2018.
[137]. White Paper: 5G Radio Access_Capabilities and Technologies, Ericsson, White Paper Uen 284 23–3204 Rev C, 2016. Accessed: Apr. 5, 2019. [Online]. Available: https:// www.ericsson.com/assets/local/publications/white-papers/wp-5g.pdf.
[138]. H. D. R. Albonda and J. Pérez-Romero, "An efficient RAN slicing strategy for a heterogeneous network with eMBB and V2X services", *IEEE Access*, vol. 7, pp. 44771–44782, 2019.
[139]. P. Popovski, K. F. Trillingsgaard, O. Simeone and G. Durisi, "5G wireless network slicing for eMBB, URLLC, and mMTC: A communication-theoretic view", *IEEE Access*, vol. 6, pp. 55765–55779, 2018.
[140]. J. Tang, B. Shim and T. Q. S. Quek, "Service multiplexing and revenue maximization in sliced C-RAN incorporated with URLLC and multicast eMBB", *IEEE J. Sel. Areas Commun.*, vol. 37, no. 4, pp. 881–895, April 2019.
[141]. C. Chang and N. Nikaein, "RAN runtime slicing system for flexible and dynamic service execution environment", *IEEE Access*, vol. 6, pp. 34018–34042, 2018.
[142]. F. Bahlke, O. D. Ramos-Cantor, S. Henneberger, and M. Pesavento, "Optimized cell planning for network slicing in heterogeneous wireless communication networks", *IEEE Commun. Lett.*, vol. 22, no. 8, pp. 1676–1679, Aug. 2018.
[143]. J. García-Morales, M. C. Lucas-Estañ and J. Gozalvez, "Latency-sensitive 5G RAN slicing for industry 4.0", *IEEE Access*, vol. 7, pp. 143139–143159, 2019.

[144]. Z. C. Ming Zhu and J. Cao, "Providing flexible services for heterogeneous vehicles: An NFV-based approach", *IEEE Netw.*, vol. 30, no. 3, pp. 64–71, May/ Jun. 2016.

[145]. K. Yang and S. Ou, "A multihop peer-communication protocol with fairness guarantee for IEEE 802.16-based vehicular networks", *IEEE Trans. Veh. Technol.*, vol. 56, no. 6, pp. 3358–3370, Nov. 2007.

[146]. K. Wang and K. Yang, "Joint energy minimization and resource allocation in C-RAN with mobile cloud", *IEEE Trans. Cloud Comput.*, vol. 6, no. 3, pp. 760–770, Jul./Sep. 2018.

[147]. G. Qiao and S. Leng, "Collaborative task offloading in vehicular edge multi-access networks", *IEEE Commun. Mag.*, vol. 56, no. 8, pp. 48–54, Aug. 2018.

[148]. K. Xiong, S. Leng, J. Hu, X. Chen and K. Yang, "Smart network slicing for vehicular fog-RANs", *IEEE Trans. Veh. Technol.*, vol. 68, no. 4, pp. 3075–3085, April 2019.

[149]. V. Chamola and B. Sikdar, "Solar powered cellular base stations: Current scenario, issues and proposed solutions", *IEEE Commun. Mag.*, vol. 54, no. 5, pp. 108–114, May 2016.

[150]. Y. Xiao and M. Krunz, "Dynamic network slicing for scalable fog computing systems with energy harvesting", *IEEE J. Sel. Areas Commun.*, vol. 36, no. 12, pp. 2640–2654, Dec. 2018.

[151]. S. Bu, F. R. Yu, and H. Yanikomeroglu, "Interference-aware energy efficient resource allocation for OFDMA-based heterogeneous networks with incomplete channel state information", *IEEE Trans. Veh. Technol.*, vol. 64, no. 3, pp. 1036–1050, Mar. 2015.

[152]. Y. Wang, X. Wang, and L. Wang, "Low-complexity Stackelberg game approach for energy-efficient resource allocation in heterogeneous networks", *IEEE Commun. Lett.*, vol. 18, no. 11, pp. 2011–2014, Nov. 2014.

[153]. S. Guruacharya, D. Niyato, D. I. Kim, and E. Hossain, "Hierarchical competition for downlink power allocation in OFDMA femtocell networks", *IEEE Trans. Wireless Commun.*, vol. 12, no. 4, pp. 1543–1553, Apr. 2013.

[154]. L. Liang, G. Feng, and Y. Jia, "Game-theoretic hierarchical resource allocation for heterogeneous relay networks", *IEEE Trans. Veh. Technol.*, vol. 64, no. 4, pp. 1480–1492, Apr. 2015.

[155]. Y. Sun, M. Peng, S. Mao and S. Yan, "Hierarchical radio resource allocation for network slicing in fog radio access networks", *IEEE Trans. Veh. Technol*, vol. 68, no. 4, pp. 3866–3881, April 2019.

[156]. N. Zhang, Y.-F. Liu, H. Farmanbar, T.-H. Chang, M. Hong, and Z.-Q. Luo, "Network slicing for service-oriented networks under resource constraints", *IEEE J. Sel. Areas Commun.*, vol. 35, no. 11, pp. 2512–2521, Nov. 2017.

[157]. Z. Xu, Y. Wang, J. Tang, J.Wang, and M. C. Gursoy, "A deep reinforcement learning based framework for power-effcient resource allocation in cloud RANs", *Proc. IEEE ICC*, Paris, France, pp. 1–6, May 2017.

[158]. Y. He, F. R. Yu, N. Zhao, V. C. M. Leung, and H. Yin, "Software-deneed networks with mobile edge computing and caching for smart cities: A big data deep reinforcement learning approach", *IEEE Commun. Mag.*, vol. 55, no. 12, pp. 3137, Dec. 2017.

[159]. Y. He et al., "Deep-reinforcement-learning-based optimization for cache enabled opportunistic interference alignment wireless networks", *IEEE Trans. Veh. Technol.*, vol. 66, no. 11, pp. 1043310445, Nov. 2017.

[160]. R. Li et al., "Deep reinforcement learning for resource management in network slicing", *IEEE Access*, vol. 6, pp. 74429–74441, 2018.

[161]. J. Koo and V. B. Mendiratta and M. R. Rahman and A. Walid, "Deep reinforcement learning for network slicing with heterogeneous resource requirements and time varying traffic dynamics", *arXiv*, 2019.

[162]. B. Han, J. Lianghai, and H. D. Schotten, "Slice as an evolutionary service: Genetic optimization for inter-slice resource management in 5g networks", *IEEE Access*, vol. 6, pp. 33137–33147, 2018.

[163]. Z. Zhao, R. Li, Q. Sun, Chi-Lin, C. Yang, X. Chen, M. Zhao, and H. Zhang, "Deep reinforcement learning for network slicing", ArXiv, 2018.

[164]. S. He, W. Huang, J. Wang, J. Ren, Y. Huang and Y. Zhang, "Cache-enabled coordinated mobile edge network: Opportunities and challenges", *IEEE Wirel. Commun.*, vol. 27, no. 2, pp. 204–211, April 2020.

[165]. H. Yang, K. Zheng, K. Zhang, J. Mei and Y. Qian, "Ultra-reliable and low-latency communications for connected vehicles: Challenges and solutions", *IEEE Network*, vol. 34, no. 3, pp. 92–100, May/June 2020.

[166]. D. Korpi et al., "Full-duplex mobile device: Pushing the limits", *IEEE Commun. Mag.*, vol. 54, no. 9, pp. 80–87, Sept. 2016.

[167]. B. Kellogg, V. Talla, and J. Smith, "Passive Wi-Fi: Bringing low power to Wi-Fi transmissions", *Get mobile Mobile Comput. Commun.*, vol. 20, no. 3, pp. 38–41, 2017.

[168]. S. Buzzi, C. I, T. E. Klein, H. V. Poor, C. Yang, and A. Zappone, "A survey of energy-efficient techniques for 5G networks and challenges ahead", *IEEE J. Sel. Areas Commun.*, vol. 34, no. 4, pp. 697–709, April 2016.

[169]. Y. Niu, Y. Li, D. Jin et al.,"A survey of millimeter wave communications (mmWave) for 5g: Opportunities and challenges". *Wirel. Netw.*, vol. 21, pp. 2657–2676, 2015.

[170]. N. Abedini, S. Tavildar, J. Li and T. Richardson, "Distributed synchronization for device-to-device communications in an LTE network", *IEEE Trans. Wirel. Commun.*, vol. 15, no. 2, pp. 1547–1561, Feb. 2016.

[171]. S. Ali, A. Ahmad, "Resource allocation, interference management, and mode selection in device-to-device communication: A survey", *Trans. Emerg. Telecommun. Technol.*, vol. 28, no. 7, pp. e3148, 2017.

[172]. O. Sallent, J. Perez-Romero, R. Ferrus and R. Agusti, "On radio access network slicing from a radio resource management perspective", *IEEE Wirel. Commun.*, vol. 24, no. 5, pp. 166–174, October 2017.

Applications of Machine Learning in Wireless Communication: 5G and Beyond

7

Rohini Devnikar[1] and Vaibhav Hendre[2]

[1]*Research Scholar at Department of Electronics and Telecommunication Engineering, G H Raisoni College of Engineering & Management, Pune, India*
[2]*Professor at Department of Electronics and Telecommunication Engineering, G H Raisoni College of Engineering & Management, Pune, India*

INTRODUCTION

The chapter presents the technological aspects that are expected to constitute advanced technologies of the future, such as eHealth applications, industry 4.0 and massive robotics, holographic telepresence, pervasive connectivity in smart environments, huge unmanned mobility in three dimensions, Augmented Reality (AR), and Virtual Reality (VR) to name a few. These next-generation technologies are expected to offer high-quality and efficient performance. All systems for optimum functioning require effective and fast data transmission without any exceptions. So, the scope for advanced technologies is high. User demands are not fulfilled with cellular technologies from First Generation (1G) to Fourth Generation (4G), they require high data rate with very low latency to obtain high dimensional image and video quality. 1G was based on analog communication and it does not provide any security to data. The Second Generation (2G) has more data rate than 1G and is based on digital communications. It provides security to data but does not fulfil user demands in terms of speed. Low-quality dimension data is stored in 2G. The Third Generation (3G) has a higher data rate than 1G and 2G. It gives high-quality images and videos but lacks in privacy and security and does not support Internet on Things (IoT) devices. The Fourth Generation (4G) can give higher quality services to each base station by using Long-Term Evolution (LTE) and Multiple Inputs Multiple Outputs (MIMO) technology with a higher data rate than 3G. But 4G does not fully provide services to IoT devices and consumes more energy. Because of these limitations, users are eagerly anticipating the next generation of cellular network technology. The Fifth Generation (5G) technology fulfils user demands with regard to all the aspects. 5G completely supports IoT devices and massive MIMO technology. 5G gives more data rate than 1G to 4G. 5G has very low latency and higher throughput, which makes it very useful in ultra-high data communication and advanced spectrum uses management. These are connected to unauthorized and licenced group [1]. However, 5G still has limitations with regard to privacy and security of data. Moreover, 5G does not fully support higher intelligence technologies such as Super IoT, VR and Artificial Intelligence (AI) [2]. Machine Learning (ML) techniques are supported by 5G, but in case of complex protocols and network of wireless communication 5G does not support the system. AI has a very complex and massive data set so it requires a higher data rate for efficient communication. Due to this problem, some scientists and researchers are focusing on next-generation technologies, i.e., Sixth Generation (6G). 6G can provide even higher data rate, supermassive access and very low latency than 5G [3]. Compared with 5G, privacy and security provided by 6G cellular technology is expected to be better [4,5]. Highly complex Device-to-Device (D2D) communication will also be introduced with emergence of 6G [6]. Millimeter-Wave (mm-wave) is a key enabling technique and innovation for AI in 6G technology [7], which gives a higher frequency range of up to 300 GHz. ML gives optimal solutions for complex problems in wireless communication for 6G. 6G has the ability to support services such as massive Machine Type Communication (mMTC), ultra-High Speed with Low Latency Communications (uHSLLC), and ultra-High Data Density (uHDD) services [8]. This chapter is an overview that starts with the evolution of cellular network technologies from 1G to 6G, how 5G was realized, 5G at a glance, 6G vision, research challenges in 6G, moving from 5G to 6G compared with performance metrics, AI in wireless communication, what is AI and ML, AI-enabled wireless communication, ML for wireless communication, key steps in ML and then ends with types of ML categories for wireless communication.

7.1 THE EVOLUTION OF CELLULAR NETWORK TECHNOLOGIES

Technological evolution, revolution, and creation was started in the 1970s in the mobile wireless industry. The evolution of cellular network technologies from 1G to 6G is represented in Figure 7.1.

7.1.1 1G

1G was introduced in the year 1980. This generation was analog-based, a very simple network. The preceding Wireless Fidelity (Wi-Fi) communication technology had been used in push-to-communicate systems and moderately in Wi-Fi telephones in the army and maritime applications. The major difference between the prevailing wireless communication structures and 1G became the creation of the cellular generation. The land was divided into a small cell network known as cells. Each cellular network had a base station, which used a radio signal and a transceiver for communicating with cell devices. Base stations have been linked to telephone networks. 1G provided a data rate of 2.4 kbps (kilobits per second) and the 1G network could be used only for voice calling. Frequency modulation strategies were used for voice calls with the aid of base stations. 1G used analog signal for communication so additional noise could be experienced as well as the network posed considerable security issues.

7 • 5G and Beyond   151

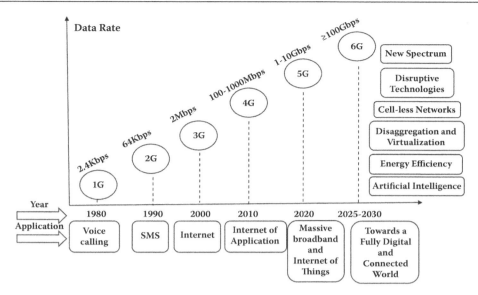

FIGURE 7.1   Evolution of cellular network technologies.

7.1.2 2G

2G generation was a digital cellular technology. It was launched in the year 1990. Voice signal was digitized in 2G and hence it addressed the disadvantages experienced in 1G. This technology was used for voice calling as well as Short Message Services and it provided speed of up to 64kbps. Security issues also solved. The 2G cellular technology was developed for Digital Advanced Mobile Phone System (D-AMPS) used for multiple call system as well as for Time Division Multiplexing (TDM) techniques. The 2G was launched on the Global System for Mobile (GSM) standard. This technology used TDM and Frequency Division Multiplexing (FDM) for managing multiple calls simultaneously. Code Division Multiple Access (CDMA) was employed in 2G and this technology was standardized by International Standards (IS-95).

7.1.3 3G

3G-internet mobile cellular network was launched in the year 2000 to give high-quality multimedia services along with quality voice transmission. This 3G technology overcomes the disadvantages associated with the previous generation by giving a higher data rate of more than the 1G and 2G generations. 3G has a 2 Mbps (megabits per second) data rate and provides higher security. 3G can access all advanced internet services. Third-Generation Partnership Project (3GPP) and 3GPP2 groups categorized the specifications for 3G. Wideband CDMA technology was used in 3G to gives Quality of Services (QoS). The 3G can be used for video calls, telemedicine, location-based services, mobile internet access, and wireless voice telephony, etc.

7.1.4 4G

The broadband cellular network services were provided by 4G-enabled mobile phone. 4G was introduced in the year 2010 and the data rate is 100–1000 Mbps, more than that provided by 1G to 3G networks. This technology enables high-quality voice transmission with high QoS than 3G. 4G is widely used in

Internet of Application. MIMO technology was introduced in 4G. LTE and Worldwide Interoperability for Microwave Access (WiMAX) standards were used in 4G to get a higher peak data rate for smooth communication. For a complex system, 4G is hard to implement. This technology is majorly utilized in high-definition mobile TV (television) and internet protocol (IP) telephony as well as streamed multimedia and data, 3D TV, and high-speed high-resolution gaming systems.

7.1.5 5G

The cellular networks from 1G to 4G compared with respect to the data rates provide very less speed than 5G. 5G provides 1–10 Gbps data rate as well as higher QoS and Quality of Access (QoA). Massive broadband and IoT deployed are using 5G technology to get higher transmission rates and throughput with low latency as compared with previous cellular network technologies. This technology was introduced in the year 2020 and it provides large bandwidth for ultra-high data communication.

7.1.6 6G

Beyond 5G, the 6G technology's expected data rate is up to 100 Gbps with very low latency and higher throughput; it is highly energy efficient and provides a large spectrum of network that can be deployed in AI and ML applications.

7.2 5G: AT A GLANCE

The 5G technology supports uHSLLC, uHDD, and mMTC services. Mm-wave massive MIMO technology was introduced in 5G for ultra-high data transmission and increased quality of performance in wireless communication. 5G is required to deliver a latency as low as 1 ms (millisecond). Connection density is expected to be 10^6 devices/km^2, which enables highly efficient signalling for IoT connectivity. 5G can achieve 30 bps/Hz spectrum efficiency with advanced antenna techniques. Positioning accuracy is 1 ms for 5G with 99.99% reliability. With a larger small cell network, the expected area traffic capacity for 5G is 10 Mbit/s/m^2. User experienced data rate is 100 Mbps with 500 kmph (kilometer per hour) mobility support in 5G technology.

7.3 HOW TO REALIZE 5G?

The key requirement of 5G is 1000-fold capacity. This goal realizes from Shannon capacity using equation 3.1 to obtain the required channel capacity (C) as shown in Figure 7.2.

$$C = D *W *M *log(1 + SINR) \tag{3.1}$$

Where C is the channel capacity of 5G. D is the number of small cells. By increasing the small cells network, we can increase the network density and the large area covered. W is the bandwidth, where to get higher frequency band range, mm-wave is deployed for ultra-high-speed data communication. M is the number of antennas. Large antenna arrays construct massive MIMO technology. To increase, antenna arrays provide high services simultaneously to each base station without any interferences. The Signal

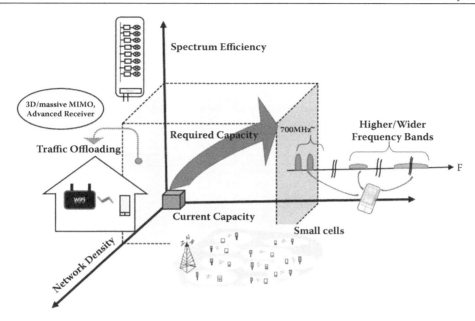

FIGURE 7.2 Three technical direction for 5G.

Interference Noise Ratio (SINR) can be determined using the logarithmic function to increase the channel capacity.

7.4 6G: VISION

Non-Terrestrial Networks (NTN) architecture was proposed for the integration of satellite communication network and cellular network. But still it is difficult to devise this interface mechanism. 5G does not fully support VR devices and AI applications. 6G is expected to give a data rate 100 Gbps with low latency and higher energy efficiency. Spectrum efficiency of 6G is 5–10x more than that of 5G. Also, 6G can deploy AI/ ML applications (Figure 7.3).

6G supports ultra-massive access and ultra-reliable services. Intelligent Connectivity is the combination of AI and ultra-high-speed connection anywhere and everywhere with a large number of connected IoT devices. Intelligent connectivity has three different connectivity's: Deep connectivity, holographic connectivity, and ubiquitous connectivity. Deep learning (DL), AI, deep sensing, and deep mind applications were developed with deep connectivity for 5G and Beyond. AR and VR technologies required higher speed of more than 10 Gbps, therefore 6G uses holographic connectivity to get high fidelity for VR/AR. Ubiquitous connectivity is being used anywhere and everywhere for applications such as air, space, ground, and sea communication.

7.4.1 Research Challenges in 6G

1. The 6G communication requires a wideband of frequency, i.e., terahertz (THz) band. The THz range is 0.1 THz–10 THz and the frequency range is expected to be 340 GHz. This frequency lies in the visible light frequency range in the electromagnetic spectrum. By the use of this

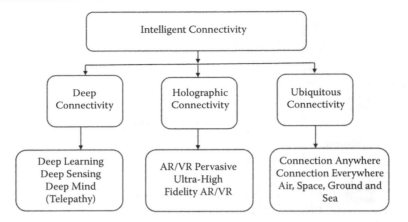

FIGURE 7.3 Intelligent connectivity for 6G.

wideband frequency range, ultra-fast feed transmission is possible and it will use a multiband range of frequency for wireless communication [9].
2. The term Super IoT is derived from 6G. Super IoT devices need more power or energy to communicate with massive data and telecommunication transmission. For efficient communication, energy harvesting techniques come into consideration [10]. Thus, these devices will work efficiently and communication will be smooth.
3. AI is easily deployed in 6G. 6G has a large frequency range and the ability to provide an optimal solution to complex algorithms and protocols is used in AI. The scope of AI research is broad.
4. Blockchain technology is constructed using 6G, so that it provides high security and privacy.

7.5 MOVING FROM 5G TO 6G

We compare 5G to 6G technology with regard to some of the performance metrics. This showed that the 5G technology has small fewer peak data rate, latency, spectral efficiency, energy efficiency. Thus it is less efficient technology compared with 6G, refer Table 7.1.

TABLE 7.1 Performance metrics: Moving from 5G to 6G wireless systems.

PERFORMANCE METRICS	5G	6G
Peak Data Rate (Gbps)	20	1000
User Experienced Data Rate (Mbps)	100	1000
Connection Density (Device/km^2)	10^6	10^7
Mobility Support (kmph)	500	1000
Area Traffic Capacity (Mbit/s/m^2)	10	1000
Latency (ms)	1	0.1
Reliability (%)	99.99	99.999
Positioning Accuracy (m)	0.01	0.0001
Spectral Efficiency (bps/Hz)	30	60
Network Energy Efficiency (J/bit)[1]	0.01	0.001

7.6 AI IN WIRELESS COMMUNICATION

Wireless communication can help deal with more applications based on AI in a dispensed manner to address the needs with regard to security and privacy along with low latency. This can be done in a very high intensity and computational manner. At the edge of wireless network system, applications based on intelligent or advanced technology would be deployed in the future. Due to this, intelligent applications with small cellular terminals, cost limits or IoT devices, generally, cannot offer sufficient computational functions. To resolve this problem, ensure that the terminals and edge servers are active in the wireless communication, also by considering bottleneck terminology the latency and capacity of all wireless connectivity are required. With this consideration, the dispensed information storage and computation with security support the various applications of AI. For examples, for privacy, federated learning is considered in every device that are ON condition and this is permitted to be distributed with training between multiple cellular devices. Bandwidth distribution, resource allocation, and scheduling between all cellular devices need to be carefully designed. The second example for dispensed inference system communication capability with ultra-low latency is necessary. So that every dispensed terminal selection can be obtained in time. To consider the various directions for a wireless communication system, it is key to maintain the overall performance on advancing. For example, with decrease in latency to zero at least 1Tb/s data rate is required to be increased. A number of devices could be linked to take part in more emerging applications consisting of autonomous driving, smart cities, VR, AR, linked infrastructure, and ground-air-space integrated networks [7].

7.6.1 Introduction to AI and ML

7.6.1.1 AI

AI is also called Machine Intelligence (MI). With respect to machine, intelligence is considered in comparison with Natural Intelligence (NI) of the animals and humans.

7.6.1.2 ML

ML considers the present and past data and accordingly predicts the outputs. Many algorithms fall under ML, categorized by each type of ML. The use of ML in AI gives high reliability, high accuracy, and solutions for complex problems. AI with ML finds applications in pattern recognition, voice recognition, statistics, and data mining process.

7.6.2 AI-Enabled Wireless Communication

In wireless communication, to facilitate intelligence, AI is the crucial technology. According to the classical Shannon communication concept, communication is the message generated at the source bred by communication objective at the destination. Shannon defined the terms "entropy" to measure the data by considering objectives to prove that separation of source coding and channel coding is as appropriate as joint processing. For realistic implementation in communication device designs within the past decades these conclusions and assumptions provided high-degree recommendations. A wireless communication system can evolve at the least in three directions with AI technology.

1. Based on the AI technologies, wireless big data analytics may be implemented in the future wireless communication systems. To help with system layout, performance tuning and fault

detection, future states may be predicted from historical data from which beneficial feature can be mined.
2. To optimize many physical layer modules AI technologies may be used for wireless link. To address network optimization issues with nonconvexity and massive scale which are tough to solve via the conventional techniques so the latest advances in AI provide many useful tools for these issues.
3. In refining, the end-to-end chain AI technology plays an essential role in wireless communications.

7.6.2.1 Big Data Analytics

Big data analytics is one of the applications of AI. There are four different types of analytics. Predictive analytics: The use of data to expect future events such as traffic patterns, consumer locations, known content and resource availability, consumer behavior. Descriptive analytics: For traffic profile, channel conditions, user views or historic data are used to get insights into a network, and so on. The situational awareness of network operators and carrier providers will significantly increase. Diagnostic analytics: Diagnostic analytics increases the reliability and security of 6G wireless systems and allows independent detection of network faults and basic reasons for network anomalies and service impairments. Prescriptive analytics: To indicate autonomous driving, network reducing and virtualization, cache placement, edge computing, selection alternatives for resource allocation, and so on to take benefit from predictions. For example, using wireless center network to significantly reduce peak traffic loads by proactive caching, which has been recently emerged and is helping in predicting future consumer needs through big data analytics [11].

7.6.2.2 Intelligence in Wireless Communication

From the transmitter to the receiver, in the end-to-end optimization of the entire chain of the physical layer signal processing, the AI technology will play an important role. A number of mismanagement issues in end-to-end communication devices include hardware complications such as amplifier distortion, quadrature imbalance, nearby oscillator and clock harmonic leakage, and the channel impairments consisting of fading and interference. To increase, the wide variety of things and parameters to be controlled will maintain. Nowadays, the wireless systems are never realistic for use in end-to-end optimization with this level of complexity. Rather, the capabilities of real-global systems are not appropriately or holistically captured because the existing methods divide the overall chain into multiple unbiased blocks, each with a simplified model. Over combinations of hardware and channel outcomes, AI technology gives opportunities to analyze the high-quality ways to communicate. By the use of combining advanced sensing and information collection of AI technology and domain-unique signal processing techniques in a future wireless communication system including an "intelligent PHY (Physical) layer" standard in 6G, it is possible for end-to-end devices to self-optimize and self-learn to know the system [11] (Figure 7.5).

7.7 ML MODEL FOR WIRELESS COMMUNICATION

7.7.1 Key Steps in ML

7.7.1.1 Training Data

The training data is categorized into two parts: Labelled data and unlabelled data. Labelled data is a set of unlabelled data sorted into a meaningful and desirable information set of data. Unlabelled data contains

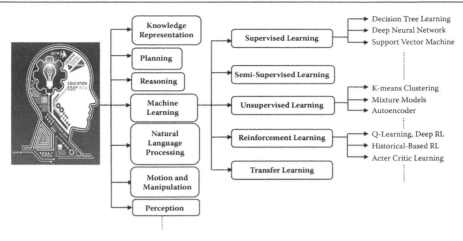

FIGURE 7.4 Brief classification of ML for AI.

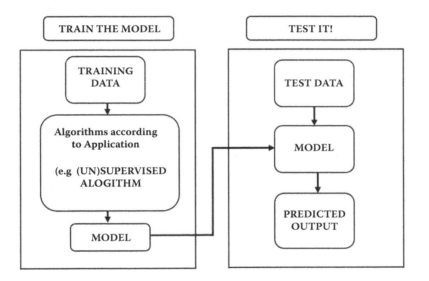

FIGURE 7.5 ML training model.

only data but no information. It does not give any desired information from a taken set of data. Training data take both labelled data and unlabelled data and predict the desired outputs from it.

7.7.1.2 Feature Extraction

Feature extraction methods describe a massive set of data and comprise a combination of different parameters or variables that describe a set of data with sufficient accuracy. The effective model construction is enabled by the appropriately optimized feature extraction method.

7.7.1.3 Learning Function and Prediction

The learning function uses a gradient descent algorithm to deal with individual threshold and weight of a set of labelled data and unlabelled data and accordingly how it would be manipulated to take a decision. By going through these key steps, the final predicted output is taken from the system.

7.7.2 ML for Wireless Communication

ML is an essential technique widely used in wireless communication because it gives efficient and quick solution for complicated problems in wireless networks [12,13]. ML techniques/algorithms first train the model and then give predicted and accurate output to the system.

ML is used at different layers of wireless communication network as explained below:

7.7.2.1 Physical Layer with ML

The mathematical models used in the physical layer and some of the several major modules in physical layers are modelled and optimized separately. DL is mostly suitable for the physical layer [14]. ML at the physical layer includes channel coding, channel modulation, beamforming, synchronization, positioning, and channel estimation. Both channel coding and modulation have some adaptive techniques that are useful for a complex system in wireless communication.

7.7.2.1.1 Channel Coding and Modulation
The error detected on a channel is corrected by channel coding and modulation. To solve channel coding problems in wireless networks, DL, ML, and Reinforcement Learning (RL) techniques are used. RL is the most widely used technique to provide an optimal solution for the channel problem of complex wireless networks.

7.7.2.1.2 Channel Estimation
The transmitted signal distorts through the channel and this information is determined through channel estimation. Through the training of the NNs, even with complicated channel environments, channel estimation can be optimized based on DL.

7.7.2.1.3 Synchronization
All devices must undergo a time/frequency and cellular synchronization technique without exception. To synchronize signal separately from signal detection/decoding Deep Neural Network (DNN) technologies are widely used in wireless communication.

7.7.2.1.4 Positioning
To adapt the actual environment with adaptive technology for positioning, the DNN location-based technology is found to be most suitable in solving the positional problem. The location of the user is recognized by advanced mobile positioning. The signals are indoor or outdoor based on wireless channels that are based on a mathematical approach or numerous signals obtained from users' mobile device. High positional errors occur if there are Non-Line-of-Sight (NLoS) communication multi-paths with mathematical approach technique.

7.7.2.1.5 Beamforming
To reduce energy consumption, extend coverage and enable advanced mobile application these features given by smart antenna, and intelligent beamforming solution reduces the sensitivity of interferences and increases the stability of the throughput. For real-time massive MIMO beamforming supported with adaptive technique represented by DL techniques gives an optimal solution for this type of beamforming.

7.7.2.2 Medium Access Control Layer with ML

To proactively determine the user's orientations and mobility, the Echo State Network (ESN) prediction algorithm is used. ML-based predictive resource allocation is very useful in systems such as Non-

Orthogonal Multiple Access (NOMA), massive MIMO, and cell-free massive MIMO system, possibly by using the so-called fast uplink grant. It is the most effective due to direct management of radio which consumes the most energy strength conservation using ML techniques only at Medium Access Control (MAC) layer. ML-based algorithms increase system performance based on network action data and traffic pattern at the MAC layer.

7.7.2.3 Application Layer with ML

Performance management automation in 6G network could improve the QoS prediction, power control, power-saving, maintenance, automatic network configuration, the performance of coverage, operation, throughput, beam management, and fault management. Unmanned Arial Vehicle (UAV) control is enhanced with energy consumption, reliability, and safety using ML techniques. By the use of ML, techniques improve traffic safety and efficiency in vehicular networks. Engineering standards change from process towards an extract and learn data, data-driven monitor, predict cycles in the development of the system, driven project, services, and deterministic such problem/challenges reduce the use of ML technique in software development.

7.7.3 The applications of ML in wireless communication

1. Image and video, e.g., Google, YouTube
2. Text & language, e.g., Wikipedia, Associated Press
3. Relational data/social network, e.g., Facebook and Twitter
4. Product recommendation, e.g., Amazon, Netflix
5. And many more

7.8 TYPES OF ML CATEGORIES FOR WIRELESS COMMUNICATION

As mentioned in Figure 7.4. a brief classification of ML is explained as follow:

7.8.1 Supervised Learning

Supervised learning (SL) takes labelled data and trains the machine by using the data and get desired known output. SL is used in wireless communication for channel estimation, signal parameter estimation, and user location prediction [15,16]. SL has three different algorithms: Decision Tree Learning, DNN, and Support Vector Machine (SVM). DNN algorithm is used in channel estimation and symbol detection. SVM algorithm is used for user location predictions. SL is categorized into regression and classification method.

1. Regression method is used in weather forecasting, market forecasting, population on growth prediction, advertise popularity prediction, and estimate life expectancy and get predicted data as output [17].
2. Classification method is used in image classification, identity formed detection, customer retention, and diagnostics as well as to get predicted outputs from it.

7.8.2 Unsupervised Learning

Learns patterns from the untagged data types algorithms are used in unsupervised learning (USL). USL is used in wireless communication for spectrum sensing, signal detection, and base stations clustering [18]. USL is used in K-mean clustering, mixture models and autoencoder algorithms. The K-mean model is used for cooperative spectrum sensing in cognitive radio networks. Autoencoder algorithm is widely used in signal detection in MIMO-Orthogonal Frequency Division Multiplexing (OFDM) networks [19]. The unsupervised algorithm uses two techniques—Clustering and Dimensionality Reduction.

1. The clustering technique divides data into several groups. This technique is used in target marketing, customer segmentation, and recommender systems.
2. Dimensionality reduction is to reduce the input variable data in training data. This method is used in big data visualization, structure discovery, and meaningful compression.

7.8.3 Transfer Learning

Transfer learning (TL) is to gain knowledge and focus on storing it while solving one problem and applying gained knowledge to a different related problem [20]. TL is used in wireless communication in Green Reynolds Averaged Navier–Stokes (RANs), base station switching, spectrum management, and cache placement.

7.8.4 Reinforcement Learning

Reinforcement learning (RL) mainly focuses on making a selection that might be generated by mapping the situations to action and comparing which moves are needed to be considered for maximizing an extended time reward. RL is used in making real-time decisions, robot navigation, game AI, and skill acquisition. RL is also used in wireless communication using historical-based RL algorithm in vehicular network [21], deep RL algorithm used in cloud RANs and D2D network in communication.

7.9 DEEP LEARNING

DL is based on Artificial Neural Network (ANN) consisting of one or more hidden layers. DL has access to large data set and computational power. Conventional Neural Network (CNN) is used in image recognition tasks to reduce the input size, which is frequently utilized. Recurrent Neural Network (RNN) is most suitable for learning tasks that require sequential models. For dimension reduction, the autoencoder-based techniques on DL are used. To generate samples similar to the available data set, Generative Adversarial Network (GAN) is used. In 6G networks, DL will not be of much help in obtaining an optimal solution for all data analysis tasks [22].

7.10 CONCLUSION

This chapter presents the importance of ML in wireless communication for 5G and Beyond. ML works efficiently at the different layers of a wireless communication system, such as the physical layer, MAC layer and application layer. At the physical layer, ML is the most suitable technique for optimizing channel coding, beamforming, channel estimation, and signal detection. AI in wireless networks gives high QoS with regard to end-to-end communication. Big data analytics for 6G has a wide scope in exploring advanced technologies. AI with ML is a key enabling technology for 6G networks. This technology is used in complex D2D communication to provide an optimal solution for ultra-high speed communication problems.

REFERENCES

[1]. M. Giordani, M. Polese, M. Mezzavilla, et al., "Towards 6G networks: Use cases and technologies". *IEEE Comm. Mag.*, vol. 58. pp. 55–61, 2020. doi: 10.1109/MCOM.001.1900411

[2]. S. Mumtaz, J. Miquel Jornet, J. Aulin, et al., "Terahertz communication for vehicular networks". *IEEE Trans. Veh. Technol.*, vol. 66. pp. 5617–5625, 2017. doi: 10.1109/TVT.2017.2712878

[3]. Z. Zhang, Y. Xiao, z Ma, et al., "6G wireless networks: Vision, requirements, architecture, and key technologies". *IEEE Veh. Technol. Mag.*, vol. 14. pp. 28–41, 2019. doi: 10.1109/MVT.2019.2921208

[4]. M. Chowdhury, Md. Shahjalal, S. Ahmed, et al., "6G Wireless communication systems: Applications, requirements, technologies, challenges, and research directions". *IEEE Open J. Commun. Soc.*, vol. 1. pp. 957–975, 2020. doi: 10.1109/OJCOMS.2020.3010270

[5]. C. Wang, F. Haider, X. Gao, et al., "Cellular architecture and key technologies for 5G wireless communication networks". *IEEE Commun. Mag.*, vol. 52. pp. 122–130, 2014. doi: 10.1109/MCOM.2014.6736752

[6]. P. Botsinis, Z. Babar, D. Alanis, et al., "Quantum search algorithms for wireless communications". *IEEE Commun. Surv. Tut.*, vol. 21. pp. 1209–1242, 2019. doi: 10.1109/COMST.2018.2882385

[7]. T. Rappaport, S. Sun, R. Mayzus, et al., "Millimetre-wave mobile communications for 5G cellular: It will work!" *IEEE Access*, vol. 1. pp. 335–349, 2013. doi: 10.1109/ACCESS.2013.2260813

[8]. H. Gao, Y. Su, S. Zhang, et al., "Antenna selection and power allocation design for 5G massive MIMO uplink networks". *China Commun.*, vol. 16. pp. 1–15, 2019. doi: 10.12676/j.cc.2019.04.001

[9]. L. Khan, I. Yaqoob, Z. Han, et al., "6G wireless systems: A vision, architectural elements, and future directions". *IEEE Access*, vol. 8. pp. 147029–147044, 2020. doi: 10.1109/ACCESS.2020.3015289

[10]. J. Du, c Jiang, J. Wang, et al., "Machine learning for 6G wireless networks: Carrying forward enhanced bandwidth, massive access, and ultrareliable/low-latency service". *IEEE Veh. Technol. Mag.*, vol. 15. pp. 122–134, 2020. doi: 10.1109/MVT.2020.3019650

[11]. B. Khaled, W. Chen, J. Zhang, et al., "The roadmap to 6G: AI empowered wireless networks". *IEEE Commun. Mag.*, vol. 57. pp. 84–90, 2019. doi: 10.1109/MCOM.2019.1900271

[12]. J. Kaur, M. Khan, M. Iftikhar, et al., "Machine learning techniques for 5G and beyond". *IEEE Access*. 2021, doi: 10.1109/ACCESS.2021.3051557

[13]. C. Andrieu, N. De Freitas, A. Doucet, et al., "An introduction to MCMC for machine learning". *Springer, Mach. Learn.*, vol. 50. pp. 5–43, 2003. doi: 10.1023/A:1020281327116

[14]. T. O'Shea and J. Hoydis, "An introduction to deep learning for the physical layer". *IEEE Trans. Cogn Commun. Netw.*, vol. 3. pp. 563–575, 2017. doi: 10.1109/TCCN.2017.2758370

[15]. S. Kotsiantis, I. Zaharakis, and P. Pintelas, "Supervised machine learning: A review of classification techniques", *Informatica*, vol. 31. pp. 249–268, 2007. Available: https://datajobs.com/data-science-repo/Supervised-Learning-[SB-Kotsiantis].pdf

[16]. O. Chapelle, B. Scholkopf, and A. Zien, *Semi-supervised learning (adaptive computation and machine learning series)*. The MIT Press, 2006.

[17]. J. Friedman, T. Hastie, and R. Tibshirani, *The elements of statistical learning*. Springer, 2001. doi: 10.1007/978-0-387-84858-7

[18]. D. Heckerman, *A tutorial on learning with Bayesian networks*. Springer, pp. 301–354, 2008. doi: 10.1007/978-3-540-85066-3_3

[19]. R. Andrei and G. Thadeu, *6G: The wireless communications network for collaborative and AI applications*. Cannel University, 2019. doi: arXiv:1904.03413v1.

[20]. M. Sheraz, M. Ahmed, X. Hou, et al., "Artificial intelligence for wireless caching: Schemes, performance, and challenges". *IEEE Commun. Surv. Tut.*, vol. 23, pp. 631–661, 2021. doi: 10.1109/COMST.2020.3008362

[21]. F. Tang, N. Kato, J. Liu, et al., "Future intelligent and secure vehicular network toward 6G: machine-learning approaches". *Proc. IEEE*, vol. 108. pp. 292–307, 2020. doi: 10.1109/JPROC.2019.2954595

[22]. A. Zappone, M. Renzo, and M. Debbah, "Wireless networks design in the era of deep learning: Model-based, AI-based, or both?" *IEEE Trans. Commun.*, vol. 67. pp. 7331–7376, 2019. doi: 10.1109/TCOMM.2019.2924010

GREEN-Cloud Computing (G-CC) Data Center and Its Architecture toward Efficient Usage of Energy

Devasis Pradhan[1], K C Priyanka[2], and Rajeswari[3]

[1] *Assistant Professor, Department of Electronics & Communication Engineering, AIT, Acharya, Bangalore*
[2] *Assistant Professor, Department of Electronics & Communication Engineering, AIT, Acharya, Bangalore*
[3] *Professor, Department of Electronics & Communication Engineering, AIT, Acharya, Bangalore*

8.1 INTRODUCTION

Cloud computing is a group of computing services that puts together administrations that furnish with respect to the request, universal, advantageous organization access to a common platform for configurable registering assets that can be quickly provisioned and included with negligible change in the existing administration. Cloud computing can be quickly developed through various stages that involve lattice utilization of computing and application-oriented services [1]. These administrations can be scaled all over relying upon the customer's variable activity requirements, guaranteeing highest expense proficiency. Appropriation of cloud-based administrations empowers organizations to stay abreast with the quick advancing and dynamic business environment, thereby gaining better unwavering quality, less support and higher availability.

In the 21st century, "green" has advanced to identify with natural issues [2]. Undeniable degrees of CO_2 emanation is hazardous and can mess with the wellbeing of future generation. As registering turns out to be progressively unavoidable, the energy utilization because of processing continues to increase. Green computing is growing as a basic data correspondence innovation to alter the course. It utilizes figuring of assets in a climate in well-disposed way while keeping up by and large figuring execution.

The main objective of this chapter is to discuss the techniques that help decrease the utilization of unsafe gadgets or they must be used within the stipulated lifetime to make the cloud energy more efficient. Green-cloud computing (GCC) is an efficient step toward the feasible usage of data center with minimal usage of energy to make the earth green. It is a significant step in registering frameworks, from handheld cell phones to data centers, which are hefty shoppers of energy. Minimizing the energy utilization in data centers is a difficult and complex task since figuring applications and information are developing so rapidly that inexorably bigger workers and plates are expected to handle them rapidly within the required time. GCC is imagined to accomplish not only effective preparing and use of processing foundation, but also limit energy utilization. To address this challenge, server farm assets should be overseen in an energy-effective way to drive GCC [1,3–5].

8.1.1 The Need for GCC

Cloud computing is inconceivably demanding nowadays on account of its simple openness and value-saving nature to framework and administrations on interest. Numerous organizations are working with extremely colossal server farms around the world. On account of the force of distributed computing, it provides capacity to PC to measure the huge amount of data created every day. Virtual data centers require high-energy utilization for all activities. To minimize the utilization, GCC is implemented. By utilizing GCC, energy utilization and carbon dioxide discharge by cloud computing will actually be an extraordinary ecological area of concern [6,7].

There are various reasons for utilizing green computing. Some of them are listed below.

a. Personal PC and electronic gadget burn-through a lot of power which destructively affects our current circumstance. It gives rise to diverse types of toxins which contaminate the air, land and water. Power is produced utilizing the petroleum derivatives.
b. Most of the PCs and electronic gadgets produce a great deal of heat which causes CO_2 emission. CO_2 is the greenhouse gas; with increase in CO_2 in the air the general temperature likewise increases, which leads to global warming [5].
c. While receiving PCs and its assets create a ton of risky squander, which can in all ways harm our current climate circumstance. It likewise leads to emission of heavy metals like lead, mercury, cadmium into air.
d. Assembling of PC and its item discharge vigorously utilization of poisonous compounds for electrical protection, binding and fire creation.

8.2 HISTORY OF CLOUD COMPUTING

Cloud computing is just the availability of planning power, amassing and applications passed on interest to the customers over the web. For example, when we say dealing with power, that is the high-end laborers, which are made open for encouraging your regions, running your endeavor applications and that is just a hint of something larger. Figure 8.1(a) and (b) discusses the general structure of cloud computing and the service provider.

(a)

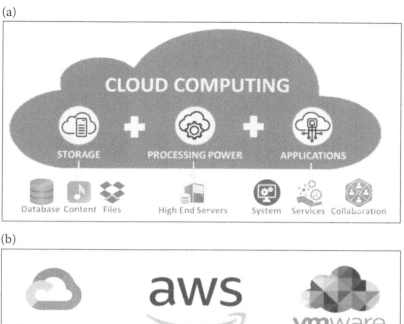

(b)

FIGURE 8.1 (a) Cloud computing (Source: Simply Coding). (b) Companies like Google, Amazon, Microsoft, IBM and many others provide such Cloud Platforms (Source: Simply Coding).

In the past, in 1960s, when PCs were considered a luxury, IBM and DEC used to share their PCs. In 1972, IBM developed the principle of Virtual Machine. The cloud picture was used way back in 1977. At the point when the website came into picture, many telecom associations offered VPN and Compaq started offering it on the web circle storage to keep records. The word cloud itself was conceived by Ramesh Chellappa in 1997. The word cloud was used as a portrayal for the Internet and cloud picture was used to address association of enrolling gear. In 1999, cloud associations, for instance, Salesforce and VM Ware, were combined. In 2002, Amazon launched the AWS. In the 2006, Hadoop was released. In 2008, Google launched the Google App Engine. In 2010, Microsoft released Microsoft Azure and in a couple of years nearly everybody bounced onto spread handling design. Figure 8.2(a) discusses the history of cloud computing.

451 Research predicts that around 33% of large business IT spending will be for encouraging and cloud benefits this year "showing a creating reliance on external wellsprings of establishment, application, and the heads and security organizations". Inspector Gartner predicts that part of the overall undertakings using the cloud right now will have wagered everything on it by 2021. As demonstrated by Gartner, overall spending on cloud organizations will reach $260 bn this year up from $219.6 bn [8]. It is moreover being created at a faster rate than the specialists expected. Regardless, it is not totally clear whether the measure of that solicitation is coming from associations that truly need to move to the cloud and what sum is being made by dealers who presently offer cloud variations of their things. Figure 8.2(b) discuss about the service revenue created from cloud computing.

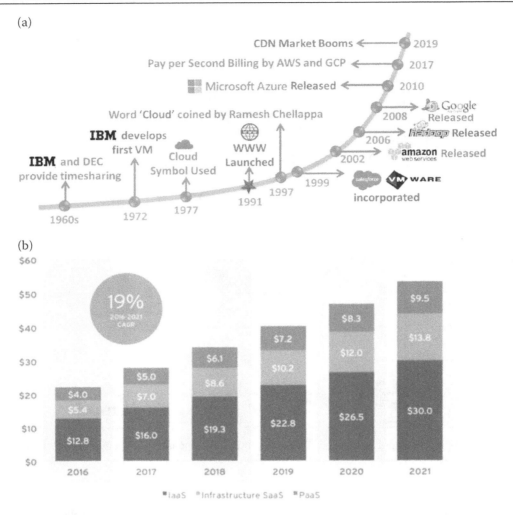

FIGURE 8.2 (a) History of cloud computing (Source: Simply Coding). (b) Cloud computing as a service revenue ($bn) (Source: Predictions for cloud computing revenues to 2021 from 451 Research).

8.3 OVERVIEW OF GREEN CLOUD COMPUTING

Presently, one of the well-known green computing services is the strategic gradual. This service applies and utilizes green figuring methods of reasoning, basically to set aside on expenses as opposed to save the climate. This green computing idea came into existence normally as organizations end up constrained to augment assets to contend viably on the lookout. This development emerged primarily from monetary suppositions as opposed to political pressing factor. Key leaders consider the social and natural effects of new and arising innovations. Besides limiting expenses, this specific development additionally considers different factors like advertising and marketing. Dissimilar to the position held by strategic steady, essential pioneers perceive the need to upgrade some current arrangements or underlying cosmetics of the association [8–10].

Perhaps the greatest test confronting the climate today is an unnatural weather change because of fossil fuel byproducts. The expression "green" is utilized to allude to naturally feasible exercises. It implies utilizing PCs in manners that save the climate, energy and economy. GC was started in 1992 by the USA.

To be considered green the following is required:

a. Decreasing CO_2 level and making efficient use of energy in data centers
b. Minimizing e-waste
c. Enabling lifestyle changes that lower influence on the environment

GCC may be seen as the demonstration of using enrolling resources for achieve most prominent benefit with no damaging effect on the environment.

The vital goals of GC incorporate:

a. Limiting energy consumption
b. Buying efficient power energy
c. Decreasing the utilization of paper and other consumables
d. Limiting gear evacuation necessities
e. Decreasing travel prerequisites for representatives/clients

8.3.1 Features of Green Cloud Computing

Lower petroleum derivative byproduct (CO_2) is ordinary in cloud listing due to an incredibly energy-capable system and decreasing establishment of software farms. The basic concept behind the development of energy-capable cloud is "Virtualization", which grants basic improvement in energy adequacy of service providers. GCC basically incorporates four key zones outlined in Figure 8.3 and clarified as follows [8]:

a. **Green Use:** Using assets in a naturally stable way while decreasing their energy utilization.
b. **Green Design:** Planning energy profitable and earth sound things and organizations.
c. **Green Scrapping:** Reusing e-waste with irrelevant or no effect on the climate.
d. **Green Fabrication:** Amassing electronic gadgets with insignificant effect or no effect on the climate.

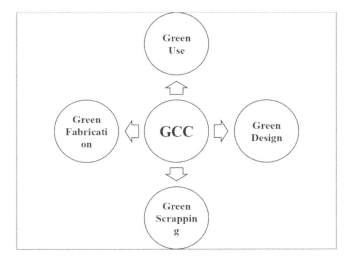

FIGURE 8.3 Basic component of GCC.

8.3.2 Approaches

By union of underutilized workers as various VMs sharing same actual worker at higher usage, organizations can acquire high investment funds for space, the board and energy. Figure 8.4 discusses the basic approaches for GCC.

8.3.2.1 Dynamic Provisioning

There are different purposes:

  a. Hard to expect leisure activities at a time; that is especially broad for web programs.
  b. Ensure accessibility of associations and to keep up explicit degree of association quality to end users. The construction provisioned with a moderate framework achieves nonuse of assets [9].

Cloud service providers screen and foresee the request and consequently apportion assets as indicated by request. Along these strains, server farms reliably maintain up the powerful worker's consistent with current hobby, which achieves the utilization of energy in a better manner.

8.3.2.2 Multi-Occupancy

With the help of this methodology, basic requirement of cloud foundation diminishes by and large energy utilization and related fossil fuel byproducts. The SAAS suppliers serve various organizations on same foundation and programming. This methodology is clearly more energy effective than numerous duplicates of programming introduced on various foundation. Besides, organizations have exceptional factor requests for designs by and large, and consequently multi-tenure on a similar worker permits the smoothing of the general pinnacle request which can limit the need for additional foundation [10].

8.3.2.3 Server Utility

On-premise system run with low utilization, a portion of the time it goes down depend upon 5% to 10% of ordinary use. Utilizing virtualization movements, different applications can be supported and executed on a relative worker in separation, in similar way lead to go through levels to 70%. Regardless of the way

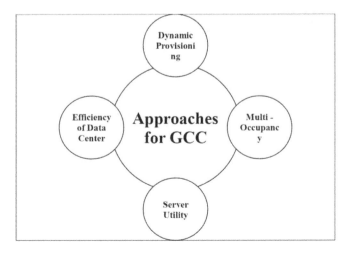

FIGURE 8.4 Approaches for green cloud computing (GCC).

that high utilization of laborers achieves extra force use, workers running at higher use can deal with greater obligation with equivalent power use [10,11].

8.3.2.4 Efficiency of Data Center

As previously discussed, the force productivity of data centers significantly affects the complete energy utilization of cloud processing. By utilizing the most energy fit advances, cloud providers can fundamentally improve the PUE of their worker ranches. Gigantic cloud master networks can achieve Power Usage Adequacy (PUE) levels as low as 1.1 to 1.2, which is about 40% more force ability than the standard worker ranches. The specialist plan as indicated by isolated compartments, water- or air-based cooling, or progressed power the board through power supply streamlining, are by and large advances toward that have basically improved PUE in worker ranches [12].

8.4 BENEFITS OF GOING GREEN

The basic establishments of these thoughts can be followed back to 1992 when the United States Environmental Protection Agency launched a program to reduce ozone-harming substance transmissions—Energy Star. This program was intended to distinguish and advance items that followed energy-productive rules.

8.4.1 Decrease in Green House Gases Emission (GHG-E)

Cloud computing cuts the measure of GHG produced from server farms. A customary on location server farm produces GHG for the duration of its lifecycle including:

   a. Delivering crude materials for the hardware
   b. Collecting the hardware
   c. Moving the hardware to the on location server farm
   d. Utilizing the hardware
   e. Arranging off the hardware when its lifecycle closes

Cloud diminishes the pace of fossil fuel byproducts by reducing the energy necessity and utilization. A study directed by Accenture, Microsoft and WSP uncovered that distributed computing substantially affects fossil fuel byproducts. It confirmed that by utilizing distributed computing, enormous organizations can reduce per-client carbon impression by 30% and up to 90% for little endeavors [13–15].

8.4.2 Dematerialization

Cloud computing makes a positive commitment to maintainability through dematerialization.
   Dematerialization alludes to the substitution of high-carbon actual items with virtual counterparts. This assists with decreasing energy use and carbon impression. Cloud administrations urge individuals to utilize virtual administrations like web-based video rather than asset substantial actual items [15].
   Relocating to the cloud implies that you utilize less machine and equipment that utilize less energy and affect the climate. Thus, organizations have lower energy charges and opened up funding to designate to productive undertakings. Cloud administrations help firms center their time and exertion around different undertakings rather than every day IT assignments and issues.

8.4.3 Move to Renewable Energy Sources

Some cloud server farms have moved to sustainable wellsprings of energy to control their tasks and thus lessen their carbon impression. They are utilizing sustainable wellsprings of energy like sun-powered, geothermal, hydropower assets and wind. A great representation of this is Agile IT cooperating with Arcadia Power to completely control its base camp and organization activities focusing on wind power [16].

Cloud computing offers organizations something other than an approach to securely store information, increment proficiency and decrease costs. It has significant natural advantages that are beyond any reasonable amount to overlook in a period where a dangerous atmospheric deviation and environmental changes are major problems [16,17]. Distributed computing can save billions of dollars in energy costs and decrease fossil fuel byproducts by a huge number of metric tons. It gives organizations the capacity to diminish their energy utilization, cut their carbon impression and move to a greener and more intelligent future.

8.4.4 Redesign of Cooling Framework

The cooling framework should consolidate outside air and working temperature that diminishes the energy use up to an extraordinary setting. Warm profiling, which is an unpredictable arrangement of time-temperature information, ought to be utilized to distinguish problem areas and overcooling in the data center.

8.4.5 Risk Management

Risk management is the greatest challenge for any business. Brand and popularity that requires a very long time to build can be destroyed within a couple of hours through practices that are against climate. Green practices can ensure organizations against administrative changes, rising fuel costs, advertising liabilities and the changing necessities of clients [17].

8.4.6 Improved Social and Corporate Image

Environmental manageability can be an amazing Corporate Social Responsibility (CSR) activity for organizations that help them assemble client devotion, solid brand mindfulness and separation from contenders. This better connects with workers and they can turn out to be more dedicated to the organization by taking an interest in exercises that advance green practices.

8.5 RECENT TRENDS IN GREEN CLOUD COMPUTING

The advantages of distributed computing are driving more organizations and undertakings across industry spaces and districts to put resources into cloud-fueled applications and administrations that are very mainstream as of now [16].

8.5.1 The Emergence of Digital Natives in the Workforce

In the following decade, the labor force will consist of millennial experts who are innovation experts and are open to working with web advancements. The new innovation laborers, called advanced locals, are likewise knowledgeable in cloud computing future patterns and will utilize them consistently for work and correspondence. The incorporation of computerized locals in the cutting-edge workspace is among the most energizing patterns in the cloud computing space. With their insights into web-based media and specialized apparatuses, computerized locals can be a resource for any innovation empowered association. Associations actually need to incorporate computerized locals with the current colleagues to build up a more assorted and hearty labor force. They are likewise embracing late patterns in distributed computing, just as computerized rehearses like opposite coaching to prepare the more established labor force in cloud-based administrations [17,18].

8.5.2 Artificial Intelligence (AI)

As indicated by industry statistics delivered by Statista, the worldwide market for Artificial Intelligence or AI is projected to cross USD 89 billion by 2025. The rise and reception of AI-related advancements lately are likewise fuelling cloud innovation patterns. The utilization of self-learning calculations through AI is fueling the computerization of business measures, in obvious sense. Moreover, with AI getting more brilliant, associations should construct clever cycles inside their framework—that is quicker and precise. Be it through abilities like machine learning or facial recognition, AI frameworks are creating huge volumes of big data that must be scaled up through cloud computing arrangements [19].

8.5.3 Hybrid Cloud Computing

With cloud applications getting more convenient, undertakings are sending their jobs across various IAAS stages, including Amazon Web Services (AWS), Google Cloud and Microsoft
    Azure. Thus, hybrid cloud computing that coordinates nearby workers with private and outsider public cloud administrations are among the most recent patterns in cloud computing. Among the arising patterns in GCC, application holders have gotten very well-known in the past as they give simplicity of overseeing and in any event, moving application code. Since its launch, Kubernetes has become the favored stage for making holders and is projected to be among the best future innovation patterns in GCC [19,20].

8.5.4 Quantum Computing

Projected among the growing patterns in cloud computing, quantum computing has caught the consideration of driving IAAS suppliers, including Amazon and Microsoft. In engineers' language, quantum computing is the figuring of different factors simultaneously—that can take care of issues rapidly and precisely. Quantum computing has become one of the top patterns in cloud computing fundamentally because of the progressions in PC equipment and design that permits information to be handled quicker than previously. Different advantages of quantum computing incorporate diminishing energy utilization, in this way positively affecting the climate. As quantum computing is as yet in the beginning stage of improvement, it has some more years to disturb enterprises, similarly, various advancements like AI and machine learning have accomplished. Because of quick development, quantum computing is among the future innovation patterns in cloud computing and can take registering capacity to the next level [20].

8.5.5 High Performance Computing in Public Cloud Storage

Public cloud storage is a model that empowers you to store, oversee and alter the information. To perform computations and interaction these information at high velocity is the thing that we call high-performance computing (HPC) in the realm of the cloud. Be that as it may, as smooth it sounds, High-execution registering is really costly because of the great finished assets, necessities and popularity. This is the reason associations have been hesitant about utilizing it in earlier years. In 2021, 60% of the ventures have concluded as of now to run the HPC responsibility in their public cloud. While, in 2016, just 26% of them knew about its genuine advantages. This ascent is supported by new sorts of public cloud occurrences like Azure H-/NC-/ND-/NV-arrangement, AWS C5/P2/P3/G3 occasions and Google V100/P100/K80. In the coming years, we will likewise see a rising number of HPC-centered arrangement suppliers [20,21].

8.6 TECHNIQUES OR METHODS TO MAKE CLOUD GREEN

These methods could change the future of green cloud computing in 5G era:

a. **Nano Data Centers:** Newly created processing stage which utilizes network access suppliers (ISP) controlled home doors to offer registering and capacity administrations. Nano data centers are more energy-proficient than traditional server farms. They help diminish the expense of warmth dissemination, they have high assistance vicinity, and have the limit with respect to self-variation or self-versatility.
b. **Dynamic Voltage Frequency Scaling:** A technique that lessens the force and energy utilization measures utilized with recurrence scaling. Executing this strategy will decrease energy utilization and influence the usage of the asset. This technique depends upon a clock being identified with electronic circuits; its working rehash is synchronized with the stock voltage at any rate, its force hypothesis saves are low stood apart from different frameworks. DVFS draws in processors to run at various blends of frequencies with voltage to reduce the utilization of the processor's force [21].
c. **Virtualization:** A strategy that improves machine, board and energy proficiency through sharing a solitary actual example of an asset/application with numerous clients or associations simultaneously. Virtualization augments the quantity of accessible framework assets in an eco-accommodating way. It empowers better checking and the executives of the asset portion, and it helps the worker bunch in amplifying their capacity to share assets.
d. **Optimal Server Utilization:** Customarily, numerous servers stay inactive of 85%–95% of the time utilizing almost as much force as they do when they are dynamic. Virtualization innovation empowers facilitating of numerous applications through one worker. The quantity of dynamic workers is decreased and the force utilization is lower.
e. **Energy-Efficient Client Gadgets:** The public cloud model decreases the quantity of energy devouring customers through little energy-effective gadgets (e.g., slender customers).

8.7 GREEN CLOUD COMPUTING ARCHITECTURE

Cloud service providers, being benefit situated, are searching for arrangements which can diminish the force utilization and in this way, fossil fuel byproduct without harming their market. Consequently, we give a bound together answer for empower GCC. Green cloud systems consider these objectives of suppliers while controlling the energy utilization of clouds. Figure 8.5 discuss about the GCA. In the

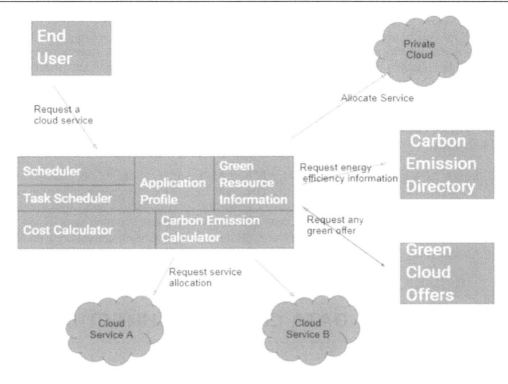

FIGURE 8.5 Green cloud architecture.

GCA, customers present their cloud supplier requests through another middleware green broker that manages the assurance of the greenest cloud provider to serve the customer's sales. A customer organization requesting can be of three sorts, that is, programming, stage and system [21,22].

The green cloud contains green organizations, assessing and time when it should be gotten to for least petroleum product side-effect. Green broker gets the current status of energy limits for using diverse cloud network from CER. The CER keeps up all the information identified with energy effectiveness of cloud distribution. This data may fuse PUE and cooling viability of cloud server farm which is offering the help, the association cost and non-renewable energy source result in speed of force, green broker figures the petroleum derivative side-effects of all the cloud providers who are offering the referenced cloud organization. By then, it picks the course of action of organizations that will achieve least petroleum product side-effect and buy these organizations for customer advantage.

Assuming a pivotal part in observing and choosing the cloud services dependent on the client QoS prerequisites, and guaranteeing least fossil fuel byproduct for serving a client. By and large, a client can utilize cloud to get to any of these three sorts of services, and in this way interaction of serving them ought to likewise be energy proficient [23,24].

a. **SAAS Level**

SAAS provider fundamentally offer programming presented in their own worker homestead or resources from IAAS providers, the SAAS providers need to show and check energy productivity of their thing, that is, plan, execute and send. For serving clients, the SAAS supplier picks the worker ranches which energy are convincing similarly as close to clients. The base number of pantomimes of clients confined information ought to be kept up utilizing energy-reasonable breaking point [25].

b. **PAAS Level**

PAAS providers offer by and large the stage associations for application progress. The stage engages the movement of businesses which guarantees framework-wide energy ability. This should be possible by breaker of different energy profiling contraptions, for example, JouleSort [5]. Other than application movement, cloud orchestrates in addition permit the relationship of client applications on hybrid cloud. For the current condition, to accomplish most unmistakable energy ampleness, the stages profile the application and pick what part of use or information ought to be dealt with in house and in cloud [24–26].

c. **IAAS Level**

They go through most to-date advancements for IT and cooling frameworks to guarantee most energy productive foundation. By utilizing virtualization conjointly, affiliation, the energy utilization is to boot cut by murdering un-utilized laborers. Absolutely interesting energy meters and sensors region unit familiar with work out the current energy capacity of each IAA suppliers and their protests. This data is pitched methodologically by cloud providers in CER totally extraordinary unpracticed booking and quality provisioning approaches can ensure least energy use [26,27].

8.8 GREEN DATA CENTER ARCHITECTURE

Figure 8.6. portrays the engineering of the green data center fueled by both sustainable power and conventional energy from the utility network. The framework utility and sustainable power are consolidated together through the programmed move switch to give capacity to the data center. The IT gadgets incorporate workers, stockpiling and systems administration switches that help applications and administrations facilitated in the data center [28]. The cooling gadgets convey the cooling assets to dispatch the warmth created by IT hardware. In this chapter, the cooling limit is conveyed to the data center through the PC room cooling units from the cooling micro-grid that comprises conventional chiller plant. The design does not consider the energy stockpiling gear, in light of the fact that the energy stockpiling hardware has accompanying inadequacies [29]:

a. The interior opposition and self-discharge of the battery can bring about loss of energy
b. The battery-related costs prevail in sunlight based fueled frameworks
c. The synthetic substances in the battery can harm the climate in any way

Considering the utilization of environmentally friendly power as a piece of the energy supply in the maintainable data center and the way that the sun-oriented force age is obscure and flimsy, the measure of sun powered energy created ahead of time to all the more likely timetable the batch-type occupations. Along these lines, the effect of unsteady sun powered energy supply on data center booking occupations can be somewhat dodged. There are numerous specialists who utilize an assortment of techniques to anticipate the measure of force age.

8.9 ENERGY TO BE SAVED

The goal is to reduce power utilization of the laborer farm. It offers highlights like electronic checking, live virtual machine movement and progress in distribution of VMs. Through this proposed GCA, up to 27% energy can be saved. The need to oversee different applications in a specialist farm makes the

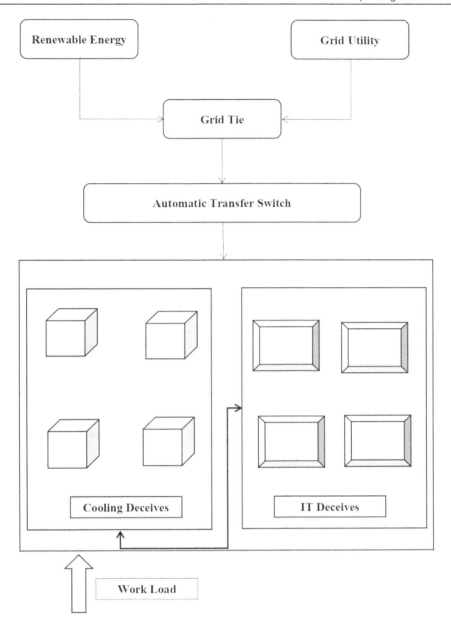

FIGURE 8.6 Green data center architecture.

preliminary of on-request asset provisioning and task because of time-moving positions. Specialist farm assets are statically expected to applications, considering the zenith load qualities, to give execution ensures and to keep up the withdrawal [29–32].

Energy-capable resource the load up has been first given respect to battery dealt with phones, where energy usage should be diminished. The principal framework is to mix the dynamic voltage frequency scaling method in a suitable manner to help the asset use and to limit energy utilization of the worker ranches which will accomplish the diminished carbon impressions and accordingly will help more in accomplishing the green computing. The green metrics for energy savings are discussed in Table 8.1.

TABLE 8.1 Green metrics-energy/power measurement for data center (Source).

SL. NO	METRICS	FORMULA	EXPLANATION
1	Power Usage Efficiency (PUE)	$PUE = \dfrac{\text{Total Facility Energy}}{\text{IT Equipment Energy}}$	It is the small amount of aggregate energy devoured by the assistance of a server farm to the complete energy devoured by IT types of gear.
2	Carbon Usage Efficiency	$CUE = \dfrac{\text{Total CO}_2 \text{ emmitted}}{\text{Total energy consumed by IT Equipment}}$	It is a calculation of greenhouse gases (CO_2, CH_4) release in atmosphere by the data center.
3	Water usage efficiency	$WUE = \dfrac{\text{Average usage of water}}{\text{Total energy consumed by IT Equipment}}$	It is estimation of yearly water utilized by server farm like for cooling, energy Creation.
4	Energy reuse factor	$ERF = \dfrac{\text{Used of reuse of energy}}{\text{Total energy consumed by IT Equipment}}$	It computes the reusable energy Like hydro power, sunlight based force and so forth utilized by data center
5	Energy reuse efficiency	$ERE = \dfrac{\text{Total energy} - \text{reused energy}}{\text{Total energy consumed by IT Equipment}}$	It is a boundary for estimating the benefit of reuse energy from a server farm.
6	Data center infrastructure efficiency	$DCiE = \dfrac{\text{Total IT Equipment Power utilization}}{\text{Total power facilitated}}$	This factor is utilized to figure the energy effectiveness of a data center.
7	Data center productivity	$DCP = \dfrac{\text{Total Useful Work}}{\text{Total resource used for the work}}$	It computes the measure of valuable work done by data center.
8	Compute power efficiency	$CPE = \dfrac{\text{IT Equipment Utilization Energy}}{\text{PUE}}$	It decides the aggregate sum of force is genuinely utilized for processing
9	Green energy coefficient	$GEC = \dfrac{\text{Green Energy Used}}{\text{Total Energy Consumed}}$	It measure the measure of efficient power energy used to give administrations to server farm.
10	Data center energy productivity	$DCeP = \dfrac{\text{Total Useful Work Done}}{\text{Total Energy used to do the work}}$	It ascertains the amount of valuable work done by data center as contrast with all out energy burned-through to make this work.

8.10 OPTIMIZATION OF ENERGY EFFICIENCY

Energy is quite possibly the most significant and scant assets accessible to the world, an incredible bit of which is currently being devoured all together catalyst PCs and registering foundation. Fundamentally, a large portion of the superior equal machines and dispersed processing framework, including server farms, supercomputers, groups, constant frameworks and matrices burn-through extensive measures of force as well as require cooling to keep the frameworks cool. The unexpected development in registering is quickly expanding the utilization of valuable normal assets like oil and coal, reinforcing the disturbing peril of energy deficiency. The scientists raise this issue every now and then and the potential measures are being taken to survive and to diminish it [29].

8.10.1 Improvement Techniques in Performance-Energy-Temperature Aware Computing

The dramatic development in figuring movement and the rizing worry for energy protection have made energy productivity in PCs an innovative issue of prime significance. The tradeoff between Performance-Energy-Temperature must be made for so the most extreme advantages can be gotten. Planning methods that are ideal regarding execution, energy and temperature are most extreme necessity taking everything into account.

8.10.2 Data Resource Tier Optimization

The data asset level addresses significant information base administration frameworks in the worldwide calculation world. General ideal models incorporate information bases, registries, record frameworks and level documents. It moreover incorporates the coordination of various information base constructions with the goal that various data sets can be investigated regardless of their putting away instruments and information structure.

8.10.3 Exceptionally Productivity Data Center Plan

Bigger data centers can be made substantially more energy effective than more modest server farms. Norms are arising for estimating this, like the idea of PUE. PUE is characterized as the proportion of all out office power separated by IT hardware power [13]. Consequently, it is a proportion of the amount of the force being devoured by the office, which is really being utilized to control the IT gear itself instead of the multitude of different things. Subsequently it will be a test to make the greater server farms power productive.

8.10.4 Creating Green Maturity Model

Full hardware lifecycle is the primary territory for green development model, with energy decrease as the best proportion of greenness. The need for development models for equipment, IT associations and registering procedures is an issue which has been tended to by a few analysts yet is restricted to explicit regions. Green development model for virtualization [14] portrays that each level portrays the level of green attributes.

8.10.5 Green Software

Recently, green programming development has become an examination subject for the majority of the product engineer's organizations as a result of need for manageable improvement [16]. The majority of the exploration has been done on the portrayal, measurements and specialized response for green programming; however, few have tended to green programming from the business point of view. Business associations are moving toward green programming and still some extensive advances should be taken.

8.10.6 Wireless Sensor Network for Data Center Cooling

Data center cooling is a serious issue taking everything into account. Server farms are spine of any registering association and should be dependable and accessible at each purpose of time. Estimating the server farm viability and keeping up the pattern is an issue. Remote sensors could assume a major part for overseeing server farms power the executives.

8.11 CHALLENGES IN IMPLEMENTATION AND FUTURE SCOPE

In this segment, we distinguish key open issues that can be tended to at the degree of the board of framework assets. Virtualization advances, which cloud registering conditions intensely depend on, give the capacity to move VMs between actual hubs utilizing live of disconnected movement. This empowers the method of dynamic combination of VMs to a negligible number of hubs as per current asset necessities. Table 8.2 describes the future challenges in 5G-GCC [33–35].

8.11.1 Energy-Aware Dynamic Resource Allocation

Ongoing improvements in virtualization have brought about its expansion of use across server farms. By supporting the development of VMs between actual hubs, it empowers dynamic relocation of VMs as indicated by QoS prerequisites. At the point when VMs do not utilize all given assets, they can be consistently re-sized and united on an insignificant number of actual hubs, while inactive hubs can be turned off. As of now, asset allotment in a cloud server farm means to give superior while meeting SLA, without a zero in on allotting VMs to limit energy utilization. To investigate both execution and energy effectiveness, three vital issues should be tended to. To start with, exorbitant force cycling of a worker could lessen its unwavering quality. Second, turning assets off in a unique climate is hazardous from a

TABLE 8.2 Application domains research challenges and future directions toward 5G-Green Cloud Computing.

SL. NO.	SUB-AREA	CHALLENGES & FUTURE DIRECTION
1	IoT	i. Security ii. Real-time data processing iii. Standardization iv. Dynamic deployment in cloud-fog v. Cross-cloud IoT services vi. Application oriented resource modeling
2	Cyber physical system	i. Need for standard ii. Privacy and security iii. Programming abstraction
3	E-Science	i. Data lock-in ii. Real-time processing of scientific data
4	Mobile cloud computing	i. Low bandwidth ii. Heterogeneous network iii. Offloading in the static and dynamic environment iv. Efficiency of data centers
5	Cloud-based application model	i. Unified modeling technique ii. Modeling QoS requirement iii. Modeling dynamic aspect iv. Simulation support to deployment model

QoS planned. Because of the fluctuation of the responsibility and forceful solidification, some VMs may not acquire needed assets under top burden, so neglecting to meet the ideal QoS. Third, guaranteeing SLA carries difficulties to exact application execution the board in virtualized conditions [31].

8.11.2 QoS-Based Resource Selection and Provisioning

Server farm assets may convey various degrees of execution to their customers; subsequently, QoS-aware asset determination plays a significant job in cloud computing. Furthermore, cloud applications can introduce changing jobs. It is accordingly vital for do an investigation of cloud administrations and their jobs to recognize regular practices, designs, and investigate load anticipating approaches that can conceivably prompt more productive asset provisioning and resulting energy productivity. In this specific circumstance, we will explore test applications and connections among jobs, and endeavor to construct execution models that can help investigate the compromizes among QoS and energy saving. Further, we will explore another online way to deal with the combination procedure of a server farm that permits a decrease in the quantity of dynamic hubs required to handle a variable responsibility without debasing the offered administration level. The online technique will naturally choose a VM arrangement while limiting the quantity of actual hosts expected to help it. Additionally, another objective is to give the agent (or shoppers) with asset choice and responsibility combination approaches that misuse the compromises between execution and energy saving [31,32].

8.11.3 Effective Consolidation of VMs for Managing Heterogenous Workloads

Cloud framework administrations give clients the capacity to arrange virtual machines and dispense any sort of applications on them. This prompts the way that various sorts of utilization (e.g., endeavor, logical and interpersonal organization applications) can be dispensed on one actual PC hub. Nonetheless, it is not clear how these applications can impact one another, as they can be information, organize or process escalated consequently making variable or static burden on the assets. The issue is to figure out what sort of utilization can be apportioned to a solitary host that will give the most proficient by and large use of the assets. Current ways to deal with energy productive union of VMs in server farms do not research the issue of consolidating various kinds of responsibility. These methodologies typically center around one specific responsibility type or do not consider various types of uses expecting uniform responsibility. As opposed to the past work, we propose a savvy solidification of VMs with various responsibility types [29].

8.11.4 Advancements in Virtual Network Topologies

In virtualized server farms VMs frequently convey between one another, setting up virtual organization geographies. Be that as it may, because of VM relocations or non-upgraded designation, the conveying VMs may wind up facilitated on legitimately removed physical hubs giving expensive information move between one another. On the off chance that the imparting VMs are dispensed to the hosts in various racks or fenced in areas, the organization correspondence may include network switches that devour critical measure of force. To dispense with this information move overhead and limit power utilization, it is important to notice the correspondence among VMs and spot them on the equivalent or firmly found hubs. To give viable re-distributions, create power utilization models of the organization gadgets and gauge the expense of information move contingent upon the traffic volume. As movements burn-through extra energy and they adversely affect the presentation, prior to starting the relocation, the redistribution regulator needs to guarantee that the expense of relocation does not surpass the advantage [29,33].

8.11.5 Security

Quite possibly the main zones of conversation around distributed computing innovation would be security. It is significant for specialist organizations to guarantee that the information is put away both securely and safely. This calls for more expertise and information around distributed computing. For sure, this is one of the significant reasons why fate of distributed computing occupations is splendid. Organizations will require gifted experts who can guarantee security on the whole phases of cloud administrations. It is likewise significant for specialist organizations to guarantee that digital assaults are kept under control. Indeed, even little organizations that do not focus on security need to change their plan of action. Studies and innovations that pressure the improvement of things to come extent of distributed computing security will discover greater and better stages to demonstrate their speculations in the impending days [33,34].

8.12 CONCLUSION

Cloud computing business potential and commitment to previously disturbing fossil fuel byproduct from ICT have led to a discussion on whether cloud figuring is truly green. It is determined that the natural impression from server farms will significantly increase somewhere in the years 2002–2020, which is at present 7.8 billion tons of CO_2 each year. There are reports on green IT investigation of clouds and data centers that show that cloud processing is "Green", while others show that it will prompt disturbing expansion in carbon outflow. Hence, we previously dissected the advantages offered by cloud processing by contemplating its central definitions and advantages, the administrations it offers to end clients, and its sending model. At that point, we talked about the parts of cloud that add to fossil fuel byproduct and the highlights of clouds that make it "Green". A few exploration endeavors and advances that expansion the energy productivity of different parts of clouds can be utilized. The proposed green cloud framework introduced a few outcomes for its approval. Despite the fact that our green cloud structure installs different highlights to make cloud registering significantly more green, there are as yet numerous innovative arrangements needed to make it a reality.

Taking everything into account, by basically improving the proficiency of gear, cloud figuring cannot be professed to be green. What is significant is to make its utilization more carbon-productive, both from clients' and suppliers' point of view. Cloud providers need to decrease the power usage of clouds and make significant strides in utilizing environmentally friendly power sources as opposed to simply seeking cost minimization.

REFERENCES

[1]. G. Greco, C. Lucianaz, S. Bertoldo, and M. Allegretti, "A solution for monitoring operations in harsh environment: A RFID reader for small UAV", *2015 Int. Conf. Electromagn. Adv. Appl. (ICEAA)*, IEEE, pp. 859–862, 2015.

[2]. L. Wang and G. Von Laszewski, "Scientific cloud computing: Early definition and experience", *Proc. 10th IEEE Int. Conf. High Perform. Comput. Commun*, 2008.

[3]. S. Niles and P. Donovan, "Virtualization and cloud computing: Optimized power, cooling, and management maximizes benefits", White paper 118. Revision 3, Schneider Electric, 2011.

[4]. A. Beloglazov, R. Buyya, Y. C. Lee, and A. Zomaya, "A Taxonomy and survey of energy-efficient data centers and cloud computing systems", *Adv. Comput.*, vol. 82, 2011.

[5]. I. Chih-Lin, C. Rowell, S. Han, Z. Xu, G. Li, and Z. Pan, "Toward green and soft: A 5G perspective," *IEEE Commun. Mag.*, vol. 52, no. 2, pp. 66–73, 2014.

[6]. A. Ashraf and I. Porres, "Multi-objective dynamic virtual machine consolidation in the cloud using ant colony system", arXiv preprint rXiv:1701.00383, 2017.

[7]. P. Matre, S. Silakari, and U. Chourasia, "Ant colony optimization (aco) based dynamic vm consolidation for energy efficient cloud computing", *Int. J. Comput. Sci. Inf. Secur.*, vol. 14, no. 345, 2016.

[8]. D. Amendola, N. Cordeschi, and E. Baccarelli, "Bandwidth management VMs live migration in wireless fog computing for 5G networks, Cloud Networking (Cloud net)", *2016 5th IEEE Int. Conf.*, IEEE, pp. 21–26, 2016.

[9]. P. K. Paul, et al., "Green and environmental friendly domain and discipline: Emerging trends and future possibilities", *Int. J. Appl. Sci. Eng.*, vol. 2, no. 1, pp. 55–62, 2014.

[10]. X. Wang, C. Xu, G. Zhao, and S. Yu, "Tuna: An efficient and practical scheme for wireless access point in 5G networks virtualization", *IEEE Commun. Lett.*, 2017.

[11]. A. T. Al-Hammouri, Z. Al-Ali, and B. Al-Duwairi, "ReCAP: A distributed captcha service at the edge of the network to handle server overload", *Tran. Emerg. Telecommun. Technol.*, 2018.

[12]. J. Shuja, A. Gani, K. Ko, K. So, S. Mustafa, S. A. Madani, and M. K. Khan, "SIMDOM: A framework for SIMD instruction translationd instruction translation and offloading in heterogeneous mobile architectures", *Trans. Emerg. Telecommun. Technol.*, 2017.

[13]. P. Arroba, J. M. Moya, J. L. Ayala, and R. Buyya, "Dynamic voltage and frequency scaling-aware dynamic consolidation of virtual machines for energy efficient cloud data centers", *Concurr. Comput. Pract. Exp.*, vol. 29, no. 10, 2017.

[14]. N. Kandavel and A. Kumaravel, "Offloading computation for efficient energy in mobile cloud computing", *Int. J. Innov. Technol. Explor. Eng.*, vol. 8, pp. 4317–4320, 2019, doi:10.35940/ijitee.J9842.0881019.

[15]. S. Guo, J. Liu, Y. Yang, B. Xiao, and Z. Li, "Energy-efficient dynamic computation offloading and co-operative task scheduling in mobile cloud computing", *IEEE Trans. Mob. Comput.* vol. 18, pp. 319–333, 2019, doi:10.1109/TMC.2018.2831230.

[16]. N. Khattar, J. Sidhu, and J. Singh, "Toward energy-efficient cloud computing: a survey of dynamic power management and heuristics-based optimization techniques", *J. Supercomput*, vol. 75, pp. 4750–4810, 2019, doi:10.1007/s11227-019-02764-2.

[17]. M. Askarizade Haghighi, M. Maeen, and M. Haghparast, "An energy-efficient dynamic resource management approach based on clustering and meta-heuristic algorithms in cloud computing IaaS platforms: Energy efficient dynamic cloud resource management", *Wirel. Pers. Commun.* vol. 104, pp. 1367–1391, 2019, doi:10.1007/s11277-018-6089-3.

[18]. M. S. Mekala and P. Viswanathan, "Energy-efficient virtual machine selection based on resource ranking and utilization factor approach in cloud computing for IoT", *Comput. Electr. Eng.* vol. 73, pp. 227–244, 2019, doi:10.1016/j.compeleceng.2018.11.021.

[19]. A. F. S. Devaraj, M. Elhoseny, S. Dhanasekaran, E. L. Lydia, and K. Shankar, "Hybridization of firefly and improved multi-objective particle swarm optimization algorithm for energy efficient load balancing in cloud computing environments", *J. Parallel Distrib. Comput.* vol. 142, pp. 36–45, 2020, doi:10.1016/j.jpdc.2020.03.022.

[20]. R. Khorsand and M. Ramezanpour, "An energy-efficient task-scheduling algorithm based on a multi-criteria decision-making method in cloud computing", *Int. J. Commun. Syst.* vol. 33, art. e4379, 2020. doi:10.1002/dac.4379.

[21]. D. Ding, X. Fan, Y. Zhao, K. Kang, Q. Yin, and J. Zeng, "Q-learning based dynamic task scheduling for energy-efficient cloud computing". *Futur. Gener. Comput. Syst*, vol. 108, pp. 361–371, 2020, doi:10.1016/j.future.2020.02.018.

[22]. N. Mc Donnell, E. Howley, and J. Duggan, "Dynamic virtual machine consolidation using a multi-agent system to optimise energy efficiency in cloud computing", *Futur. Gener. Comput. Syst.*, vol. 108, pp. 288–301, 2020, doi:10.1016/j.future.2020.02.036.

[23]. X. Hu, L. Wang, K.-K. Wong, M. Tao, Y. Zhang, and Z. Zheng, "Edge and central cloud computing: A perfect pairing for high energy efficiency and low-latency", *IEEE Trans. Wirel. Commun*, vol. 19, pp. 1070–1083, 2020, doi:10.1109/TWC.2019.2950632.

[24]. X. Wu, H. Wang, D. Wei, and M. Shi, "ANFIS with natural language processing and gray relational analysis based cloud computing framework for real time energy efficient resource allocation", *Comput. Commun.* vol. 150, pp. 122–130, 2020, doi:10.1016/j.comcom.2019.11.015.

[25]. S. Elashri and A. Azim, "Energy-efficient offloading of real-time tasks using cloud computing", *Cluster. Comput.* 2020, doi:10.1007/s10586-020-03086-2.

[26]. P. Sharma, R. Kumari, and I. K. Aulakh, "Task-aware energy-efficient framework for mobile cloud computing", in M. Tuba, S. Akashe, A. Joshi, Eds., *ICT Systems and Sustainability. Advances in Intelligent Systems and Computing*, Springer: Singapore, Vol. 1077, pp. 387–395, 2020, ISBN 9789811509353.

[27]. X. Li, Y. Dang, M. Aazam, X. Peng, T. Chen, C. Chen, "Energy-efficient computation offloading in vehicular edge cloud computing", *IEEE Access*, vol. 8, pp. 37632–37644, 2020, doi:10.1109/ACCESS.2020.2975310.

[28]. M. Rajabzadeh, A. Toroghi Haghighat, and A. M. Rahmani, "New comprehensive model based on virtual clusters and absorbing Markov chains for energy-efficient virtual machine management in cloud computing", *J. Supercomput.* 2020, doi:10.1007/s11227-020-03169-2.

[29]. T. Sun, Y. Tao, and R. Tang, "An algorithm towards energy-efficient scheduling for real-time tasks under cloud computing environment", in Geo-Spatial Knowledge and Intelligence. GSKI 2017. Communications in Computer and Information Science; H. Yuan, J. Geng, C. Liu, F. Bian, T. Surapunt, Eds., Springer: Singapore, Vol. 848, pp. 578–591, 2018, ISBN 9789811308925.

[30]. S. R. Jayasimha, J. Usha, and S. G. Srivani Iyengar, "A comprehensive review of energy efficiency in cloud computing environment", *Int. J. Eng. Technol.*, vol. 7, pp. 249–252, 2018.

[31]. V. K. Saroha, S. Rana, and R. Malik, "Study for the enhancement of energy efficiency mechanism in cloud computing network: A review", *J. Adv. Res. Dyn. Control Syst.*, vol. 10, pp. 1357–1362, 2018.

[32]. J. Kavitha and C. V. Phani Krishna, "A survey on energy-efficient useful resource allocation based on various optimization techniques for cloud computing". *J. Adv. Res. Dyn. Control Syst*, vol. 10, pp. 550–559, 2018.

[33]. N. M. N. Pham, V. S. Le, and H. H. C. Nguyen, "Energy-efficient resource allocation for virtual service in cloud computing environment", in *Information Systems Design and Intelligent Applications: Advances in Intelligent Systems and Computing*; V. Bhateja, B. Nguyen, N. Nguyen, S. Satapathy, D. Le, Eds.; Springer: Singapore, vol. 672, pp. 126–136, 2018, ISBN 9789811075117.

[34]. R. K. Verma, B. Pati, C. R. Panigrahi, J. L. Sarkar, and S. D. Mohapatra, "M2C: An energy-efficient mechanism for computation in mobile cloud computing", in *Progress in Advanced Computing and Intelligent Engineering. Advances in Intelligent Systems and Computing*; K. Saeed, N.Chaki, B.Pati, S.Bakshi, D.Mohapatra, Eds.; Springer: Singapore; Vol. 563, pp. 697–703, 2018, ISBN 9789811068713.

[35]. N. Rehani and R. Garg, "Energy efficient reliability aware workflow scheduling in cloud computing", *Int. J. Sensors, Wirel. Commun. Control*, vol. 7, pp. 198–210, 2018, doi:10.2174/2210327908666180123162717.

9
SDR Network & Network Function Virtualization for 5G Green Communication (5G-GC)

Devasis Pradhan[1], K C Priyanka[2], and Rajeswari[3]

[1]Assistant Professor, Department of Electronics & Communication Engineering, AIT, Acharya, Bangalore
[2]Assistant Professor, Department of Electronics & Communication Engineering, AIT, Acharya, Bangalore
[3]Professor, Department of Electronics & Communication Engineering, AIT, Acharya, Bangalore

9.1 INTRODUCTION

With the increase in the number of the most recent applications past private interchanges, versatile devices will likely achieve heaps of billions until the business arrangement of 5G network. The significant aspects in general execution markers of 5G are foreseen to include: Better, pervasive and raised inclusion of practically 100% inclusion for "each time anyplace" network, 10 to 100 cases higher customer realities expenses, above 90% force investment funds, a total help unwavering quality and accessibility of 99.99%, an offer up-to-surrender over-the-air inactivity of under 1 ms and decreased electromagnetic territory levels in contrast with LTE.

Fifth generation networks characterized by using heterogeneity because of mixed usage of notably varied get admission to technologies. Therefore, community operators endorse precise necessities to equipment carriers for fee-green and energy-saving answers. The few new paradigms to cope with the above problems consist of NFV, SDR and SDN, toward 5G Green Communication. NFV utilizes the standard IT virtualization age to several local area gadget types onto venture stylish serious degree workers, switches and capacity contraptions. Thus, overseers can create networks toward conveying local area administrations onto vogue devices. NFV allows turning local area highlights without introducing the equipment or gadget for each new transporter, making it substantially less subsidizing and considerably less consuming with regard to organization control and activity. It empowers the standard, worn out organization gear to proceed, starting with one equipment stage then onto the next [1–3].

The primary objective of SDN, as the name infers, is to authorize numerous modes through redesigning the radio with explicit programming. The product program can be pre-stacked inside the gadget or downloaded through fixed insights or over-the-air (OTA). SDR has been majorly utilized in military communication structures and as of late brought to the customer hardware commercial center. Notwithstanding the way SDN and NFV were in the first place progressed as fair systems administration standards, advancements in the two technologies has demonstrated their robust integration. SDN and NFV rate ordinary longings and tantamount specialized musings, and are reciprocal to each other. Incorporating SDN and NFV in future systems administration may furthermore trigger reformist organization plans that may totally misuse the advantages of the two standards [4,5].

9.2 5G—GREEN NETWORK

The term Green Communication is integrated for energy harvesting of the network, to produce minimal quantity of CO_2 in the surroundings. The small cellular network (SCN) is the technique that renders the 5G community as an inexperienced community. The performance of the mobile networks might be expected to be improved considerably through implementation of 5G. It will be beneficial in supplying electricity for billions of devices. This energy efficiency can be made possible through various strength harvesting strategies such as spectrum sharing, millimeter wave, D2D communique, extremely dense network, network-centric strategies, big MIMO, IoT permit network, femto cells and Cloud RAN. Figure 9.1 shows the green communication network.

Extra fact of green verbal exchange is energy harvesting which supports energy transmission at remote access. The strength harvesting provisions using sun electricity, wind strength and other environmental energy resources for charging the energy station. In different way, the radio frequency indicators interference that is an unfavorable phenomenon, also the above mentioned complex interference within the cell area is taken into consideration as a supply of green electricity. It could be used as power harvesting gadgets to improve life of power supply and efficiency of the community. Figure 9.2 shows utilization of energy or power with temperature distribution in 5G green network by a smart phones or smart gadgets [6,7].

9.2.1 Classification of 5G Green Communication Network

The 5G green network can be categorized on foundation of base station electricity assets. The strength station can be charged through the renewable power supply by means of using off grid or on grid technique.

9.2.1.1 Off-Grid Base Stations (OFF-GBS)

In the off grid network the bottom stations are not linked to the grid. The opposite renewable assets are used to provide the electricity to base station. They may be no longer geared up to impart energy supply to every other, and the possible collaboration includes supporting every different remotely via changing their transmitted energy or probably offloading Clients. The aim of the inexperienced network is optimizing of the electricity resource on this manner which can be preserve the offerings to the quit users [7].

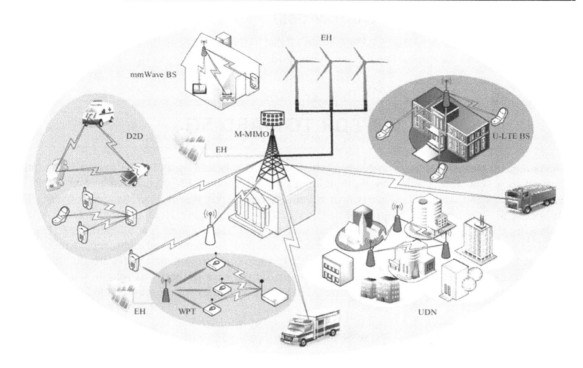

FIGURE 9.1 5G Green communication network ecosystem.

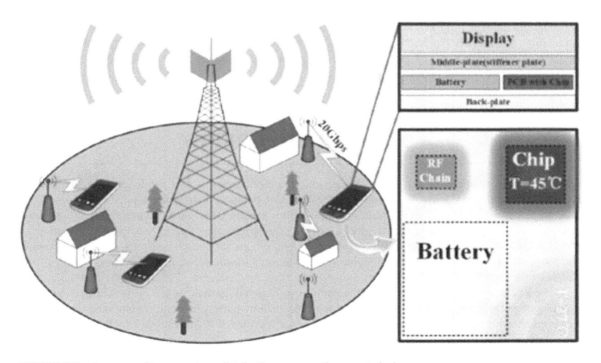

FIGURE 9.2 Power and temperature distribution among the smart devices.

9.2.1.2 On-Grid Base Stations (ON-GBS)

Inside the on grid community the base stations are related to the grid. The bottom station is connected to the strength grid and the ones related to the clever grid.

9.3 SDR TOWARD 5G

SDR is a radio correspondence machine in which added substances by and large executed in equipment are as a substitute applied by programming on an individual PC or inserted gadget. SDR innovation brings the adaptability and worth execution crucial for power interchanges forward. The appropriate SDR has different groups and modes with open engineering. Remote capacities are refined by stacking the product program providing an assortment of radio dispatch administrations. The essential SDR stage involves a gathering mechanical assembly, a multi-band radio repeat module, a broadband A/(D/A) converter, DSP processors, and various developments. Software defined radio moves A/D and D/A nearer to the radio frequency, from the base to the focus rehash or even to the radio frequency, and replaces submitted virtual circuits by strategies for programmable DSP or FPGA gadgets [8,9].

Software portray network control in SDR is decoupled from sending and promptly programmable, which allows the fundamental system to be distracted for packs and association commitments and treats the association as an insightful or virtual component. Network insight is unified in programming based absolutely SDN controllers. With SDN, enterprises and suppliers oversee the entire organization from a solitary intelligent factor, because they not have any desire to perceive and method stacks of convention prerequisites anyway just take conveyance of guidelines from the SDN regulators. SDN makes the local area less programming mindful but rather more utility-hand crafted, and programs less organization cognizant but rather more organization usefulness conscious [10]. Subsequently, processing, stockpiling, and network sources might be streamlined. Accordingly, NFV can uphold SDN by means of giving the foundation whereupon the SDN programming can run. NFV adjusts eagerly with SDN focuses by methods for the utilization of product workers and switches. Moreover, SDR also gives include virtualization help to versatile organizations.

9.4 SDR NETWORKING

Radio exists everywhere, at any rate does not exist in a dumb manner. It is far an entire system, where each base station approaches radio limits; it has issues of booking, help apportioning, examining, load changing, and protection. Without thinking about the information, the most prompt dating mong radio, clients, and suppliers is to permit providers to use negligible expense to permit customers to experience range resources. As there ought to be contention among clients, and carrier carriers need to respond to the longings of customers to make genuine rules, which can moreover contain business practices and perplex this issue [11,12].

The advancement of SDR conquers the dangers of radio equipment. When all is said in done, radio equipment gadgets are expensive, and comprise of the ensuing parts: clear out, regulation demodulation, and converter for managing the one of a kind recurrence ascribes of organizations which have distinctive organization conventions. Equipment gadget may moreover come from particular equipment producers, consequently dispensing possible similarity issues.

9.4.1 Architecture

SDR is an allowing age, significant all through a wide scope of regions inside the remote business that offers green and nearly less expensive responses to several of the issues inborn in extra customary radio designs. Unmistakably positioned, programming program portrayed radio is the term used to depict radio innovation where a couple or all the remote actual layer highlights are programming program characterized. In this manner, for the fifth generation network vision of the fate, network sellers pick more prominent straightforwardness, in correlation with the past multi-band. All the above will start from recurrence range sharing or reuse to a solitary range portion issue, and the utilization of programming program-characterized attributes will powerfully recognize obstruction and crash [13–15]. Ordinary areas of standardization, all of which might be fairly inter-associated, are as follows:

a. **Utility frameworks:** A software framework provides the same old framework for putting in place, tearing down, configuring and controlling waveform packages going for walks on the SDR platform. Those frameworks normally specify the software program running surroundings, which includes the base software interfaces, essential to support the waveform packages.
b. **Hardware abstraction layer:** A hardware abstraction layer, as described through the SDR discussion board's hardware Abstraction Layer working group, "assists with the portability of excessive velocity sign processing code (FPGA, DSP, or different) within the signal processing subsystem (SPS)".
c. **Radio service:** Radio provider are described by the SDR discussion board's device Interface operating group as "an settlement of offerings provided and required behavior amongst associated software and/or hardware modules" (Figure 9.3).

9.4.2 Design

Range of frequency spectrum is an undetectable component. Step by step instructions to accurately control such undetectable substance is an absolutely imposing mission. Thus, we should have some of the accessible stuff and procedures. On the off chance that SDR is a critical thinking gadget for range access, Cognitive Radio (CR) [16] is a real methodology to a difficulty that the got range information are outfitted to customer arranged system to designate range valuable asset based absolutely at the technique for self-insight and learning, CR characterizes the range use inclusion. The inclusion may also go in sync with the detected range utilization prospects and ecological boundaries. Be that as it may, the objective is to locate the most extreme appropriate organization and setup rebuilding. Figure 9.4 shows the design principle of SDR network which is divided into certain layers as per the hardware and software requirement.

The primary distinction among SDR and equipment radio is the utilization of the client quit to try accessible recurrence groups, and incorporate extraordinary association innovation in a solitary interface. Which implies a client can utilize just one device to obtain signs of WiMAX, Wi-Fi, GSM, or LTE. Introducing the principles to be utilized are inserted inside the chip, the purchaser can utilize a recurrence band in accordance with the standards. Albeit each sign has stand-out coding and examination methodologies, it requires a down heap of the front line popular modules, along with bringing in a Library to far and wide compilers, for you to ordinarily procure and interpret radio. Thus, it could lessen costs and keep space. From the administration point, the control of radio equipment is bendy since it could best control associations and substitute TCP/IP and applications. On the opposite, the range of control of SDR can cowl the bodily layer, as shown above. Figure 9.4 illustrates the simple SDR type [17,18].

188 Future Trends in 5G and 6G

(a)
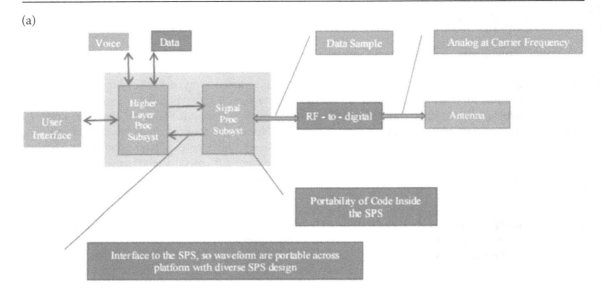

(b)
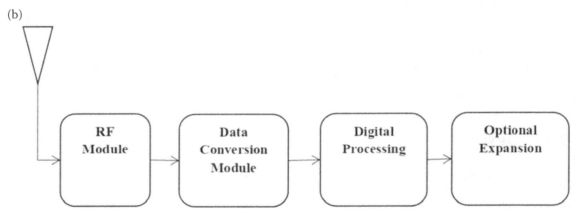

FIGURE 9.3 (a) SDR architecture model (Source: Lee Pucker, Spectrum Signal Processing). (b) Basic block diagram of SDR model.

9.5 SDN

SDN is a brand new structure that that has been designed to permit greater agile and value- effective networks. SDN allows dynamic reconfiguration of the community by means of taking a new method to the community architecture. In a conventional community device, like a router or transfer, it includes each of the statistics aircraft. The control plane determines the path that site visitors will take via the community, while the statistics plane is the part of the network that surely includes the visitors. With the aid of isolating the control and statistics aircraft, community equipment can be configured externally via seller unbiased management software and has the capacity to convert the network from a closed machine to an open machine. SDN also allows to permit centralization of network management for special entities inside a mobile community. There's still work occurring inside academia and industry to apply SDN standards to mobile networks [19].

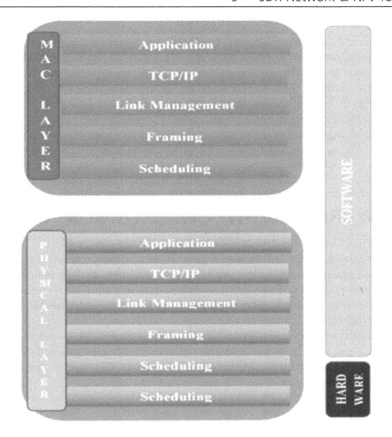

FIGURE 9.4 Design principle of SDR architecture.

9.5.1 Architecture

The whole architecture is divided into three basic layer:

a. **Application layer**: This layer helps the hosts for SDN programs and talk with the SDN enabled controller thru standardized software programming interface. The program developers can write the programs for configuring the network and do now not have to worry approximately the information of underlying network. The SDN programs can be network packages, cloud orchestration or business applications.
b. **Manage layer**: SDN decouples manage data plane. The SDN controller is placed on top of things layer and translates application layer necessities and controls the SDN data paths. The SDN controller constructs and gives a logical map of the network for efficient selection making by means of SDN programs which might be sitting inside the utility layer.
c. **Infrastructure layer**: This layer is called the real network hardware (middle community, base stations, switches, routers, and so forth) that implement SDN data paths and forward the real site visitors. The infrastructure layer is needed to put into effect open standards-primarily based programmatic get admission to infrastructure to allow programming from SDN controllers. Figure 9.5 shows the architecture of SDN.

9.5.2 Design

The future network site guests, as foreseen by utilizing primary undertakings and specialists, might be various occasions the cutting edge traffic. In simple expressions, the essential commitment of SDN is that the local area might be remade. The best circumstance is to get a completely robotized the executives without the format and changes of the chairman intercession inclusion. Inclusion is the general guideline for recreation, and SDN has a helpful idea, which demonstrates effortlessly oversaw network issues with the guide of characterizing a decent inclusion set, with a self-acquiring information on inclusion exchange system [20]. The simple design of SDN architecture diagram is as shown in Figure 9.6.

9.6 HYBRID ARCHITECTURE OF SDR AND SDR

SDN and SDR have come to be a warm theme on the grounds that the portable local area improves every day. Since the raised clients conveyed the outstanding guests that this environmental factors should require a hearty control plan to facilitate any records and demands from all features. Each machine has explicit gifts and that they regularly talk, facilitate or even affect with one another. Yet, in certain examples, it can advance into shared resistance among structures if there are lopsided characteristics marvel since a framework grabbed the machine assets. Alongside SDR contraption need to plan the client's channel need, and SDN gadget should uncover and appropriates guests. Every one of those two constructions are have own structures, elements and association. Constantly, they are additionally involved with a unified control in the event that they need to mix [20,21].

In a 5G Green Network, the association among range and drift turns out to be increasingly more evident as range reuse can soothe traffic among exceptional recurrence groups. In various words, if the coast is appropriately prepared, the range will avoid resistance, and over-burdening of transfer speed additionally can be turned away. In this manner, it is anticipated that SDR can review following results while seeing the recurrence range or exchanging groups, while SDN can allude to the SDR recurrence range circumstances while changing the strategy. To understand the anticipated outcomes, this investigation

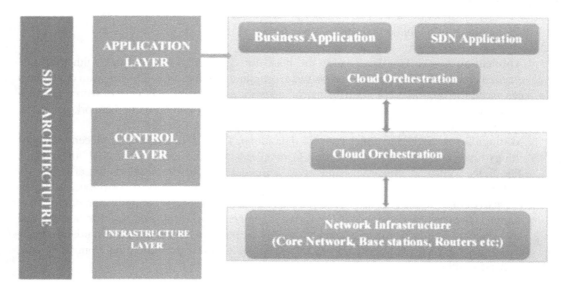

FIGURE 9.5 Software defined networking (SDN) architecture (Source: openairinterface.org).

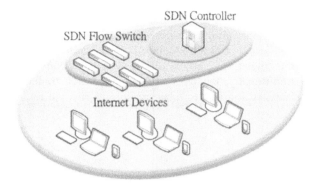

FIGURE 9.6 Design of SDN architecture.

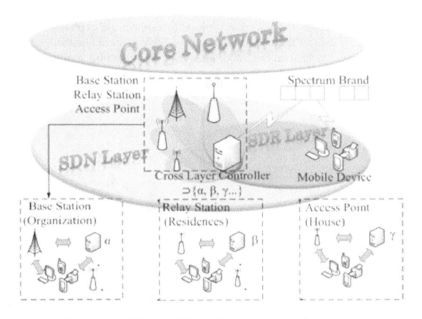

FIGURE 9.7 Hybrid architecture of SDN and SDR for 5G green network.

shows that the regulator in the bona fide SDN engineering should cross the SDN Layer and SDR Layer, as demonstrated in Figure 9.7. Normal, while stood up to with a 5G network, it genuinely approach permitting additional contraptions to appreciate high best administrations at any rate esteem. SDR or SDN all alone is not sufficient, and there must be cross-layer engineering.

9.7 NETWORK FUNCTION VIRTUALIZATION (NFV)

Remote virtualization is a rising age that is dealing with the creating request of differs organizing administrations. It permits more than one virtual network operators (VNOs) to serve over and rate a similar foundation while improving in reverse similarity get right of passage to networks (by utilizing applying Ethernet and combination stage as IP convention). NFV tends to these inconveniences by means of favored IT virtualization period uniting numerous local area gear sorts onto industry favored high amount workers, switches, and carport devices situated in records places, network hubs, and quit individual premises [22]. The people group trademark is actualized in a product pack taking strolls in virtual machines. NFV permits the state of affairs of late systems to convey and work organization and foundation contributions. Additionally, the possibility of local area trademark virtualization NFV of radio sources is to utilize framework assets to meet transporter necessities.

9.7.1 Architecture

NFV is a time principally based at the idea of control/information sending partition, Virtualization, SDN regulator, and data focus. Its prevalent trademark is to make the elements of the organization component and Offer virtual sources to decouple the equipment and programming project of the customary media transmission gadget. The functions of network devices at this point do not depend on novel equipment, and might be disconnected into virtual sources like actual resources through virtualization age, all together that they can be utilized for upper degree projects to achieve the explanation of virtualization of organization capabilities [22,23]. The essential element of NFV is to acknowledge computerized local area components and virtual organizations, and the local area creation and association shape have changed, anyway the general execution of the organization is not changed. The architecture is shown in Figure 9.8.

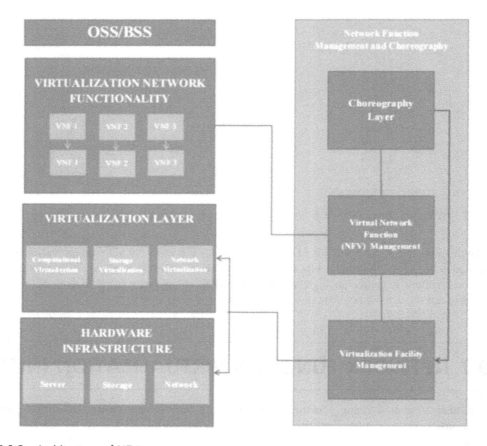

FIGURE 9.8 Architecture of NFV.

9.7.2 Design

The design area incorporates the format and arrangement of business stock related with local area decreasing, and the improvement of transporter degree arrangements based at the type of administration and the inhabitant's requirements. Cut help administration (SSS) plays an essential capacity inside the plan of web cuts. The SSS uncommonly comprises of two element blocks: network Slice format style fashioner and cross-space Slice administrator. One of the network cut layout creators is to plan the relating cut format with regards to the assistance kind, the organization abilities of each specialized control of delivering the local area cut formats, and the component and in general execution prerequisites of the occupants. The space cut director (DSM) in DSS is chargeable for planning the stock of organization cuts for subnets in an area, runtime SLA assurance and cut disconnected. DSS guarantees a continuous assurance of the ability to break down SLAs in genuine time in each space.

9.8 CURRENT STANDARDIZATION FOR SDR, SDN & NFV ENABLED

Sequentially, the idea of software defined radio and software defined network were each conveyed during the 1990s based absolutely at the longings of uses and that they turned out to be warm examinations subjects inside the new century. In examination with the previous innovation, NFV is an unmistakably new idea proposed by utilizing net supplier transporters to manage the pattern of framework virtualization. SDR age is advanced by methods for the wi-fi innovation forum, which become recently called the SDR conversation board [11]. The conversation board is working cooperatively with various necessities of bodies, along with ITU, IEEE, ETSI, etc., and administrative bodies which incorporate the FCC and NTIA, to finish reports, rules, and specialized determinations for SDR and related innovation which incorporate programming program correspondence structure, intelligent radio, and dynamic spectrum access (DSA) [24].

The Open Networking Function (ONF), shaped via web foundations and transporter organizations [12], characterizes the design of SDN, proceeds with the Open accept the way things are stylish, and sorts out the conformance leaving SDN empowered gadgets. In the design of SDN, network gadgets are streamlined by means of programming, oversee, and framework layers. The highlights of the abatement layers are alluded to as the Open Flow with the stream regulator and the open flow switch, which identify with the control plane and information plane of customary IP/MPLS network switches and switches (Figure 9.9).

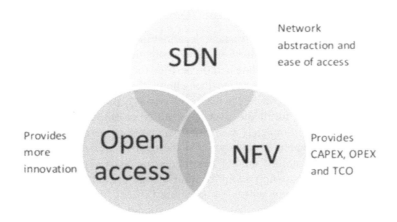

FIGURE 9.9 Inter-relationship between Open Access—NVF and SDN.

9.8.1 SDR

The exorbitant realities throughput and portability of 5G are refined generally because of the utility of MIMO innovation and dynamic range usage, furthermore, these updates should be embedded in SDR. IMT- predominant, the ITU-T specialized necessities for 4G designs, conscious to expand the transfer speed from 20–40 Mbs, the heterogeneous local area shape will accentuate the prerequisite that gadgets must be modified to adjust to the fundamental organization plans, which incorporate remote or cell, utilizing range assets either affirmed or no longer [24,25]. To acquire the capacities above, more impressive models and proficient calculations should be depicted and brought to the SDR structure.

9.8.2 SDN

SDN has discovered its lovely practices in grounds organizations with server farms. In future portable telecom organizations, the intense applications will lie in cell the executives, between ISP handoff, and control plane protection. Since the control airplane is decoupled from the records airplane, the exchanging of multi-homing individual gear between heterogeneous organizations, which incorporate wi-fi and versatile, might be without trouble overseen absolutely at a more significant level without surprising absence of network or administration interference. To acquire consistent handoff among the organizations of various transporter suppliers, a procedure called portable IP (MIP) changed into proposed through IETF to permit wandering contraptions to ship from one organization to some other with a perpetual IP manage appended.[25]

9.8.3 NFV

The movability and interoperability of virtual machines for network gear will encourage the state of NFV. Versatility will convey the freedom to advance assistance and local area arrangement. The interoperability decouples the computerized home gear from the actual gadget given by utilizing unique organizations [26] (Table 9.1).

9.9 INTEGRATION OF SDN WITH NFV

The general objectives of 5G subject performance, dependability, electricity efficiency, and reasonably-priced value reduction. The duties to recollect on the coordination competencies of the orchestrator and of the controller are as follows: to manipulate and orchestrate the distinctive SDN and NFV technologies deployed; to enforce a network cutting scheme that allows the efficient consciousness of the unique predicted 5G verticals; to allocate the wireless resources wanted; subsequently to monitor the different additives of the 5G Network. To fulfill the advancing different help prerequisites, SDN information airplane gadgets need to perform completely standard float coordinating and bundle sending, which may furthermore extensively development of intricacy. On the oversee plane, current SDN structure needs sufficient assistance of interoperability among heterogeneous SDN regulators, and thus restricts its ability to arrangement bendy stop-to-stop administrations across self-sufficient area names. Figure 9.10 shows the incorporated SDN regulator with NFV structure [26,27].

Alternatively, many technical demanding situations should be tended to for knowing the NFV worldview. The executives and orchestration has been perceived as a vital perspective inside the ETSI NFV structure. A decent arrangement extra modern to oversee and the executives instruments for both

TABLE 9.1 Standard for integrating NFV, SDR and SDN.

SL. NO.	TECHNOLOGY	STANDARDIZATION	RELATED STANDARD AND ACTIVITIES	USABILITY
1	NFV (Network function virtualization)	ETSI (*European Telecommunications Standards Institute*)	ETSI GS NFV-PER 001 002	Execution and ideas
			ETSI GS NFV 001 004	Examples and structural requirements
			ETSI TR 103 062 064 ETSI TR 102 681/ 803/839/944/945	Use cases and business
			ETSI TS 102 969/ ETSI EN 302 969	Mobile device standards
			ETSI TS 103 095/ 146-1	
			ITU-R SM.2152	Definition
2	SDR (Software Defined Radio)	ITU (*International Telecommunication Union*)	ITU-R M.2117-1 ITU-R M.2063/2064	Specific applications standards
			ITU- R Resolution 805/956	
		WINNF (*Wireless Innovation Forum*)	SDRF-01-P-0006 SDRF-02-P-0002 SDRF-02-P-0002	Framework design and safety
3	SDN (Software Defined Network)	ONF (*Open Network Foundation*)	"Software Defined Networking:	Definition and interoperability
			Open Flow series	Design and the executives and switch detail
		ITU (*International Telecommunication Union*)	ITU-T Resolution 77	Normalization for SDN
		IETF (*Internet Engineering Task Force*)	IETF RFC 7149	Perspective

virtual and actual assets are needed by the uncommonly powerful systems administration environmental factors empowered by means of NFV, in which automatic local area control is basic. Utilizing the SDN statute—decoupling control insight from the controlled assets to allow a sensibly unified programmable control/control airplane—in the NFV engineering may likewise considerably encourage acknowledgment of NFV.

9.9.1 Migration of Network

A period to be important for 5G should furnish a perfect relocation way with legitimate similarity with the inheritance frameworks. The blending of SDN and NFV as proposed innovations for 5G need to diminish the alterations in local area factors, subsequently providing a proceeding with movement dependent on administrator wants. This permits the gradual updates of local area components in specific parts of the local area simultaneously as keeping up inheritance factors in different components of the local area network [27].

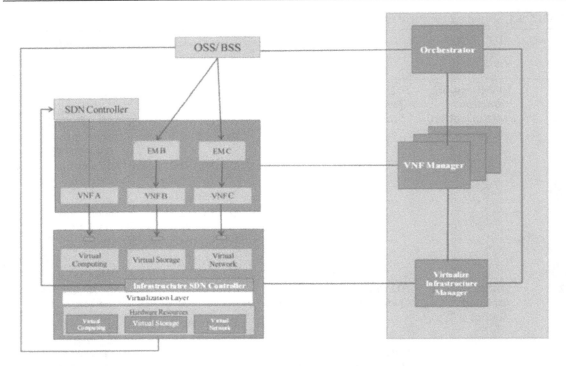

FIGURE 9.10 Integrating SDN controller into NFV architectural framework.

9.9.2 Network Monitoring

Some other significant suspicion for a ground-breaking sending is having legitimate organization observing. Other than surveying security, local area following aides confirmation, SLA, QoS and QoE, investigating, and the appraisal of advancements and utilization of resources. On the solitary hand, network virtualization sets new prerequisites for cell network following anyway on the other and it furthermore gives implies for actualizing progressed local area following solutions [28]. NFV/SDN permits the mix of cloud foundation that gives higher phases of opportunity concerning the area of estimation focuses and adaptable oversee of traffic streams. A convoluted and ground-breaking QoS following arrangement need to contain both a dispensed (SDN/NF-based absolutely) QoS estimation gadget and a unified assessment gadget.

9.9.3 Service & Optimization

Administration provisioning and advancement is some other presumption needed to ensure asset accessibility. This could be accomplished by a solitary coordination element. This suspicion might be sent in SDN networks via the utilization of oversee applications which have total perspective on local area arrangement. This along with notoriety data gave through local area following and data arrangement structures license cell local area orchestrator programming to upgrade administration or potentially help use less muddled than customary organizations that need to depend on flagging. The orchestrator can deal with various local area factors through the control programs, surely from numerous organizations. This grants to present new contributions through composition or adjusting the orchestrator while in regular organizations all of the framework require to be moved up to help the new supplier kind.[29] It is accepted that 5G organizations will presently do not be applied after a fresh start procedure; as a substitute, inheritance and new SDN based absolutely 5G arrangements should exist together over the long haul. In any case, so you can make the most the capability of SDN, participation between both of these arrangements is required, e.g., with the guide of presenting reflection and robotization layer for the heritage network part.

9.9.4 Cost Reduction

Cost decrease is a top notch necessity. Virtualization of the 5G organization is foreseen to contribute in such manner, taking advantage of normalized network factors and better guide use with SDN. However, virtualized network elements may also expand the requirement for seriously figuring power, more muddled organization control, and make more perplexing expense organizations. The web advantage of SDN in5G networks must be tried.

9.9.5 Requirements

Inside the innovation of 5G, the remote correspondence network is prepared to do gathering gigantic measures of clients' different supplier needs of different records costs. As an answer, the possibility of insightful heterogeneous organizations (HetNets) is proposed as a Hierarchical people group asset usage plot abusing each the versatile and WLAN innovation. In HetNets with the help of NFV, SDN, and SDR to improve the degree of explicit. The application of SDR is to monitor the interface between heterogeneous hubs and gadgets, through which it can manage the actual layer cautions and concentrate data bundles for upper layers. In order to stop the bundle transportation, SDN is employed to deftly build the start to finish blockage tough transmission channels by methods for misusing of various predefined protocols [29,30].

9.10 NETWORK MANAGEMENT & OPERATION IN 5G ENVIRONMENT

Incorporating SDN with network virtualization finishes in decoupling of measurements and oversee/the executives planes on both framework layer and administration layer, thus calling for discrete interfaces for controlling and taking care of real foundation sources and virtual transporter capacities, separately. Such interface on the foundation layer is the actual SB interface among regulators and switches in every framework territory, which has been particularly all around concentrated inside the setting of SDN (e.g., OpenFlow and ForCES). Be that as it may, oversee/the executives interface on the transporter layer between computerized networks and their regulators (i.e., the advanced SB interface) has obtained little consideration and merits more examination inside what is to come. Appropriate models for abstracting advanced sources and transporter highlights are needed by utilizing this interface. Also, such interface ought to disengage the control/control for extraordinary individual VNs to help more than one VNs with redid conventions. Moreover, versatile assistance provisioning calls for adaptable components for scaling-up/down VN oversee capacity and powerfully sending and relocating VN controllers [27,28].

Helpful asset control and range sharing based at the deliberation of sources in a heterogeneous organization climate has come to be an essential trouble, specifically in remote organization virtualization. Wherein SDN control airplane used to control resources in virtualized networks effectively and an approach to relegate the detached assets to the assorted assembly of various remote organizations and administrators. As a final product, it requires more prominent green and first in class helpful asset the executives and composed control procedures in the heterogeneous stage way [16]. Huge examination has demonstrated, uncommonly, the CRRM plans on the apex of asset allotment through right setups. The product portrayed plan can adjust to different correspondence standards and presents transmission unwavering quality concerning QoS ensures through the most fitting control of the format and the improvement of the CRRM for the 5G versatile organization.

Innovation way inside the plan for a 5G organization searching for to procure virtual remote hyperlinks and accomplish start to finish (E2E) virtual organizations through amassing the entirety of the actual assets from unmistakable radio get passage to procedures, by means of planning computerized highlights to actual sources. Besides, on the grounds that the cloud thought is expanding, the SDN empowered cloud based entomb activity structure regularly diminishes inertness and jitter when contrasted with the independent cloud answer the utilization of decentralized control hub systems. Mist registering gives on-request contributions and applications near contraptions, thick topographical allocated and low-idleness reactions, bringing about additional purchaser experience.

9.11 SECURING NFV-BASED SDN

The assessed security structure pursuits at providing self-wellbeing, self-recovery, and self-reestablish abilities through novel empowering influences and parts. It is far intended to control security guidelines and characterize applicable security controls to be organized over heterogeneous organizations. The ideal insurance moves can be upheld in explicit kinds of substantial/virtual apparatuses, comprising of each IoT organizations and programming based organizations. The arrangement airplane plays a vital situation in interpreting the individual approaches in well-being instruments and presents dynamic reconfiguration and release if there should arise an occurrence of deviation from the normal conduct. The well-being implementation plane deals with the guide usage and real time activity of the administrations and gives local area availability parts to the security empowering influences [15].

9.11.1 Client Plane Policy

The client plane joins interfaces, organizations, and contraptions to end-customers for methodology definition, structure noticing and organization the chiefs. Its methodology administrator gives an intuitive and simple to utilize gadget to orchestrate security approaches managing the arrangement of the system and association, similar to affirmation, endorsement, filtering, channel protection, and sending. The overall courses of action fill in as commitment to the methodology middle person portion of the security association plane to support the plan of security enabling impacts expected to satisfy the customer techniques [19].

9.11.2 Security Orchestration plane

The Security Orchestration plane maintains methodology based security parts and gives run-time reconfiguration and variety of security engaging impacts, thusly giving the design with clever and dynamic direct.

It includes:

   a. Monitoring segment,
   b. Reaction segment,
   c. Security Orchestrator,
   d. Policy mediator, and
   e. Security Enablers Provider.

The Policy Interpreter module gets as data the plans decided in an evident level language and recognizes the limits expected to execute such methodologies (capacity matching). Then, the Interpreter helps out the Security Enablers Provider to perceive the SDN/NFV-based engaging specialists that are prepared to approve the ideal capacities. The Policy Interpreter plays out a first refinement measure translating the overall security methodologies into a lot of plans described in a medium-level security language [24].

This medium-level security technique language licenses to extricate the game plan communication, therefore captivating interoperability among different security engaging impacts, which may use dealer express instruments to deliver low-level security courses of action to be enforceable in NFV and SDN associations. Ensuing to tolerating these medium-level security draws near, the Security Orchestrator picks the enabling impacts to be effectively passed on, addressing the security necessities, the available resources in the crucial structure, and headway measures. By then, it requests a second course of action refinement measure, which is finished by the Policy Interpreter to unravel the medium-level security systems into express low-level [24–26].

The Monitoring segment gathers security-zeroed progressively data identified with the framework conduct from physical/virtual apparatuses. Its principle objective is to give cautions to the response module on the off chance that something is getting out of hand. Security tests are sent in the foundation space to help the checking administrations. At that point, the Reaction segment is accountable for giving fitting countermeasures, by powerfully characterizing reconfiguration of the security empowering influences as per the conditions. The response results are at that point investigated by the Security Orchestrator, which implements the relating empowering agents' countermeasures. Along these lines, the general system can promise self-recuperating and versatility capacities, by continually guaranteeing the fulfillment of the security prerequisites characterized in the end-client approaches. Designs as per the chose empowering agents.

9.11.3 Security Enforcement Plane

The Security Enforcement Plane joins both the envisioned security enabling specialists and the sections required for their organization.

It includes:

a. Control and management domain
b. Infrastructure and Virtualization domain
c. VNF domain

   A. **Control and Management Domain:** It is a module that regulate the utilization of assets and duration of activities toward security empowering experts sent through schedule based and 5G Network. A great deal of circumnavigated SDN controller acknowledges responsibility for conversing with the SDN- based association portions to direct organize in the under virtual and genuine design. NFV ETSI MANO- unsurprising modules keep up secure situation and the heads of virtual security limits over the virtualized system. These SDN/NFV integrated controller are normally deployed to the network (e.g., entryways) to uphold security capacities in heterogeneous 5G Green Communication [27].
   B. **Infrastructure and Virtualization Domain:** This zone incorporates all the real machines fit for giving enlisting, storing, and frameworks organization abilities to gather an Infrastructures as a Service (IaaS) layer by using appropriate virtualization developments. This plane also joins the association parts liable for traffic sending, noticing the guidelines of SDN/NFV regulator, and a passed on arrangement of security tests for data combination to help the checking organizations.
   C. **VNF Domain:** The VNF region addresses the VNFs conveyed over the virtualization establishment to approve security inside association organizations. Express instruments will be made to affirm the unwavering quality of VNFs and to reliably screen their key limits. Express

contemplation will be directed to the provisioning of bleeding edge security VNFs (like virtual firewall, Intrusion Detection/Prevention System (IDS/IPS), channel protection, etc), fit to give the defend instruments and peril countermeasures referenced by security draws near.

9.12 CHALLENGES

There are numerous difficulties are related with usage of green organization. The enormous number of clients and the base stations are expanded inside the organization hence the adaptability is vital for the organization to deal with these diverse base stations or sub organizations. At the point when the clients are expanded then the traffic and huge game plan of little cells complete a principle challenge for sub organizations to advance enormous traffic to the primary organizations in an ease and exceptionally energy-proficient way. The association with sub organization and center organization is creating the way delay yet it requires the lower transfer speed as contrast with fundamental way subsequently the enormous course of action of little cells would be restricted if backhaul limitations are estimated.

9.12.1 Virtualization for Infrastructure Abstraction

Virtualization of certifiable designs for layer-assessment reflection expects an essential part in future structures association with SDN-NFV coordination. Foundation virtualization is if all else fails comprehensively pressed in appropriated figuring and structures association, yet repeating design research revolves all the more energetically around information plane framework. The SDNV structure shows that virtualization on the control/the heap up plane to accomplish decoupled control/the bosses for physical plus, virtual affiliations is moreover an evaluation subject that benefits concentrated appraisal. Another new test is to empower joined discussion of heterogeneous designs (e.g., association, register, and cutoff) through a standard stage for supporting composite associations across structures association and figuring zones [29,30].

9.12.2 Virtual Network Construction

To meet customer necessities of building VNs is a middle limit with respect to future assistance provisioning, which might be colossally supported by mix of SDN and NFV following the SDNV framework. In this framework, the control/the load up plane on the help layer picks and makes reasonable data plane VSFs to outline VNs for meeting organization necessities. The best strategy to give exceptional depictions of VSF credits, how to make VSFs open what is more, discovered, and how to pick and make the ideal set out of VSFs are altogether relevant issues that require more escalated examination. Cloud organization creation has been broadly inspected and may offer some supportive techniques for VSF piece to create VNs. Regardless, cloud organization plan research fundamentally based on handling organizations rather than frameworks organization organizations; likewise, further assessment on VSF association in the SDNV setting, especially piece of VNFs and VCFs across frameworks organization and figuring spaces [30].

9.12.3 Service Quality Assurance in Virtual Network Environments

Virtualization-based systems administration climate acquires new difficulties to support quality confirmation. How could programming based virtual capacities accomplish similar degree of administration quality as what committed equipment ensures is a significant issue that should be intended to? The SDNV system demonstrates that more different useful jobs, like infrastructure suppliers, VSF providers, VN administrators, and composite organization cloud specialist co-ops, may be empowered by SDN-NFV reconciliation in future organizations. These significant parts in the new help organic framework, who may have conflicting income, ought to partake for meeting execution necessities of organization provisioning. The example toward network-cloud organization association particularly calls for better approaches to manage offering beginning to end QoS guarantees for 5G Green network [16,30].

9.12.4 Energy-Aware Network Design

Building harmless to the ecosystem network framework by decreasing energy utilization is a vital part of future organization plan. Organization asset virtualization along with adaptable SDN control and the board gives incredible potential to accomplish energy-productive systems administration; nonetheless, such favorable position is yet to be completely abused. A test to energy-careful NFV-SDN coordination lies in the grouping of bury weaved in 5G organization segments that ought to be considered in this issue, remembering the two structures and organization capacities with respect to both data and control/the heads planes.

9.12.5 Millimeter-Wave

5G Green Communication (5G-GC) system can accomplish very wide data transfer capacity from mm-wave, as its recurrence goes from 26.5–300 Gbps. Moreover, mm-wave belong to much smaller radio wire shaft size contrasted with the microwave, so it can all the more accurately point the objective. The standard test of mm-wave to network function virtualization is the essential for high accuracy. It can pass on organization limits without presenting gear equipment for each new help, and thus offer a chance of low stuff cost and operational cost. It makes virtualized examples relocating starting with one equipment then onto the next, which precisely meets the order of distributed computing conditions. The mm-Wave innovation, be that as it may, requires more exact gadgets for transmission rate, which will extend the planning of NFV [30].

9.12.6 Massive MIMO

The Massive MIMO upgrade the working principle of basic MIMO technique in order to get all the compensations in critical situations, which consolidates:

    a. Inexpensive, low-power portions
    b. An immense abatement of idleness
    c. Effective to maintain both the man-made impedance as well as air impedance in order to maximum power delivered
    d. Decreasing the limitations on exactness and linearity of every individual enhancer of radio frequency chain

As a rule, colossal MIMO is an engaging specialist for upcoming broad band network, which is capable of handling the energy consumption. The potential of huge MIMO is to addresses a enormous test to execute the joining of NFV, SDR, and SDN. Tremendous MIMO groups produce gigantic proportions of base band data logically, through raise the computational [16,28–30].

9.12.7 Heterogeneous Networks

In HetNets, distinctive eNodeBs with various transmission force can establish diverse cell sizes or layers, like large scale, miniature, pico, and femto. Heterogeneous organizations exploit of the reciprocal qualities of various organization levels, and accordingly become an inescapable pattern for future advancement of data organizations. Not with standing, various levels of systems administration access innovation also, different assistance prerequisites progression and difficult to implement in HetNets. Specifically, with SDR improvement, clients may switch rehash oftentimes for high information rate, causing the relationship to maintain and the load up pushed [29].

9.12.8 C-RAN

This forward jump of RAN configuration can support range capability through the co-employable radio with appropriated accepting wires arranged in radio heads. It is a fundamental development to create virtual base stations all through a progressing cloud structure of C-RAN. Network function virtualization is surely a legitimate response for meet this need. Regardless, it needs to face various challenges, since far off base stations have unbending objectives of nonstop and tip top. Regardless, if various standards are allotted to a comparative reach, RRH can maintain various standards just for the most part.

9.13 CONCLUSION

This chapter discuss about the planned designing of discussed title versatile associations while considering the improvement of rules and technological enhancement. The difficult issue of coordinating SDN and NFV in future organizations by introducing a compositional structure that consolidates the critical standards of both paradigms. SDN and NFV may profit by one another and introduced a two-dimensional model to show that both SDN and NFV rely upon consideration yet focusing in on the plane and layer estimations, independently. The Software-Defined Network Virtualization (SDNV) design to give a sensible extensive vision of fusing the SDN and NFV principles into united association designing, which grants inventive association plans to totally abuse the advantages of the two ideal models. The SDNV system offers helpful rules that may encourage orchestrating research endeavors from different viewpoints toward the normal target of coordinating SDN and NFV in future networks.

REFERENCES

[1]. D. Kreutz, F. Ramos, P. Verissimo, C. E. Rothenberg, S. Azodolmolky, and S. Uhlig, "Software-defined networking: A comprehensive survey," *Proc. IEEE*, vol. 103, no. 1, pp. 14–76, Jan. 2003.

[2]. N. M. M. K. Chowdhury and R. Boutaba, "A survey of network virtualization," *Elsevier Comp. Netw. J.*, vol. 54, no. 5, pp. 862–876, Apr. 2010.

[3]. ETSI NFV ISG, "Network function virtualization – Introduction white paper," *Proc. SDN OpenFlow World Congress*, Oct. 2012.

[4]. J. Liu, S. Zhang, N. Kato, H. Ujikawa, and K. Suzuki, "Device-to-device communications for enhancing quality of experience in software defined multi-tier LTE-A networks," *IEEE Netw. Mag.*, vol. 29, no. 4, pp. 46–52, Jul. 2015.

[5]. M. Casado, T. Koponen, S. Shenker, and A. Tootoonchian, "Fabric: A retrospective on evolving SDN," *Proc. 1st Workshop on Hop Topics in Software-Defined Networks (HotSDN'12)*, Aug. 2012.

[6]. M. Casado, T. Koponen, S. Shenker, and A. Tootoonchian, "Software-defined internet architecture: Decoupling architecture from infrastructure," *Proc. of the 11th ACM Workshop on Hot Topics in Networks (HotNet'12)*, Oct. 2012.

[7]. R. Sherwood, G. Gibb, K.-K. Yap, G. Appenzeller, M. Casado, N. McKeown, and G. Parulkar, "FlowVisor: A network virtualization layer," *OpenFlow Switch Consortium, Tech. Rep*, Oct. 2009.

[8]. D. Drutskoy, E. Keller, and J. Rexford, "Scalable network virtualization in software-Defined networks," *IEEE Int. Comp.*, vol. 17, no. 2, pp. 20–27, Mar. 2013.

[9]. T. Wood, K. Ramakrishnan, J. Hwang, G. Liu, and W. Zhang, "Toward a software-based network: Integrating software defined networking and network function virtualization," *IEEE Netw. Mag.*, vol. 29, no. 3, pp. 36–41, May 2015.

[10]. S. Abdelwahab, B. Hamdaoui, M. Guizani, and T. Znati, "Network function virtualization in 5g," *IEEE Commun. Mag.*, vol. 54, no. 4, pp. 84–91, April 2016,

[11]. C. Y. Chang, N. Nikaein, and T. Spyropoulos, "Impact of packetization and scheduling on C-RAN fronthaul performance," in *2016 IEEE Global Commun. Conf. (GLOBECOM)*, pp. 1–7, Dec. 2016.

[12]. A. S. Thyagaturu, Y. Dashti, and M. Reisslein, "Sdn-based smart gateways (Sm-GWs) for multi-operator small cell network management," *IEEE Trans. Netw. Service Manage.*, vol. 13, no. 4, pp. 740–753, Dec. 2016.

[13]. D. Garcia and R. Lopez, "EAP-based authentication service for CoAP," *IETF, Internet-Draft draft-marinace-wg-coap-eap-05*, work in Progress, Apr. 2017.

[14]. Y. Faqir Zarrar, M. Gramaglia, F. Vasilis, B. Gajic, D. von Hugo, B. Sayadi, V. Sciancalepore, and M. R. Crippa. "Network slicing with flexible mobility and QoS/QoE support for 5G networks," in *Proc. IEEE Int. Conf. Commun. Workshops (ICC Workshops)*. 2017.

[15]. A. Ibrahim, T. Tarik, S. Konstantinos, A. Ksentini, H. Flinck, "Network slicing & softwarization: A Survey on principles, enabling technologies & solutions," *IEEE Commun. Surv. Tutor.*, vol. 20, no. 3, pp. 2429–2453, 2018.

[16]. D. A. Temesgene, J. Nez-Martnez, and P. Dini, "Softwarization and optimization for sustainable future mobile networks: A survey," *IEEE Access*, vol. 5, pp. 25, 421–425, 436, May 2017.

[17]. U. Habiba and E. Hossain, "Auction mechanisms for virtualization in 5G cellular networks: basics, trends, and open challenges," *IEEE Trans. Netw. Serv. Manage.* vol. 20, no. 3, pp. 2264–2293, 2018.

[18]. A. Laghrissi and T. Taleb. 2018. "A survey on the placement of virtual resources and virtual network functions," *IEEE Commun. Surv. Tutor.*, vol. 21, no. 2, pp. 1409–1434.

[19]. C. Massimo and M. Toktam, "Softwarization and virtualization in 5G mobile networks: benefits, trends and challenges," *Comput. Netw.* vol. 146, pp. 65–84, 2017.

[20]. I. Afolabi, M. Bagaa, T. Taleb, and H. Flinck, "End-to-End network slicing enabled through network function virtualization," *Proc. IEEE Conf. Stand. Commun. Netw. (CSCN)*, no. 5, pp. 30–35, 2017.

[21]. "Video quality using MPTCP and segment routing in SDN/NFV," *Proc. IEEE Conf. Netw. Softwarization*, 2018.

[22]. X. Zhou, R. Li, T. Chen, and H. Zhang, "Network slicing as a service: Enabling enterprises' own software-defined cellular networks," *IEEE Commun. Mag.* vol. 54, pp. 146–146, 2016.

[23]. J. Sanchez, I. G. B. Yahia, N. Crespi, T. Rasheed, and D. Siracusa, "Softwarized 5G networks resiliency with self-Healing," *Proc. 1st Int. Conf. 5G Ubiquitous Connectivity* (5GU) pp. 229–233, 2014.

[24]. R. Ravindran, A. Chakraborti, S. O. Amin, A. Azgin, and G. Wang, "5G-ICN: Delivering ICN services over 5G using network slicing," *IEEE Commun. Mag.*, vol. 55, no. 5, pp. 101–107, 2017.

[25]. A. A. Barakabitze and T. Xiaoheng, "Caching and data routing in information centric networking (ICN): The future internet perspective," *Int. J. Adv. Res. Comput. Sci. Softw. Eng.* vol. 4, no. 11, pp. 26–36, 2014.

[26]. T. X. Tran, A. Hajisami, P. Pandey, and D. Pompili, "Collaborative mobile edge computing in 5G networks: New paradigms, scenarios, and challenges," *IEEE Commun. Mag.* vol. 55, no. 4, pp. 54–61, 2017.

[27]. C. Tselios and G. Tsolis, "On QoE-awareness through virtualized probes in 5G networks," *Proc. IEEE Int. Workshop Comput. Aided Model. Des. Commun. Links Netw. (CAMAD)*, pp. 159–164, 2016.

[28]. D. Pradhan and Priyanka K. C., "RF- Energy harvesting (RF-EH) for sustainable ultra dense green network (SUDGN) in 5G green communication," *Saudi J. Eng. Technol.*, June, vol. 5, no. 6, pp. 258–264, 2020.

[29]. D.Pradhan and K. C.Priyanka, "A comprehensive study of renewable energy management for 5G green communications: Energy saving techniques and its optimization," *J. Seybold Rep.*, vol. 25, no. 10, pp. 270–284, 2020.

[30]. P. Devasis, "Massive MIMO technique used for 5th generation system with smart antenna," *Int. J. Elect. Electron. Data Commun. (IJEEDC)*, vol. 6, no. 7, pp. 81–87, 2017.

An Intensive Study of Dual Patch Antennas with Improved Isolation for 5G Mobile Communication Systems

T. Prabhu[1], E. Suganya[2], J. Ajayan[3], and P. Satheesh Kumar[4]

[1]Department of ECESNS College of Technology, Tamil Nadu
[2]Department of ECESri Eshwar College of Engineering, Tamil Nadu
[3]Department of ECESR University, Telangana
[4]Department of ECECoimbatore Institute of Technology, Tamil Nadu

10.1 INTRODUCTION

The era of cellular technology began from the zero generation networks in which only one channel was used for push and talk communication which led to development of first generation in the next decade that includes voice services. Second generation purely focused on digital services and the third generation supported better voice and data quality. In the next few decades, the fourth generation flourished with high data rate and high channel capacity. The advanced wireless communication technology is required to fulfill the demand for features like high data rate, low latency, maximum throughput, reliable connection and low power consumption for 5G communication systems that are suitable for SMART city and SMART transportation with the help of Internet of Things (IoT). In recent years, many researchers have proposed some innovative antenna designs to meet these challenges. Among these innovative antenna designs, massive MIMO gains tremendous attention to fulfill these demands by proper deployment of multiple antennas at the transmitting as well as receiving end of 5G wireless mobile communication systems [1–6]. This MIMO antenna technology empowers the caption by sending the data parallel on multiple channels to maximize the channel capacity. Accommodation of multiple antennas on a single device (i.e., within compact space) is a challenging task that results in isolation problems. Several antenna designs have been reported for the enhancement of isolation between radiating elements. Some of the techniques are complementary split ring resonator (CSSR) [7–9], neutralization lines (NL) [10–13], electromagnetic band gap (EBG) structure [14–16], metamaterials [17,18], decoupling network [12–21] and defected ground structure (DGS) [22–27]. CSRR is found to be effective in minimizing the mutual coupling up to -22 dB, NL can reduce the mutual coupling up to -23 dB, EBG can reduce the mutual coupling up to -28 dB and metamaterials can reduce mutual coupling up to -42 dB. Also, the decoupling networks, DGS and DRA (dielectric resonator antenna), can effectively minimize the mutual coupling up to -20 dB, -18 dB and -25 dB, respectively. For improving the isolation between two patch elements, some of the antenna designs still find it difficult to retain the MIMO antenna features like superior gain, low ECC and maximum DG. In this chapter, we investigated the influence of inter-element spacing between two antennas to reduce the mutual coupling, isolation improvement and also to obtain low ECC and maximum gain. Initially a single element antenna that resonates at 2.32 GHz is designed, and then a dual MIMO antenna is designed with an elemental separation of λ/4 which also resonates at 2.32 GHz. Later, the mutual coupling has been reduced by increasing the elemental separation to λ/2 [28–30].

10.2 ANTENNA CONFIGURATION AND DESIGN CONCEPT

10.2.1 Single Antenna Design

Microstrip antennas are much more suitable due to their low profile, light weight features. The single rectangular patch antenna (RPA) with dimensions 29.48 mm x 38.03 mm resonates at 2.32 GHz. The rectangular patch is attached on the dielectric material FR4 substrate (ε_r = 4.4 and loss tangent tanδ = 0.0007) with the size of 48.68 mm x 54.23 mm x 1.6 mm which is further attached to the ground plane having the dimensions of 48.68 mm x 54.23 mm. The overall structure of single RPA is shown in Figure 10.1. The dimensions of single RPA have been calculated by using the below-mentioned equations and the calculated dimensions listed in Table 10.1.

Case: 1 If $\left(\frac{w}{h}\right) < 1$

$$\varepsilon_{\text{reff}} = \frac{\varepsilon_r + 1}{2} + \frac{\varepsilon_r - 1}{2}\left[1 + 12\frac{h}{w}\right]^{\frac{-1}{2}} \tag{10.1}$$

Case: 2 If $\left(\frac{w}{h}\right) > 1$

$$\varepsilon_{\text{reff}} = \frac{\varepsilon_r + 1}{2} + \frac{\varepsilon_r - 1}{2}\left[1 + 12\frac{h}{w}\right]^{\frac{-1}{2}} + 0.04\left(1 - \left(\frac{w}{h}\right)\right)^2 = 4.09 \tag{10.2}$$

Where, w refers to width of the patch and h indicates the thickness of substrate.

Equations (10.3) and (10.4), respectively, estimate the value of single patch antenna width and length. The width of antenna patch (w) can be obtained as

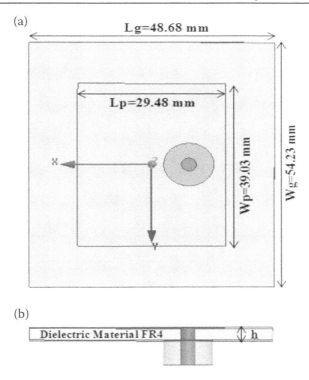

FIGURE 10.1 Configuration of single RPA (a) top view and (b) side view.

TABLE 10.1 Dimensions of the single RPA

PARAMETERS	FR (GHZ)	LP (MM)	WP (MM)	H (MM)	ER	LG (MM)
Values	2.32	29.48	38.03	1.6	4.4	48.68
Parameters	Wg (mm)	Feed type	ro (mm)	ri (mm)	ho (mm)	Operating mode
Values	57.23	Probe	5	1.5	3	TM10

$$w = \frac{c}{2f_0 \sqrt{\frac{(\varepsilon_r + 1)}{2}}} = 38.03 \text{ mm} \tag{10.3}$$

Where,
$c = (3 * 10^8$ m/s) speed of EM wave, f_0= resonant frequency= 2.32 GHz, ε_r= relative permittivity.
An effectual dielectric constant (ε_{eff}) can be obtained from the equation (2)
Effective length (L_{eff}) of the single RPA is calculated as

$$L_{eff} = \frac{c}{2f_0 \sqrt{\varepsilon_{eff}}} = 31.96 \text{mm} \tag{10.4}$$

The extension length of patch Antenna (ΔL) can be calculated as

$$\Delta L = 0.412h \frac{(\varepsilon_{eff} + 0.3)\left(\frac{w}{h} + 0.264\right)}{(\varepsilon_{eff} - 0.258)\left(\frac{w}{h} + 0.8\right)} = 1.24 \text{mm} \tag{10.5}$$

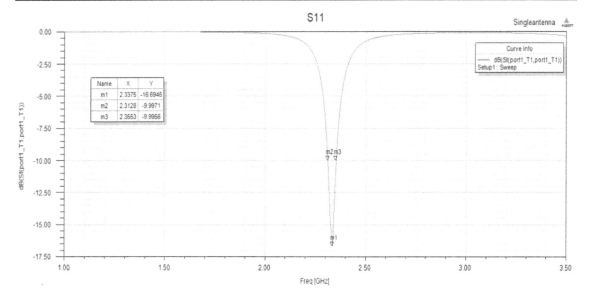

FIGURE 10.2 S_{11} parameter plot for single antenna.

Therefore, the actual length of patch is $L_{eff} - 2\Delta L = $ **29. 48mm**.

The plot of frequency versus S_{11} is shown in Figure 10.2. The single RPA achieved a bandwidth (BW) of 42.5 MHz at -10 dB. The antenna resonates at 2.33 GHz with the input reflection coefficient of -16.69 dB. Moreover, the proposed antenna is compared with the existing single antenna and the comparison is shown in Table 10.2.

10.2.2 Design of a 2 x 2 Antenna with λ/4 Distance

The main aim of the proposed antenna design is to minimize the mutual coupling of two radiating elements. For that, a mirror of single patch antenna is used with a distance of λ/4 between the elements.

TABLE 10.2 Comparison between proposed single antenna with other existing methods.

REFERENCE	SUBSTRATE (E_R)	RESONANT FREQUENCY (GHZ)	RETURN LOSS (S_{11}) (DB)	SIZE (MM^2)	GAIN (DBI)	BANDWIDTH
[7]	FR4 (4.4)	2.45	-22	14 x 18	-0.8	50 MHz
[15]	Rogers RO4350B (3.48)	2.4	NA	18.8 x 17.9	4.83	27.44 MHz
[16]	FR4 (4.35)	2.45	33.03	20.5 x 20.5	2.6	75 MHz
[17]	FR4 (4.4)	5.8	>-15	23.6 x 26	4	10 MHz
[18]	Rogers RO4003C (3.5)	2.42	-17	80 x 51.06	5.1	35 MHz
[22]	FR4 (4.4)	5.98	-20.34	30 x 30	3	1.21 GHz
Proposed	**FR4 (4.4)**	**2.33**	**-16.69**	**29.48 x 38**	**4.23**	**42.5 MHz**

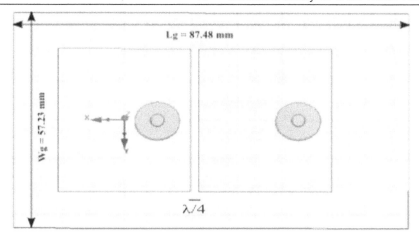

FIGURE 10.3 2 x 2 MMO antenna with λ/4 spacing.

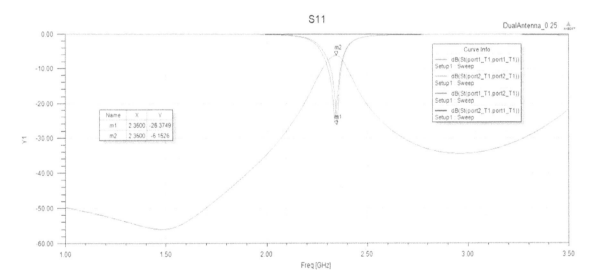

FIGURE 10.4 Scattering parameters of 2 x 2 MIMO antenna for λ/4 distance.

The geometry of 2 x 2 antennas is depicted in Figure 10.3. The plot of frequency versus S parameters is illustrated in Figure 10.4. From the Figure 10.4, it is clear that the resonant frequency of dual antenna shifted from 2.33 GHz to 2.36 GHz and also improved the gain from -16.69 dB to -26.37 dB. The mutual coupling has been reduced up to -6.15 dB using the proposed design.

10.2.3 Design of 2 x 2 Antenna with λ/3 Distance

To further improve the isolation between two patch elements, the spacing between the elements was increased from λ/4 to λ/3 and it is illustrated in Figure 10.5. The spacing is calculated at center-to-center patch. The plot of frequency versus S parameters is displayed in Figure 10.6. It was observed that when the spacing between the patch elements is increased from λ/4 to λ/3, the mutual coupling decreases from -6.152 dB to -16.64 dB.

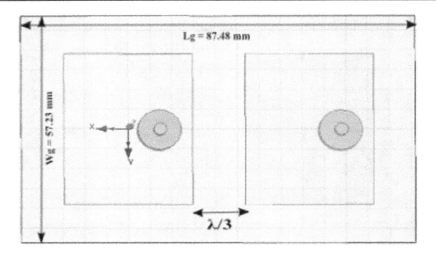

FIGURE 10.5 λ/3 spacing between two elements.

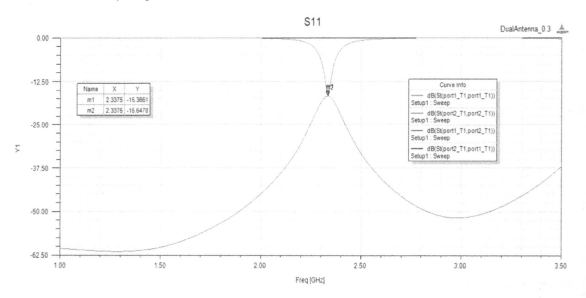

FIGURE 10.6 S_{21} parameter plot for 2 x 2 antenna.

10.2.4 Design of 2 x 2 Antenna with λ/2 Spacing

Finally, the mutual coupling value was drastically reduced by implementing λ/2 distance among the two elements. Figure 10.7 displays the configuration of the proposed 2 x 2 MIMO antenna with λ/2 spacing and their scattering parameter is showed in Figure 10.8. From Figure 10.8, it can be clearly understood that the proposed MIMO antenna resonates at 2.32 GHz with S_{11} of -17.66 dB and an isolation of -24.95 dB.

10.2.5 Electric Field Distribution

To justify the reduction of mutual coupling, the electric field distribution is simulated, which displayed in Figures 10.9 and 10.10. It can be clearly seen from the figures that the surface current distributed into

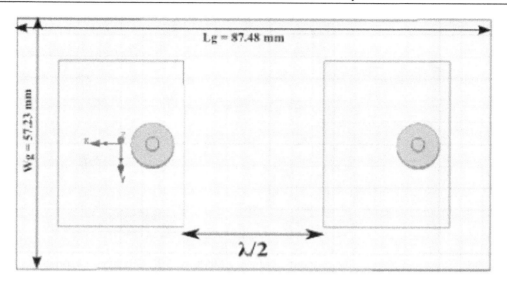

FIGURE 10.7 λ/2 spacing between two elements.

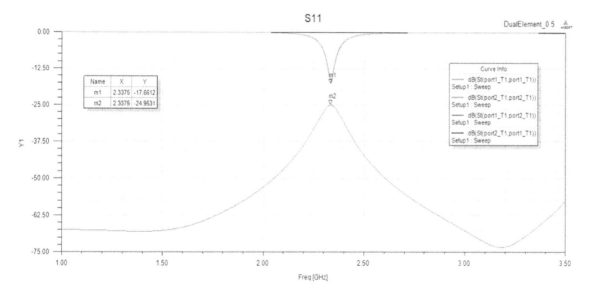

FIGURE 10.8 S_{21} parameter plot for 2 × 2 MIMO antenna.

port2 is significantly reduced by implementing λ/2 inter-element spacing. The Figure 10.9 depicts the electric field distribution by considering the excitation in port1 and termination in port2. The electric field varies from 1.138 V/m to 1.138 × 10^4 V/m and from Figure 10.9 it is noticed that electric field is extreme at the edges and smallest at the center of the patch.

Figure 10.10 depicts the electric field distribution by considering the excitation at port2 and termination at port1. The distribution of electric field is found to be not disturbing each other if any port is excited and terminated. From Figure 10.10, we can conclude that the mutual coupling is effectively minimized by enlarging the distance between two patch elements and here the improved result is obtained at λ/2 spacing.

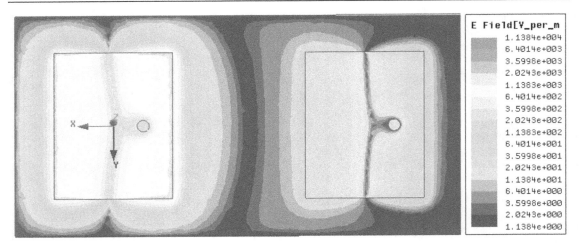

FIGURE 10.9 Simulated electric field distribution of 2 × 2 MIMO at 2.33 GHz at excitation in port1.

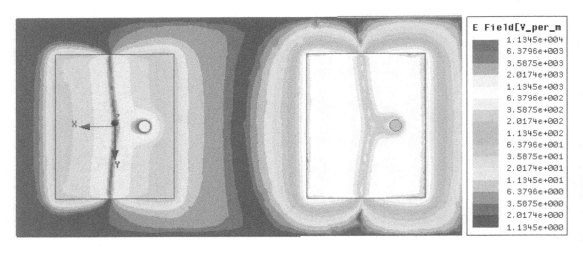

FIGURE 10.10 Simulated electric field distribution for 2 × 2 MIMO at 2.33 GHz at excitation in port2.

The gain of the proposed antenna can be obtained from its radiation pattern. The radiation patterns of single and dual patch antennas with different spacing are illustrated in Figure 10.11. Figure 10.11 demonstrates that gain of the antenna is significantly high at $\lambda/2$ distance (Figure 10.11(d)) when compared with $\lambda/4$ and $\lambda/3$ distance [Figure 10.11(b) & (c)].

10.2.6 Envelope Correlation Coefficient (ECC) and Diversity Gain (DG)

ECC, DG and TARC are the other key parameters that can be used for analyzing the performance of MIMO antennas. For a two antenna system, ECC can be computed as

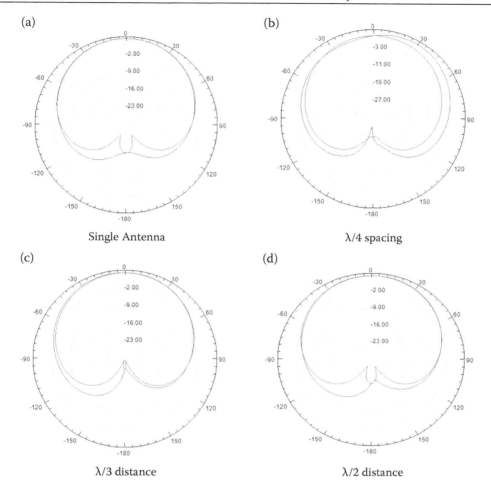

FIGURE 10.11 Radiation pattern of (a) single antenna, (b) dual patch antenna with λ/4 spacing, (c) dual patch antenna with λ/3 distance, (d) dual patch antenna with λ/2 distance.

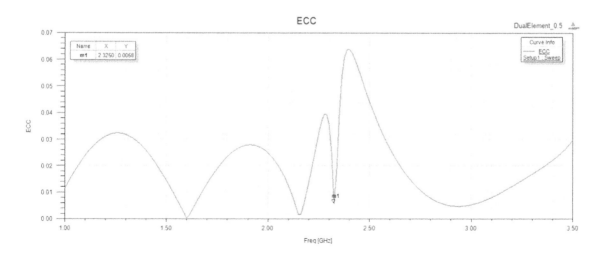

FIGURE 10.12 Simulated ECC of the proposed antenna.

$$\mathrm{ECC} = \frac{\left| S_{11}^* S_{12} + S_{12}^* S_{22} \right|^2}{[1 - (|S_{11}|^2 + |S_{21}|^2)][1 - (|S_{22}|^2 + |S_{12}|^2)]} \qquad (10.6)$$

The ECC value obtained for the proposed antenna is 0.0058, which is less than the recommended value of 0.5. The simulated result for ECC is shown in Figure 10.12. The DG of a two antenna system can be calculated as

$$\mathrm{DG} = 10(1 - |\rho|^2)^{1/2} \qquad (10.7)$$

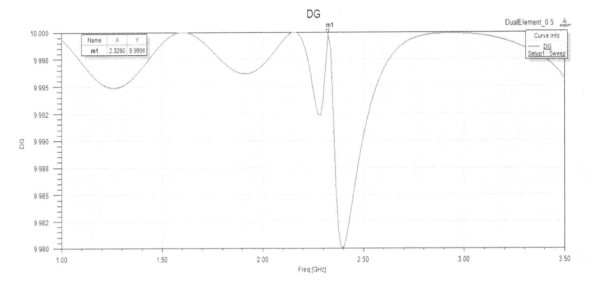

FIGURE 10.13 Simulated DG of the 2 x 2 MIMO antenna.

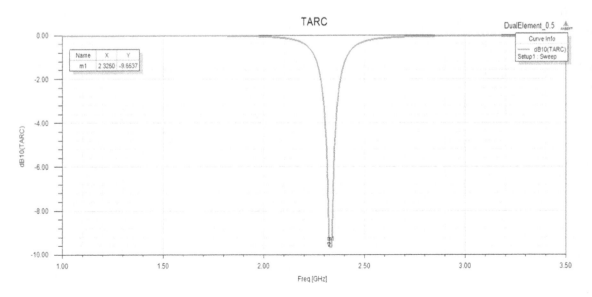

FIGURE 10.14 Simulated TARC of the MIMO antenna.

TABLE 10.3 Parametric comparison of proposed design method with previous works.

REFERENCE	MATERIAL	NO. OF PORTS	RESONANT FREQUENCY	ISOLATION (DB)	ECC	DG	GAIN	TECHNIQUES EMPLOYED
[1]	FR-4	Dual	3.5 GHz	15	<0.05	-	-	Metamaterials
[7]	FR-4	Quad	2.45 GHz	>-18	<0.3	9.53 dB	-0.8 dBi	CSRR
[8]	Rogers4003	Dual	2.45 GHz	>-20	<0.016	9.99 dB	4.025 dBi	CSRR
[9]	FR-4	Quad	2.45 GHz	>-22	0.4	9.16 dB	4 dBi	CSRR
[10]	FR-4	Dual	2.4 GHz	>-15	<0.016	9.8 dB	2.1 dBi	NL
[11]	FR-4	Dual	2.4 GHz	-19	0.006	-	2.1 dBi	NL
[12]	FR-4	Dual	2.45 GHz	>-20	0.002	9.99 dB	-	NL
[13]	FR-4	Dual	2.5 GHz	>-23	0.18	9.8 dB	-1.79 dBi	NL
[14]	FR-4	Dual	2.35 GHz	-21	0.04	-	-	EBG
[16]	FR-4	Dual	2.5 GHz	-28	0.01	9.9 dB	4.55 dBi	EBG
[18]	Rogers RO4003, RO4450B	Quad	2.4 GHz	>-42	0.02	9.89 dB	-	Metamaterials
[19]	FR-4	Three	2.4 GHz	>15	0.1	-	-	Decoupling method
[20]	FR-4	Two	2.45 GHz	-20	0.01	-	2.3	Decoupling method
[21]	FR-4	Eight	3.5GHz	>-15	0.15	9.88 dB	-	Decoupling
[22]	FR-4	Quad	0.7 GHz	>-20	0.075	9.97 dB	2 dBi	DGS
[23]	FR-4	Dual	2.7 GHz	18	0.15	9.88 dB	-	DGS
[24]	FR-4	Dual	2.75 GHz	12	0.001	9.98 dB	-	DGS
Proposed	**FR-4**	**Dual**	**2.32 GHz**	**>-24.95**	**0.0058**	**9.99 dB**	**3.98 dB**	**Inter-element spacing**

Where ρ depicts the complex cross-correlation coefficient, with $|\rho|^2 \approx$ ECC

TARC of the two antenna system can be computed as

$$\text{TARC} = \sqrt{\frac{(S(1,1) + S(1,2))^2 + (S(2,1) + S(2,2))^2}{2}} \tag{10.8}$$

The MIMO parameters ECC and TARC mainly depend on the scattering parameters. The plot of frequency versus DG is displayed in Figure 10.13. From the plot, it is noted that the proposed method achieves a high DG of 9.99 dB due to the lower ECC value of 0.0058. Lower ECC provides better diversity performance. Another important parameter for MIMO antenna is TARC and it is formulated by knowing the value of S_{11} and S_{12} and it is showed in Figure 10.14. From the Figure 10.14, it can be clearly detected that for the resonant frequency fr = 2.32 GHz, the TARC values obtained was -9.65 dB.

The performance comparison of proposed dual patch antenna system with other existing antennas is given in Table 10.3. The proposed dual patch MIMO antenna achieves significant reduction in ECC, which indicates that it has a good isolation performance.

10.3 CONCLUSION

In this chapter, we intensively studied the influence of spacing between antenna elements in a dual patch antenna system for improving the isolation between two antenna elements. First, we designed and simulated a dual patch antenna with a spacing of $\lambda/4$ and the measured isolation was -6.15 dB. Second, we increased the distance to $\lambda/3$ and measured the isolation; it was improved to -15.54 dB. Finally, a distance of $\lambda/2$ is maintained to improve the isolation, the measured value was -24.95 dB and it also attains lower ECC value and high diversity gain. The proposed MIMO antenna improves the isolation between the radiating elements and also satisfies the MIMO demands for future 5G mobile communication systems. Future studies need to be focused on the reduction of the size of antenna to find applications in IoT devices.

REFERENCES

[1]. R. N. Tiwari, P. Singh, and B. K. Kanaujia, "A compact UWB MIMO antenna with neutralization line for WLAN/ISM/mobile applications", *Int. J. RF Microw. Comput.-Aided Eng.*, vol. 29, no. 11, pp. e21907, 2019.

[2]. L. Sun, Y. Li, Z. Zhang, and Z. Feng, "Wideband 5G MIMO antenna with integrated orthogonal-mode dual-antenna pairs for metal-rimmed smartphones", *IEEE Trans. Antennas Propag.*, vol. 68, no. 4, pp. 2494–2503, 2019.

[3]. I. Elfergani, A. Iqbal, C. Zebiri, A. Basir, J. Rodriguez, M. Sajedin,... and S. Ullah, "Low-profile and closely spaced four-element mimo antenna for wireless body area networks", *Electron.*, vol. 9, no. 2, p. 258, 2020.

[4]. T. Prabhu and S. C. Pandian, "Design and development of planar antenna array for mimo application", *Wirel. Netw.*, pp. 1–8, 2020. doi: 10.1007/s11276-020-02253-y

[5]. J. Zhang, E. Björnson, M. Matthaiou, D. W. K. Ng, H. Yang, and D. J. Love, "Prospective multiple antenna technologies for beyond 5G", *IEEE J. Sel. Areas Commun*, vol. 38, no. 8, pp. 1637–1660, 2020.

[6]. J. Y. Guo, F. Liu, G. D. Jing, L. Y. Zhao, Y. Z. Yin, and G. L. Huang, "Mutual coupling reduction of multiple antenna systems", *Front. Inf. Technol. Electron. Eng.*, vol. 21, pp. 366–376, 2020.

[7]. M. S. Sharawi, M. U. Khan, A. B. Numan, and D. N. Aloi, "A CSRR loaded MIMO antenna system for ISM band operation", *IEEE Trans. Antennas Propag.*, vol. 61, no. 8, pp. 4265–4274, 2013.

[8]. D. G. Yang, D. O. Kim, and C. Y. Kim, "Design of dual-band MIMO monopole antenna with high isolation using slotted CSRR for WLAN", *Microw. Opt. Technol. Lett.*, vol. 56, no. 10, pp. 2252–2257, 2014.

[9]. A. Ramachandran, S. V. Pushpakaran, M. Pezholil, and V. Kesavath, "A four-port MIMO antenna using concentric square-ring patches loaded with CSRR for high isolation", *IEEE Antennas Wirel. Propag. Lett.*, vol. 15, pp. 1196–1199, 2015.

[10]. S. W. Su, C. T. Lee, and F. S. Chang, "Printed MIMO-antenna system using neutralization-line technique for wireless USB-dongle applications", *IEEE Trans. Antennas Propag.*, vol. 60, no. 2, pp. 456–463, 2011.

[11]. S. Zhang, and G. F. Pedersen, "Mutual coupling reduction for UWB MIMO antennas with a wideband neutralization line", *IEEE Antennas Wirel. Propag. Lett.*, vol. 15, pp. 166–169, 2015.

[12]. Y. Ou, X. Cai, and K. Qian, "Two-element compact antennas decoupled with a simple neutralization line", *Prog. Electromagn. Res.*, vol. 65, pp. 63–68, 2017.

[13]. S. Wang and Z. Du, "Decoupled dual-antenna system using crossed neutralization lines for LTE/WWAN smartphone applications", *IEEE Antennas Wirel. Propag. Lett.*, vol. 14, pp. 523–526, 2014.

[14]. L. Sun, Y. Li, Z. Zhang, and Z. Feng, "Wideband 5G MIMO antenna with integrated orthogonal-mode dual-antenna pairs for metal-rimmed smartphones", *IEEE Trans. Antennas Propag.*, vol. 68, no. 4, pp. 2494–2503, 2019.

[15]. A. Suntives, and R. Abhari, "Miniaturization and isolation improvement of a multiple-patch antenna system using electromagnetic bandgap structures", *Microw. Opt. Technol. Lett.*, vol. 55, no. 7, pp. 1609–1612, 2013.

[16]. S. Ghosh, T. N. Tran, and T. Le-Ngoc, "Dual-layer EBG-based miniaturized multi-element antenna for MIMO systems", *IEEE Trans. Antennas Propag.*, vol. 62, no. 8, pp. 3985–3997, 2014.

[17]. M. A. Abdalla, and A. A. Ibrahim, "Compact and closely spaced metamaterial MIMO antenna with high isolation for wireless applications", *IEEE Antennas Wirel. Propag. Lett.*, vol. 12, pp. 1452–1455, 2013.

[18]. G. Zhai, Z. N. Chen, and X. Qing, "Enhanced isolation of a closely spaced four-element MIMO antenna system using metamaterial mushroom", *IEEE Trans. Antennas Propag.*, vol. 63, no. 8, pp. 3362–3370, 2015.

[19]. J. Ryu, and H. Kim, "Compact MIMO antenna for application to smart glasses using T-shaped ground plane", *Microw. Opt. Technol. Lett.*, vol. 60, no. 8, pp. 2010–2013, 2018.

[20]. K. C. Lin, C. H. Wu, C. H. Lai, and T. G. Ma, "Novel dual-band decoupling network for two-element closely spaced array using synthesized microstrip lines", *IEEE Trans. Antennas Propag.*, vol. 60, no. 11, pp. 5118–5128, 2012.

[21]. C. F. Ding, X. Y. Zhang, and C. D. Xue, "Novel pattern-diversity-based decoupling method and its application to multielement MIMO antenna", *IEEE Trans. Antennas Propag.*, vol. 66, no. 10, pp. 4976–4985, 2018.

[22]. Y. S. Chen, and C. P. Chang, "Design of a four-element multiple-input–multiple-output antenna for compact long-term evolution small-cell base stations", *IET Microw. Antennas Propag.*, vol. 10, no. 4, pp. 385–392, 2016.

[23]. Y. T. Wu, and Q. X. Chu, "Dual-band multiple input multiple output antenna with slitted ground", *IET Microw. Antennas Propag.*, vol. 8, no. 13, pp. 1007–1013, 2014.

[24]. A. Dkiouak, A. Zakriti, M. El Ouahabi, A. Zugari, and M. Khalladi, "Design of a compact MIMO antenna for wireless applications", *Prog. Electromagn. Res.*, vol. 72, pp. 115–124, 2018.

[25]. S. S.Kashyap, V. Dwivedi, and Y. P. Kosta, "Electromagnetically coupled microstrip patch antennas with defective ground structure for high frequency sensing applications", *Computation and Communication Technologies*, Boston: De Gruyter, 2016.

[26]. S. Rawat, and K. K. Sharma, "A compact broadband microstrip patch antenna with defective ground structure for C-Band applications", *Cent. Eur. J. Eng.*, vol. 4, no. 3, pp. 287–292, 2014.

[27]. R. B. Chen and X. O. Ou, "A hairpin DGS resonator for application to microstrip lowpass filters", *J. Electr. Eng.*, vol. 71, no. 2, pp. 110–115, 2020.

[28]. S. Prasad Jones Christydass, S. Asha, Suraya Mubeen, B. Praveen Kitti, P. Satheesh Kumar, and V. Karthik, "Multiband Circular Monopole Metamatrial Antenna with Improved Gain", *J. Comput. Theor. Nanosci.*, vol. 18, no. 3, Mar. 2021, pp. 736–745, 2021(10), American Scientific Publishers, 10.1166/jctn.2021.9657.

[29]. S. Prasad Jones Christydass, Dr E. Kusuma Kumari, A. Sowjanya, P. Satheesh Kumar, N. Selvam, and K. Murali, "Microstrip Metamaterial Bandpass Fiter For 5G Application", *Solid State Technol.*, vol. 63, no. 3, 2020.

[30]. T. Nivethitha, S. Kumar Palanisamy, K. Mohanaprakash, and K. Jeevitha, "Comparative study of ANN and Fuzzy classifier for forecasting electrical activity of heart to diagnose Covid-19", *Mater. Today: Proc*, vol. 1, pp. 97–106. Oct. 2020.

Design of Improved Quadruple-Mode Bandpass Filter Using Cavity Resonator for 5G Mid-Band Applications

P. Satheesh Kumar[1], P. Chitra[2], and S. Sneha[3]

[1]Associate Professor, Electronics and Communication Engineering Department Coimbatore Institute of Technology, Coimbatore, Tamilnadu, India
[2]Department of ECE Coimbatore Institute of Technology, Tamil Nadu
[3]PG Scholar, Electronics and Communication Engineering Department Coimbatore Institute of Technology, Coimbatore, Tamilnadu, India

11.1 INTRODUCTION

Filters are networks that process a frequency-dependent way of processing signals. Filters are essential components of communication technology; multiple filters at various stages are used in a single transceiver. Some of them require very low losses, especially when it is positioned at the front end of the receiver [1]. In this situation, for achieving better Q-factors, the waveguide technology is used. Waveguides, unfortunately, are bulky and increase the weight of the components. Therefore, substantial efforts have been made to alleviate this problem. In this chapter, a novel quad-mode bandpass filter using a modified coaxial cavity resonator, which has four coupled conductive posts within a single cavity, is proposed. Instead of using coupling windows, as required by the conventional designs, coupling control between the inner posts is achieved by appropriate selection of the post positions and dimensions. In the proposed filter for 5G mid-band applications, the metallic walls are not required inside the cavity, thereby, the filter maximizes space utilization, thus achieving an optimal factor for a given volume. In this way, the proposed filter configuration enables compact design with an improvement in the resonator unloaded Q, compared to the conventional coaxial filters. Additionally, due to the controllable mixed electric and magnetic coupling in the cavity, the dominant sequential (inline) coupling and cross-coupling can be adjusted to inductive or capacitive according to the requirement. Consequently, multiple controllable transmission zeros and, thus, flexible stopband characteristics, are easily realized.

The proposed quad-mode bandpass filter is equivalent to a parallel-tuned circuit and is most frequently used with or without amplification for general transmitter sharpening of a single front end receiver selectivity. The requirements of size reduction and performance improvement have tended to become more and more stringent to adapt to the rapid evolution of mobile communication systems. Demands on size and performance of the filter can be achieved by dielectric loading [2,3], dual-mode and multi-mode techniques [4–6]. However, most of these methods based on waveguide technologies cannot be directly applied to coaxial cavity filters, which are preferable in base station applications. In addition to cost reduction, critical limiting factor for the wide spread of these kinds of filters is required, although they show very high-quality factor (Q-factor) and high power capability, the coaxial cavity filters based on a combline structure have been extensively studied and applied in the telecommunication industry for many decades [7], [8]. Especially in wireless base station applications, they are widely employed due to their advantages such as compact size, low production cost and relatively high factor [9], [10]. A large number of studies in the area of analysis and design of coaxial cavity filters have been conducted since the 1980s [11–15]. In recent years, most of the research interests have been focusing on the filter coupling topology to improve the stopband characteristics with a minimum resonator number [16], [17]. In [18], an adjustable blocking point at a certain frequency is realized by presetting and/or preselecting a defined capacitive and inductive coupling between two coaxial resonators, one immediately following the other on a signal path. More recently, several research studies regarding the compact inline coaxial cavity bandpass filters have been presented. By exploring controllable mixed electric and magnetic coupling, an inline coaxial cavity filter with a reduced resonator number and filter size was introduced [19]. By changing the orientation of some of its coaxial cavity resonators, the length of the filter can be reduced to a certain extent, and multiple controllable transmission zeros can be generated due to the presence of cross-coupling between the nonadjacent resonators [20]. However, generally speaking, the size-reduced bandpass filters based on the coaxial cavity resonator and the common filter topology have rarely been reported.

The resonance characteristics of the proposed compact coaxial quad-mode filter shows a high Q-factor. The filter can be designed for a specification with the adequate dimensions. The filter size is small and the bandwidth is less than 1% [21–25]. For broader bandwidths, resonator coupling is compact and are closely spaced. Coaxial cavity resonating filter is widely used in communication and radar systems with the advent of the microwave communications. The filter can be defined by the cavity structure as a standard filter and a coaxial square cavity filter. The high Q value feature of the coaxial cavity filter is easy to physically understand and is applicable to the low insertion loss, narrow band and great suppression of sidebands. Using the proposed concept, 4- and 8-pole bandpass filters are equipped with various filtering functions.

11.1.1 Cavity Filters

Cavity filters are single, dual and triple-mode bandpass filters. At present, in the satellite communications, it is common to use four-pole or six-pole dual-mode filters in output multiplexers and eight-pole dual-mode filters in input multiplexers. Thus, by this arrangement, output multiplexers have either two or three cascade waveguide cavities and input multiplexers have four cascade waveguide cavities. In the satellite communications, any weight or volume savings achieved are extremely important. Filters currently used in input and output multiplexers are generally significantly heavier and occupy a much larger volume than filters made in accordance with the present invention. Further, it is known to use two triple-mode cavities as a six-pole filter in an output multiplexer. Unfortunately, this type of six-pole filter must be launched onto a manifold of the multiplexer at a sidewall of one of the triple-mode cavities [26–28]. This side-wall launching can be much bulkier than an end-wall launching.

For some time, it has been known that if a filter can be made to produce more transmission zeros, the response of that filter will be enhanced. With previous filters, the maximum number of transmission zeros that can be produced is equal to the order of the filter minus two. For example, a six-pole prior art dual-mode filter can be made to produce four transmission zeros and such a filter is said to produce an elliptic function response.

A bandpass filter having at least one four-mode cavity with tuning screws and coupling screws arranged in the said filter so that the said four-mode cavity resonates at its resonant frequency in four independent orthogonal modes simultaneously. Any additional cavities in the four-mode cavity are immediately adjacent to one another or to the four-mode cavity. At least one of any additional cavities is immediately adjacent to the said four-mode cavity. Any immediately adjacent cavities are coupled to one another and are resonating at their resonant frequencies, said filter having an input and output.

Preferably, the filter of the present invention has at least two cavities, a first cavity being a quadruple-mode cavity and a second cavity being either a single-mode cavity, a dual-mode cavity, a triple-mode cavity or a quadruple-mode cavity. Still, more preferably, the filter of the present invention is operated in such a manner that the number of transmission zeros is equal to the order of the filter.

In particular, to a filter that has at least one cavity resonating in four independent orthogonal modes simultaneously, resonating coaxial cavity, consisting of four conductive couplings, rather than using couplings, coupling windows can be used to accomplish sufficient control between the inner posts. The proposed filter maximizes the metallic walls within the cavity. The use of space achieves an optimal filter. The suggested filter configuration enables compact configuration with resonator enhancement. Furthermore, because of the controllable electric and magnetic combination, the dominant sequential (inline) coupling in the cavity is required to change cross-coupling to inductive or capacitive effect in the 40 MHz–960 MHz spectrum range. High quality-factor, Q, improved efficiency with better stability and narrowly spaced (up to 75 kHz) frequencies is attained [8–10]. To increase the internal capacity of the resonating filter cavities, physical dimensions of traditional cavity filters can be varied. These range from 85 in the 40 MHz range to less than 11 in the 40 MHz range. The cavity filters become more realistic in the range of 900 MHz, where Q-factor is substantially higher than the resonators and filters for the lumped part [2].

11.2 FILTER STRUCTURE

Four- and eight-pole bandpass filters using this proposed concept with different filtering functions are designed, manufactured and tested. The measured results show that an excellent Q-factor can be achieved with compact size. In general, the proposed design concept of coaxial cavity resonator can provide an unloaded improvement of approximately 15% for a given volume. It means that 30%–35% volume saving can be achieved while maintaining the comparable Q-factor value with the conventional coaxial designs. It should be mentioned that a similar resonator configuration has been reported based on the substrate integrated waveguide (SIW) technique in [21]. However, only the first three resonant modes are used to design the dual-band filter, while the last mode is considered as spurious, without the investigations of multiple transmission zeros. A standard combline filter is the basic filter structure, with a number of resonators situated between two ground planes in a straight line [6] as shown in Figure 11.1.

11.2.1 Cavity Resonator

The cavity resonator is the fundamental component of waveguide filters. This is composed of a short waveguide length obstructed at both ends. Between both ends, waves trapped within the resonator are reflected back and forth. The resonance effect can be used to distinguish between those frequencies passing through. In practice, the wave requires a filter structure and it is permitted to pass out of one cavity through

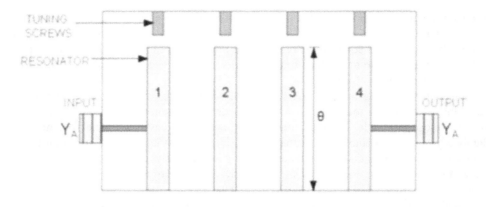

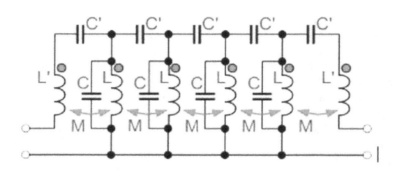

FIGURE 11.1 General combline cavity filter and equivalent circuit model.

a coupling structure into another. The architecture of the entirely closed cavity resonator and errors would be small. In different filter groups, a variety of different coupling mechanisms is used [11].

11.3 QUADRUPLE-MODE RESONATOR

Four pole bandpass filters with various filtering functions are constructed using this suggested definition. The calculated results indicate that, with a compact scale, an outstanding element can be achieved [12]. The coupling between adjacent resonators is realized through a coupling window and adjusted by the window size and/or coupling screw penetrating into the cavity from the top cover, as shown in Figure 11.1. To enhance the filter selectivity, transmission zeros are usually introduced into the stopband using cross-coupling between non-adjacent resonators. In contrast to the conventional design, the proposed modified coaxial cavity resonator eliminates the inside metallic walls within the cavity, and consequently shows new characteristics of quadruple resonant modes, as detailed below.

11.3.1 Configurations and Characteristics

The physical setup of the quad-mode resonator that has been implemented consists of four different inner conductive poles within a single resonating cavity. Its side dimensions including the length and the

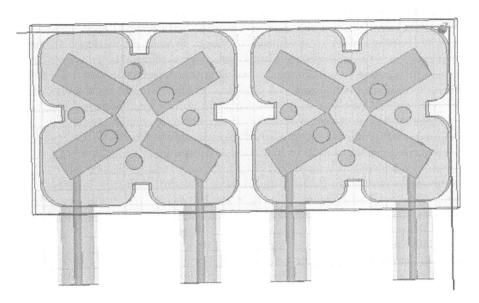

FIGURE 11.2 4-pole quad-mode resonator with proposed dimensions.

height define the size of the square cavity. The mathematical formulations for the normalized capacitances per unit element, between resonators and ground for combline filter [7], are given in (1)–(5).

Figure 11.2 depicts the physical configuration of the introduced quadruple-mode resonator, which consists of four Γ-shaped inner conductive posts within a single cavity. All four of the inner posts are short-circuited on the bottom of the metallic housing, and open-circuited at the other end. The important dimensional parameters are also illustrated in these figures. The size of the square cavity is described by its side length a and height H. The cavity corner is filleted with round R. d denotes the distance between the post center and cavity center. r and h are the radius and height of each post, respectively. Instead of the caps used in the conventional coaxial resonator, the inner posts are bended 90^0, and extended toward the cavity center to strengthen the electric coupling between them. The length and thickness of the bended part (hereinafter referred to as head) are L and T, respectively. At the open end, the two side edges of each post are chamfered with c to fine control the electric coupling. The distance between the open edge of the head and cavity center is denoted as g and, hence, the spacing between two diagonal posts is 2g.

In Figure 11.2, a four-pole elliptic filter 2 having cavity 4 resonating at its resonant frequency in the four independent orthogonal modes simultaneously is shown. The cavity 4 can be made to resonate in a first TE_{113} mode, a second TM_{110} mode, a third TM_{110} mode and a fourth TE_{113} mode. Electromagnetic energy is introduced into the cavity 4 through input coupling probe 6, which excites an electric field in the first TE_{113} mode. Energy from the first TE_{113} mode is coupled to the second TM_{110} mode by coupling screw 8. Energy is coupled from the second TM_{110} mode to the third TM_{110} mode and from the third TM_{110} mode to the fourth TE_{113} mode by coupling screws 10, 12, respectively. Energy is coupled out of the cavity 4 by means of a magnetic field transfer through aperture 14 in iris 16.

A filter in accordance with the present invention could have at least two cavities, where one cavity resonates at its resonant frequency in four independent orthogonal modes simultaneously and another cavity is either a single-mode cavity, a dual-mode cavity, a triple-mode cavity or a quadruple-mode cavity. Similarly, a two-cavity filter in accordance with the present invention could have one quadruple-mode cavity in combination with either a single-mode, dual-mode, triple-mode or quadruple-mode cavity. In any of these filters, a coupling screw can be arranged in each of the quadruple-mode cavities to create a negative feedback coupling between the first mode and the fourth mode, thereby giving rize to two transmission zeros. Where the filter has more than one quadruple-mode cavity, the negative feedback coupling is created between the first and the fourth mode of each cavity. For example, in a two-cavity filter, where each cavity resonates at its resonant frequency in four independent orthogonal modes simultaneously, the negative feedback coupling in the first cavity is M_{14} and in the second cavity is M_{58}. In other words, the fifth mode of the filter is the first mode of the second cavity and the eighth mode of the filter is the fourth mode of the second cavity. A filter in accordance with the present invention can have at least two cavities, with at least one cavity resonating at its resonant frequency in four independent orthogonal modes simultaneously and at least one of the remaining cavities being either a single-mode cavity, a dual-mode cavity, a triple-mode cavity or a quadruple-mode cavity. These filters can be operated in such a manner that the number of transmission zeros is equal to the order of the filter. Some of the transmission zeros are created in these filters by adding a resonant feedback coupling. This novel post configuration reveals new resonant modes. Simulation is conducted with the electromagnetic (EM) full-wave simulator ANSYS HFSS (eigenmode) [22]. It can be obviously seen that four different resonant modes: (a) fundamental mode, (b) differential mode and (c and d) a pair of orthogonal degenerate modes are generated in the proposed coaxial resonator. In the following explanation, the resonant modes are analyzed by simple considerations on the orientation of their magnetic fields, as the same conclusions can be derived by considering the electric fields as well. For the fundamental mode (called mode 1), the magnetic field circulating around all of the four posts has an identical orientation. As a result, it forms a large magnetic field loop in the cavity and looks like only one conductive post in the cavity center. Oppositely, as to the differential mode (called mode 2), the orientations of the magnetic field circulating around the four inner posts are different from each other. Therefore, it looks like four small independently resonated cavities, which is similar to the conventional coaxial resonator with inside metallic walls. The two degenerate modes (called modes 3 and 4, respectively) are orthogonal to each other in space. For each one, the magnetic field only concentrates in the volume around two diagonal posts accounting for the fulfillment of the boundary condition. In the case of the conventional coaxial resonator, the field is mainly confined within each cavity due to the inside metallic walls. Nevertheless, the proposed resonator shows different features. Since four inner posts are located within a single cavity, the equivalent resonant space of each post will vary according to the resonant mode and post position as long as the boundary condition can be satisfied. Therefore, the resonant frequency of each mode is determined not only by the post and cavity dimensions, but also by the post position, specifically mainly related to the parameter. Let us consider as an example a square cavity with a = 33 mm, H = 17 mm and R = 3 mm. The cavity dimensions given above remain unchanged for all the presented quadruple-mode resonators and the subsequent 4-pole filters. The other dimensions regarding the inner post are as follows: r = 2.3 mm, h = 14.9 mm, L = 11 mm, T = 3 mm, c = 1.5 mm and g = 2.3 mm as shown in Figure 11.3. The corner of the cavity has a round R fillet. Each post of its side edges are chamfered at the open end for the fine tuning of electric wave coupling. The spacing distance between two diagonal posts and distance between cavity center and open edge of head is 1 unit length.

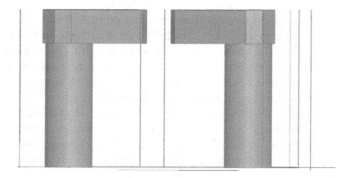

FIGURE 11.3 Projected quad-mode resonator with proposed dimensions (side view).

Let us consider a square cavity, in which the cavity dimensions are same for subsequent four-pole filters and the quadruple-mode resonators as shown in Figures 11.4 and 11.5. The dimensions of the inner poles are as follows: R = 3 mm, L = 9 mm, r = 2 mm, c = 1 mm.

The cascaded pair of degenerate orthogonal modes created in the coaxial resonator is designed. The four inner posts are placed in a single cavity, its equivalent circuit for resonant space will differ according to the resonating mode, and it is possible to satisfy the boundary condition of the post location. The resonating frequency of each mode is, therefore, defined by the cavity dimensions of the post and cavity and is primarily correlated with the parameter.

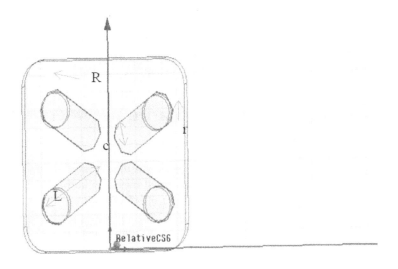

FIGURE 11.4 Proposed dimension of quadruple-mode resonator (top view).

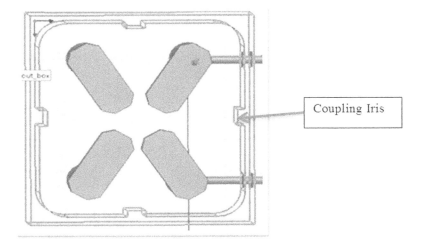

FIGURE 11.5 Perspective view of the cascaded coaxial quadruplet filter (top view).

11.4 DESIGN OF CAVITY FILTER

Resonating modes are chosen based on the passband spectrum, coupling structure, response modulation frequency tuning, band stop/passband tuning and standing wave modulation, which are fully taken into consideration. The measurement of the eigen mode resonant unit is selected on the basis of the filter's passband. The single cavity is designed having the same size as the mode by using the eigen mode solver created on the Euclidean finite element algorithm. In the frequency range of 30 MHz to 40 GHz and with bandwidth choices from less than 0.5% to over 66%, cavity filter designs are available. Cavity filters provide the consumer with very low loss of insertion, steep skirt selectivity and narrower bandwidths than discrete filters of components. The efficiency of the cavity filter is dependent on the selection of components and the physical configuration of the helical coils, resonators and the shape and size of the cavity housing. Using aluminum as the base metal, standard cavity filters are typically constructed.

11.4.1 Normalized Capacitance between Resonators and Ground

This section presents the capacitances needed for the calculation of resonator "d". The resonator is first thought to be rectangular and then transformed into a cylindrical structure with a diameter of "d". The width "w", length "L" and thickness "t" are the dimensions of the rectangular resonator. The resonator impedance is taken as 70 ohms to achieve a high unloaded "Q" for filter. Mathematical formulations are given in (11.1)–(11.6) for the standardized capacitances per unit element between resonators and ground for combline filter [7].

$$\frac{C_0}{\varepsilon} = \frac{376.7 Y_A}{\sqrt{\varepsilon_r}} \left(1 - \sqrt{\frac{G_{T1}}{Y_A}} \right) \tag{11.1}$$

$$\frac{C_n}{\varepsilon} = \frac{376.7 Y_A}{\sqrt{\varepsilon_r}} \left[\frac{Y_{an}}{Y_A} - 1 + \frac{G_{Tn}}{Y_A} - \frac{J_{n-1,n}}{Y_A} \tan \theta_0 \right] + \frac{C_{n+1}}{\varepsilon} \tag{11.2}$$

$$\frac{C_{n+1}}{\varepsilon} = \frac{376.7 Y_A}{\sqrt{\varepsilon_r}} \left(1 - \sqrt{\frac{G_{Tn}}{Y_A}} \right) \tag{11.3}$$

$$\frac{C_{01}}{\varepsilon} = \frac{376.7 Y_A}{\sqrt{\varepsilon_r}} - \frac{C_0}{\varepsilon} \tag{11.4}$$

$$\left. \frac{C_{j,j+1}}{\varepsilon} \right|_{j=1 \text{ to } n-1} = \left(\frac{376.7 Y_A}{\sqrt{\varepsilon_r}} \right) \left(\frac{J_{j,j+1}}{Y_A} \right) \tan\theta_0 \tag{11.5}$$

$$\frac{C_{n,n+1}}{\varepsilon} = \frac{376.7 Y_A}{\sqrt{\varepsilon_r}} - \frac{C_{n+1}}{\varepsilon} \tag{11.6}$$

where, $\mathcal{E}$ is the permittivity, $\mathcal{E}r = 1$, "n" is the order of filter, Y_{aj} is the resonator admittance = 1/70 and Y_A = 1/50.

where, $\theta o = 45°$ for $\lambda/8$ resonators. All the mathematical calculations for normalized capacitances are performed using SCILAB-generated codes.

11.4.2 Physical Filter Dimensions

In this work, the cavity width selected for the design is $b = 16$ mm. $\Delta C/\mathcal{E}$ [7], for $t/b = 0.4$, where $\Delta C/\mathcal{E} = C_{j,j}+1/\mathcal{E}$, derived in (11.7) gives the normalized width "w_j" [8] of rectangular resonators.

$$\left.\frac{W_j}{b}\right|_{j=1 \text{ to } n} = \frac{1}{2}\left(1 - \frac{t}{b}\right)\left[\frac{1}{2}\frac{C_j}{\varepsilon} - \frac{(C'_{fe})_{j-1,j}}{\varepsilon} - \frac{(C'_{fe})_{j,j+1}}{\varepsilon}\right] \tag{11.7}$$

where $C'_{fe'}/\mathcal{E}$ is the normalized even-mode fringing capacitance. By equating the diameter of both types of resonators, as shown in (11.8), the rectangular cavity resonators are converted to cylindrical ones.

$$\prod d_j = 2(W_j + t) \tag{11.8}$$

The diameter of each resonator is set equal to the average diameter value given by (11.9) for the simplicity and symmetry of the physical structure.

$$d = \sum_{j=1}^{n} \frac{d_j}{n} \tag{11.9}$$

Hole "$d1$" is drilled into a resonator creating a coaxial structure with a diameter "D" tuning screw inserted into this hole. Using the coaxial capacitance formula given in (11.10), the values of "$d1$" and "W" are computed.

$$C_r = W \times 0.02414 \times \frac{\varepsilon_r}{\log\left(\frac{d1}{D}\right)} \text{pF} \tag{11.10}$$

Using the distributed capacitance "Cr" built at the free end of the resonators; the resonators are tuned to resonate at f0. Capacitance Cr (11.11) is determined by means of (11.11).

$$C_r = (Y_A)\left(\frac{Y_{aj}}{Y_A}\right)\left(\frac{\cot\theta_0}{\omega_0}\right) \tag{11.11}$$

Where, ε is the permittivity, $\varepsilon = 1$, "n" is the order of filter, Y_{aj} is the resonator admittance (1/70) and $Y_A = \frac{1}{50}$. The factors $\frac{G_T}{Y_A}$ and $\frac{J_{j,j+1}}{Y_A}$ are calculated. Where, $\frac{b}{Y_A} = 0.918$ for $\lambda/8$ resonators [8], with impedance equal to 70 Ω.

Filter housings are silver plated for enhanced electrical characteristics and current flow, as most raw metals are inherently loss. To minimize frequency drift over temperature, brass, copper, aluminum or bi-metal resonators are used.

The basic circuitry of cavity and duplexer filters are highly tuned resonance circuits that can pass through only certain frequencies. A cavity filter is a resonator at the input and output inside a conducting "box" with resonating coupling loops. The widely used frequency range is from 40 MHz to 960 MHz. A better Q quality factor is obtained by increasing the stability efficiency at closely spaced frequencies (down to 75 kHz) and increasing the internal filter cavity length. Filters based on cavity resonators, in comparison to filters built on lumped-passive element LC resonators or planar cavity resonators, have better power handling capability and better insertion loss. It is popular to use cavity resonator filters in wireless and satellite applications.

11.5 COAXIAL RESONATOR CAVITY FILTERS

The coaxial resonator filters achieve magnetic coupling or capacitive coupling when they open or place the sample between two cavities. The iris or sample dimensions regulate the magnetic or capacitive coupling value. Cross-assembly between coaxial resonators measures the position and number of transfer zeros. The volume of a coaxial resonator is achieved by adjusting its internal and external drivers by taking into account the discharged Q factor and power capability of a coaxial resonator as shown in Figure 11.6. In short, the coaxial resonator cavity filter has many advantages, such as a smaller length and a slimmer bandpass, higher slopes and greater power capacity that are characteristic of Chebyshev's general function.

Tuning screws placed in the resonant cavities, which can be adjusted externally to the waveguide, allow fine tuning of the resonant frequency by adding more or less thread to the waveguide. Tuning screws in each cavity are sealed with jam nuts and thread-locking compounds. The equivalent circuit is a shutter condenser with a value rising as the screw is inserted for screws inserted at just a small distance as shown in Figures 11.7 and 11.8.

The screw has been inserted at a distance of $\lambda/4$ which is equal to the sequence of LC circuits. Inserting it further allows the impedance to change from capacitive to inductive, i.e., the arithmetic sign shifts.

Consider two separate size fusion resonators with the same resonator rods magnetically coupled with an iris in the wall. The tuner height of each other does not, therefore, create a synchronous resonance.

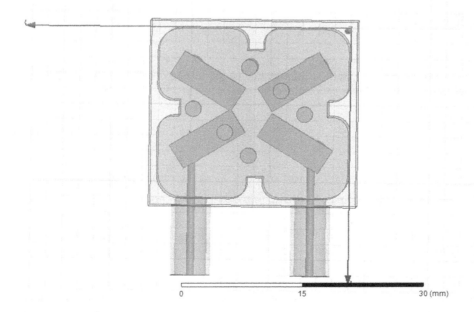

FIGURE 11.6 Design of the proposed quad-mode bandpass filter (top view).

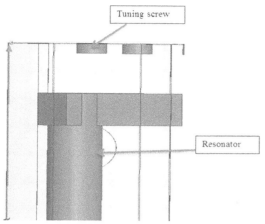

FIGURE 11.7 Conventional coaxial resonator (cross-sectional view).

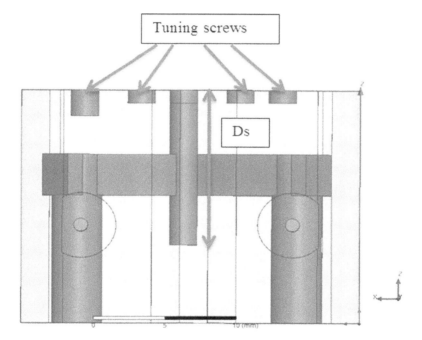

FIGURE 11.8 Penetration depth of coupling screw.

Figure 11.9 shows the proposed quadruple-mode bandpass filter including tap feeders. Figure 11.10 shows design model of eight-pole bandpass resonating cavity filter by cascading two quadruple-mode resonators.

Considering two combline cavity resonators of diverse sizes with similar resonator rods magnetically coupled to an iris in the common wall, it does not produce a synchronous resonance condition to set the tuner heights equal to other.

A synchronously tuned state can be achieved by keeping tuning screw heights (say, Th1) constant and iteratively changing the other tuning screw height dimension (say, Th2) until a minimum value is reached by the different to two individual frequencies of the entire structure. An eight-pole bandpass filter by cascading two quadrature mode cavity filters is also designed for future work.

230 Future Trends in 5G and 6G

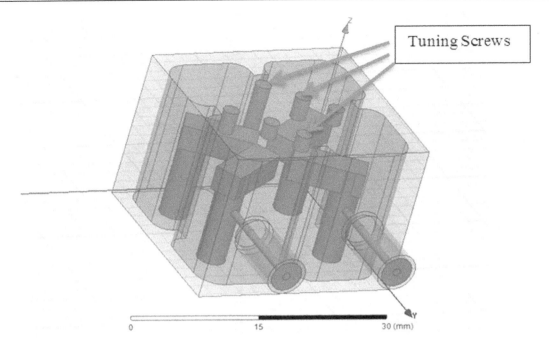

FIGURE 11.9 Proposed quadruple-mode bandpass filter including tap feeders.

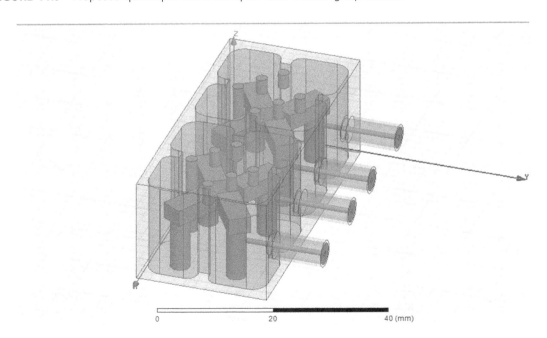

FIGURE 11.10 Design model of eight-pole bandpass resonating cavity filter by cascading two quadruple-mode resonators.

The quadrature mode cavity filter can be used as an elementary building block to create higher order resonating filters with different responses. The four-pole single cavity filter has been designed, the square with an outer width = 26 mm side length and an outer length = 20 mm height. These tuning screws used undergo small changes to compensate for manufacturing tolerances. The calculated center frequency of the filter is 4.250 GHz, with a bandwidth of 850 MHz. The configuration of the filter model is developed in the HFSS. A simulation approach is used to model the entire design process of the cavity.

11.6 RESULTS AND DISCUSSIONS

Return loss compares the power reflected to the power fed from the transmission line. It also expresses the mismatch in the signal. Return loss particularly shows the center frequency of 4.250 GHz at -23 dB as shown in Figure 11.11.

Insertion loss is nothing but a loss of signal energy/power from the transmission line and it is usually represented in decibels (dB). In S-parameter plot, it is upper portion of S_{21} plot where it touches 0 dB or near to 0 dB. As much near to 0 dB, shows less amount of insertion losses, as shown in Figure 11.12¾

The measured center frequency of the cavity resonator at 4.250 GHz and the passband bandwidth is 850 MHz, as shown in Figure 11.13. Table 11.1 shows the comparative analysis of Q-factor of conventional and proposed quadrature mode bandpass filter.

11.7 CONCLUSION

The developed quadruple-mode pass filters have four conductive posts inside a single cavity which do not have metal walls within the cavity for new resonant characteristics and high utilization for 5G midband applications. In addition, the proposed resonator will give four versatile controllers. Transmission

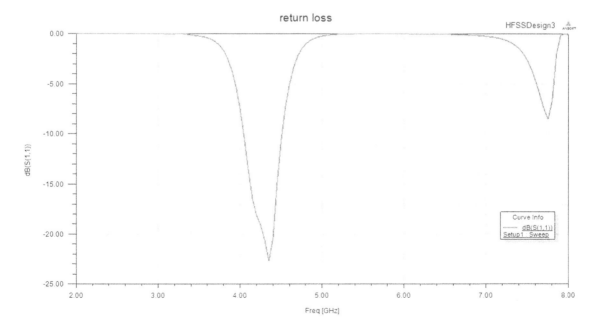

FIGURE 11.11 Return Loss of resonance cavity resonator.

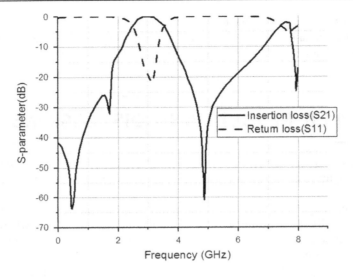

FIGURE 11.12 The insertion loss of cavity resonator.

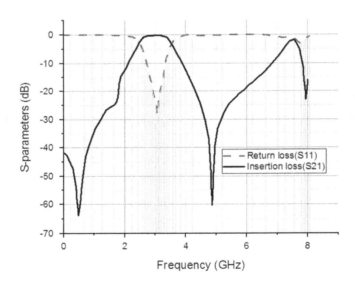

FIGURE 11.13 Filter response of S-parameter (2:1) VSWR.

TABLE 11.1 Comparison of the conventional and proposed resonators

	VOLUME	QUALITY FACTOR, Q	BANDWIDTH, BW	RESONANT FREQUENCY
Conventional	18,513 (33 mm*33 mm*17 mm)	0.865	600 MHz	2,566.8 MHz
Proposed (4-pole quadruple-mode BPF)	13,520 (26 mm*26 mm*20 mm)	2.1	850 MHz	4,250 MHz
Proposed (4-pole quadruple-mode BPF)	13,122 (27 mm*27 mm*18 mm)	4.3	880 MHz	4,200 MHz

of zeros by multiple cross-coupling will occur inside the resonating cavity. The mathematical model also highlights the capability of controlling the position of the conductive posts resonant frequencies, so that the 5G sub-band that extends along the range (3.7–4.2 GHz) can perfectly be covered with almost a flat passband. At the resonance frequency, a fractional bandwidth of 12.8% (500 MHz impedance bandwidth) has been obtained with a return loss of more than 18 dB and an insertion loss of less than 2.5 dB over the targeted bandwidth. Four-pole resonating cavity filters have been developed and the center frequency is 4.5 GHz. Insert loss at 0 dB and estimated bandwidth at 850 MHz and Q-factor of 4.3. Coefficient of reflection at 0 and return loss at -23 dB. The proposed design is appealing and ensures that low-cost, better insertion-loss, better bandwidth and Q-factor and substantially miniaturized filters for wireless base station applications are achieved for 5G mid-band applications.

REFERENCES

[1]. M. Hameed, G.B. Xiao, L. Qiu, C. Xiong, and T. Hameed, "Multiple-mode wideband bandpass filter using split ring resonators in a rectangular waveguide cavity", *Electronics*, vol. 7, p. 356, 2018.
[2]. G. Basavarajappa and R. R. Mansour, "A high-Q quadruple-mode rectangular waveguide resonator", *IEEE Microw. Wirel. Compon. Lett.*, vol. 29, pp. 324–326, 2019.
[3]. H. Campanella, Y. Qian, C. O., Romero, J. S. Wong, J. Giner, and R. Kumar, "Monolithic multiband MEMS RF front-end module for 5G mobile", *J. Microelectromechan. Syst.*, vol. 30, pp. 72–80, 2021.
[4]. A. O. Watanabe, M. Ali, S. Y. B. Sayeed, R. R. Tummala, and P. M. Raj, "A review of 5G front-end systems package integration", *IEEE Trans. Compon. Packag. Manuf. Technol.*, vol. 11, no. 1, pp. 118–133, 2020.
[5]. G. Wibisono, T. Firmansyah, H. Herudin, M. Wildan, T. Supriyanto, M. Alaydrus, and F. Ujang "Multi wideband bandpass filter based on folded quad cross-stub stepped impedance resonator", *Int. J. Antennas Propag.*, 2020. doi: 10.1155/2020/4124721
[6]. F. Teberio, I. Arregui, P. Soto, M. A. G. Laso, V. E. Boria, and M. Guglielmi, "High-performance compact diplexers for Ku/K-band satellite applications", *IEEE Trans. Microw. Theory. Tech.*, vol. 65, no. 10, Oct. 2017.
[7]. Y. I. A. Al-Yasir, N. O. Parchin, A. M. Abdulkhaleq, M. S. Bakr, and R. A. Abd-Alhameed, "A survey of differential-fed microstrip bandpass filters: Recent techniques and challenges", *Sensors*, vol. 20, p. 2356, 2020.
[8]. L. Gao, Y. Yang, and S. Gong, "Wideband hybrid monolithic lithium niobate acoustic filter in the K-Band", *IEEE Trans. Ultrason. Ferroelectr. Freq. Control.*, vol. 68, no. 4, pp. 1408–1417, 2021.
[9]. G. Jang, B. Lee, and N. Park, "Compact quad-mode bandpass filter using modified coaxial cavity resonator with improved-factor", *IEEE Trans. Microw. Theory. Tech.*, vol. 63, no. 3, Mar. 2015.
[10]. B. Yassini, M. Yu, and B. Keats, "A Ka- Band fully tunable cavity filter", *IEEE Trans. Microw. Theory. Tech.*, vol. 60, no. 12, Dec. 2012.
[11]. J. Chen and Y. Guan, "The fast design method of band tunable cavity bandpass filter", *8th Int. Symp. Comput. Intell. Des. ISC.*, Hangzhou, China, Dec. 2015.
[12]. T. Nivethitha, S. Palanisamy, K. Prakash, and K. Jeevitha, "Comparative study of ANN and fuzzy classifier for forecasting electrical activity of heart to diagnose Covid-19", *Mater. Today: Proc.* doi: 10.1016/j.matpr.2020.10.400.
[13]. A. R. Harish and J. S. K. Raj, "A Direct method to compute the coupling between nonidentical microwave cavities and techniques, vol. 52, no. 12, Dec. 2004.
[14]. S. Nassar, P. Meyer, and P. W. van der Walt, "An S-band combline filter with reduced size and increased pass-band separation", *IEEE Confer. Microw. Tech*, Pardubice, Czech Republic, 2015.
[15]. J. Lee, M. S. Uhm, and I.-B. Yom, "A dual-Passband Filter of canonical Structure for satellite applications", *IEEE Microw. Wirel. Compon. Lett.*, vol. 14, no. 6, June 2004.
[16]. S. Nam, B. Lee, C. Kwak, and J. Lee, "A New class of K-band high-Q frequency–tunable circular cavity filter", *IEEE Trans. Microw. Theory. Tech.*, vol. 66, no. 3, Mar. 2018.
[17]. H. N. Shaman, "New S-band bandpass filter (BPF) with wideband passband for wireless communication systems", *IEEE Microw. Wirel. Compon. Lett.*, vol. 22, no. 5, May 2012.

[18]. A. V. G. Subramanyam, D. Sivareddy, V. Vamsi Krishna, V. V. Srinivasan, and Y. Mehta, communication system group, ISRO satellite centre, Indian Space Research Organization, "Compact Iris-coupled Evanescent-mode Filter for Spacecraft S-band Data Transmitters", *IEEE Int. Microw. RF Conf* (IIMaRC), Dec. 2015. doi: 10.1109/IMaRC.2015.7411442

[19]. Y. H. Cho and G. M. Rebeiz, "Two- and Four-Pole Tunable 0.7–1.1-GHz Bandpass- to -Bandstop Filters with Bandwidth Control", *IEEE Trans. Microw. Theory. Tech.*, vol. 62, no. 3, Mar. 2014.

[20]. P. Wang, L. Li, and S. Wei, "Design of a tunable S-band narrow –band coaxial cavity filter", *IEEE Int. Conf. Microw. Millim. Wave Techno. ICMMT*, Shenzhen, China, May 2012.

[21]. Y. Wang and M. Yu, "True inline cross-coupled coaxial cavity filters", *IEEE Trans. Microw. Theory Techn.*, vol. 57, no.12, pp. 2958–2965, Dec. 2009.

[22]. M. Höft and F. Yousif, "Orthogonal coaxial cavity filters with distributed cross-coupling", *IEEE Microw. Wirel. Compon. Lett.*, vol. 21, no. 10, pp. 519–521, Oct. 2011.

[23]. X. Du, P. Tang and B. Chen, "Design of a C-band coaxial cavity band pass filter", *Prog. Electrm. Res. Symp. Proc.*, Guan, China, pp. 1065- 1068, Aug. 2014. doi: 10.1109/IBCAST.2018.8312328

[24]. Y. Zhan, J. X. Chen, W. Qin, J. Li, and Z. H. Bao, "Spurious-free differential bandpass filter using hybrid dielectric and coaxial resonators", *IEEE Microw. Wirel. Compon. Lett.*, vol. 26, no. 8, pp. 574–576, Aug. 2016.

[25]. S. Kurudere and V. B. Ertürk, "Novel microstrip fed mechanically tunable combline cavity filter", *IEEE Microw. Wirel. Compon. Lett.*, vol. 23, no. 11, pp. 578–580, Nov. 2013.

[26]. M. Yuceer, "A reconfigurable microwave combline filter", *IEEE Trans. Circt. Syst.*, vol. 63, no. 1, pp. 84–88, Jan. 2016.

[27]. G. L. Matthaei, L. Young, and E. M. T. Jones, *Microwave Filters, Impedance-matching Networks, and Coupling Structures*. North Bergen, NJ: Artech House, 1985.

[28]. D. Natarajan, *A practical design of lumped, semi-lumped and microwave cavity filters*. Berlin, Heidelberg: Springer-Verlag, 2013.

Wavelet Transform for OFDM-IM under Hardware Impairments Performance Enhancement

Asma Bouhlel[1], Anis Sakly[2], and Salama Said Ikki[3]

[1]*Laboratory of Electronic and Microelectronic, Faculty of Sciences MonastirUniversity of Monastir, Tunisia*
[2]*Industrial Systems study and Renewable Energy (ESIER), National Engineering School of MonastirUniversity of Monastir, Tunisia.*
[3]*Electrical Engineering Department, Faculty of Engineering Lakehead University, Thunder Bay, ON P7B 5E1, Canada*

12.1 INTRODUCTION

To meet the growing demand for high Quality of Service (QoS) in wireless communication systems, multicarrier modulation was introduced to the most popular standards. Indeed, by transmitting data over multicarrier the selective channel is divided into subchannels that could be characterized as flat fading channels. OFDM modulation was considered as a promising technique in meeting the requirements of high speed data communication [1].

Moreover, novel transmission systems such as massive multiple input multiple output (MIMO) systems [2] and non-orthogonal multicarrier communications schemes [3] were proposed as transmission techniques for 5G wireless communication. To further boost the achievable performances, a new modulation technique named index modulation (IM) has been recently introduced. Indeed, the IM includes a spatial axis to extend the constellation space of the information signal. Using this modulation, additional bits through the building blocks indices of the transmission system are implicitly conveyed. Therefore, a higher benefit with regard to spectral efficiency is obtained. To satisfy the growing demand for high speed wireless products, IM was introduced to the most popular standard. The concept of IM was adopted for transmit antennas to announce spatial modulation (SM). The SM approach was proposed by Mesleh in [4] and [5]. Unlike standard MIMO systems, in the SM technique, transmit antenna index on space axis are explored to carry bits in addition to the mapped symbols. To increase the data transfer rate, various types of SM systems have been lately introduced in [6–8].

To maximize OFDM bandwidth efficiency, IM is combined with the OFDM system [9,10]. The concept of SM was readapted to the OFDM system by extending the representation of the signal on spatial axis using the indices of active subcarriers. The system was first introduced in [11] and was called OFDM-IM. It was demonstrated that the proposed system is capable of reaching high data rate using binary phase-shift keying (BPSK) compared to the standard OFDM system regarding the indices of active subcarriers which are used to convey supplementary data.

However, although OFDM-IM was considered as a promising perspective technique in wireless communication, system performances could be affected by real implementation circumstances such as hardware impairments [12]. In analogy with classical OFDM, the orthogonality of the subcarriers is the key element in the OFDM-IM modulation. Phase noise or frequency inconsistencies between the receiver's local oscillators include orthogonality loss between subcarriers and a significant deterioration in system performance.

In addition, OFDM-IM systems are particularly sensitive to the imbalance between in-phase (I) and quadrature (Q) of transmission and receiving paths. Radio frequency (RF) components and IQ imbalance lead to detrimental effects on the performances of the system.

Moreover, the Fourier basis has long served as the unique basis for signal representation in wireless communication applications before the introduction of wavelets as an alternative base with several advantages [13–15]. The main property of the wavelet transform lies in its flexibility in the time-frequency localization. Hence, several applications of wavelets in wireless communications have been investigated in the literature, such as modeling propagation channels [16–18], antenna design and improving the cost of implementation complexity [19], signal reconstruction from noisy data [20–22], ISI and ICI interference reduction [23] and [24] and, finally, in the context of multicarrier modulations, instead of modulating the input symbols by the Fourier exponential functions, the wavelet functions can be used for data transmission. To overcome the limitations of the classic OFDM, several alternative systems have been proposed including the Wavelet-OFDM. The wide choice of the wavelet type and wavelet transform flexibility makes it a promising technique, [25] and [26]. We are interested in the following applications of wavelets in the field of multicarrier.

In this chapter, the conventional OFDM-IM and WOFDM-IM are first presented. To further boost the performance of the standard OFDM-IM system WOFDM-IM adopting wavelet transform is announced and evaluated under hardware impairments. Finally, the proposed analysis is proven by numerical results.

12.2 SYSTEM MODEL

12.2.1 Conventional OFDM-IM

In the standard OFDM, the total number of subcarriers is activated and used to transmit the m input bits which are mapped using a constellation diagram. However, in the case of OFDM-IM system, for each block we have only k active subcarriers and the rest is set to zero. Hence, the mapped symbols are transmitted using the k subcarriers and their indices are implicitly used to convey supplementary bits as illustrated in Figure 12.1.

12 • Wavelet Transform for OFDM-IM   237

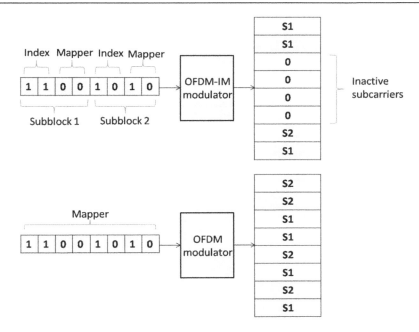

FIGURE 12.1 Example of OFDM-IM and OFDM modulators.

The activated subcarrier indices are defined based on the incoming bits and using a look up table common to the transmitter and the receiver sides. The system model of OFDM-IM is illustrated in Figure 12.2 [11].

The M input bits are first divided over G subblocks. Each subblock uses a sequence of b bits, $M = bG$. Hence, n subcarriers from a total number of N are deployed for each subblock such as $n = N/G$. From the b input bits, a sequence of b_1 bits defines the k activated subcarriers indices by referring to a look up table. On the other hand, b_2 bits are mapped using a constellation diagram as illustrated in the example of Figure 12.3. Note that $b_1 = log_2(C\,(n, k))$, $b_2 = k\,log_2(M)$, $b = b_1 + b_2$, and $C(n,k)$ is the binomial coefficient.

The activated subcarrier indices determined by b_1 bits for the jth subblock are given as:

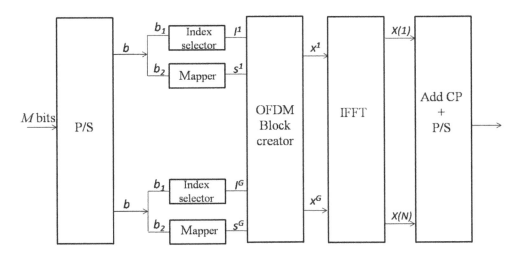

FIGURE 12.2 OFDM-IM system model.

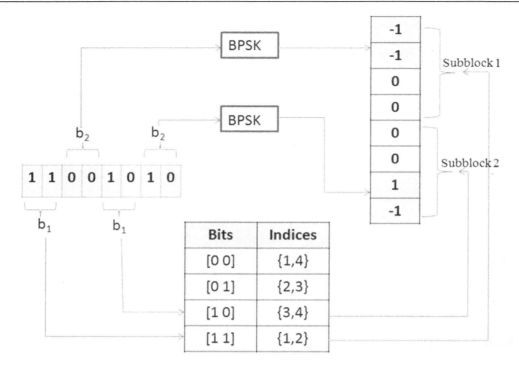

FIGURE 12.3 OFDM-IM modulator using BPSK modulation with n = 4, k = 2 and G = 2.

$$I^j = \{I^j(1), \cdots, I^j(k)\}, \tag{12.1}$$

Where $I^j(\epsilon) \in [1, \cdots, n]$ for $j = 1, \cdots, G$ and $\epsilon = 1, \cdots, k$.

Further, the complex mapped symbols from the b_2 bits sequence using a constellation diagram for the jth subblock are expressed as:

$$S^j = \{S^j(1), \cdots, S^j(k)\}, \tag{12.2}$$

with $j = 1, \cdots, G$.

Thus, the resulting signal for the jth subblock is given as

$$x^j = \begin{cases} S^j(\gamma), & \gamma \in I^j \\ 0 & \text{otherwise}. \end{cases} \tag{12.3}$$

At that stage, the inverse Fourier transform (IFFT) is applied to generate time signal from the collected symbols of the total subblocks. To protect the signal against the inter symbol interference (ISI), a CP is added at the beginning of each symbol. Then, the resulting signal is transmitted over the channel; this effect is represented using the vector h_t.

The received signal in frequency domain under ideal conditions for the jth subblock is noted as:

$$y = \sqrt{E}Xh + \eta, \tag{12.4}$$

Where $y = [y^j(1) \cdots y^j(n)]^T$, E refers to the transmitted energy, $X = \text{diag}(x^j(1) \cdots x^j(n))$, is the matrix of the conveyed signal and $h = [h^j(1) \cdots h^j(n)]^T$ represents the frequency channel response with a

Rayleigh distribution. The additive white Gaussian noise is noted as $\eta = [\eta^j(1)\cdots \eta^j(n)]^T$, which has zero mean and $N_{o,F}I_n$ variance in the frequency domain such as $N_{o,F} = (kG/N)N_{o,T}$ where $N_{o,T}$ is the noise variance in the time domain.

At the receiver, an inverse operation sequence with an equalizer is used for symbol detection.

12.2.2 WOFDM-IM

Motivated by bandwidth efficiency and performance enhancement of OFDM system, the Discrete Wavelet Transform (DWT) is applied to OFDM-IM to improve the system performance under hardware impairments.

As described in the following figure, WOFDM-IM system model is based on IDWT/DWT operations. Moreover, there is no need for addition of CP and removing blocks at the transmitter and the receiver since DWT provides signal localization in time and frequency domains. Hence, using DWT protects the transmitted signal against ISI and with a significant spectral efficiency gain (Figure 12.4).

The discrete wavelet transform decomposes a signal into detail coefficients which correspond to its projection on the wavelet functions, and approximation coefficients that correspond to its projection on the scale functions. In [27], Mallat studied the relationship between time-scale representation functions and conjugate mirror filters. In fact, wavelets and scaling functions are associated with high-pass and low-pass filters, respectively. Hence, the decomposition of the signal DWT is obtained by successively filtering the signal by a low-pass filter and a high-pass filter followed by subsampling.

Morlet proposed to construct the base of functions that satisfies

$$\Psi_{j,k}(t) = b^{-j/2}\Psi(b^{-j}t - k), \tag{12.5}$$

where b is the scale parameter and k is the translation parameter. A range of scales commonly used is the range of the dyadic scales $b = 2$, then (12.5) is expressed as

$$\phi_{j,k}(t) = 2^{j/2}\phi(2^j t - k). \tag{12.6}$$

By setting $j \to -j$ we obtain

$$\Psi_{j,k}(t) = 2^{j/2}\Psi(2^j t - k). \tag{12.7}$$

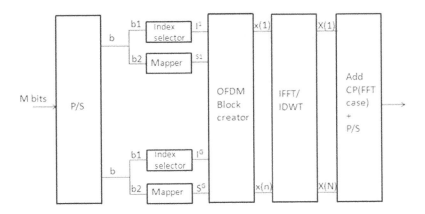

FIGURE 12.4 OFDM-IM and WOFDM-IM transmitter model.

In the same way, we define the scale functions

$$\phi_{j,k}(t) = 2^{j/2}\phi(2^j t - k). \tag{12.8}$$

The discrete wavelet transform decomposes a signal into detail coefficients which correspond to its projection on the wavelet functions, and approximation coefficients that correspond to its projection on the scale functions.

12.2.3 Hardware Impairments

The hardware effects are usually modeled as ideal. Thereby, for a wireless transmission over high frequencies and with dynamic ranges, hardware imperfections extremely affect the system performance. Accordingly, it is imperative to take into account the influence of these impairments in performance analysis. However, hardware impairments are a result of many RF components imperfections such as:

- Phase noise: In communication systems, the phase noise is a result of local oscillator's imperfections. Theoretically, the local oscillator frequency is f_{os}. But in practice, this frequency is perturbed laterally by fluctuations, thus forming the phase noise. The phase noise is generally expressed in dB/Hz at the frequency f_{el}. This corresponds to the attenuation in decibels per Hz of the oscillator spectrum, distant from the carrier frequency f_{os} with f_{el} [28]. The phase noise negatively affects OFDM subcarriers orthogonality and leads to system performance degradation, [29] and [30].
- Nonlinearities: The evaluation of the peak factor is important in the dimensioning of nonlinear components in a communication system. Indeed, before conveying the signal over the channel, a power amplifier is necessary. However, in practice, radio amplifiers have a nonlinear characteristic. To eliminate signal distortion in the amplifier, it is necessary for the signal to remain in the linear operating range and hence its maximum power is less than that corresponding to the compression point. But since OFDM modulation becomes advantageous with high number of subcarriers, a significant level of peak average power ratio is obtained and leads to OFDM performance deterioration, [31] and [32].
- IQI: The IQ imbalance is defined as the quadrature loss between I and Q channels of a transmitter or receiver. This loss due to the inevitable RF components imperfections is the principal cause of performance degradation [33,34].

12.3 PERFORMANCE ANALYSIS OF WOFDM-IM\OFDM-IM UNDER HARDWARE IMPAIRMENTS

A real implementation condition which takes into account the presence of hardware impairments is presented in this section and evaluated for both OFDM-IM and WOFDM-IM systems. Thus, a new expression of jth subblock-received signal is proposed as

$$y = [\sqrt{E}X + \omega_t]h + \omega_r + \eta, \tag{12.9}$$

Where the independent distortion noise $\omega_t = \text{diag}([\omega_t^j(1), \cdots, \omega_t^j(n)])$ and $\omega_r = \text{diag}([\omega_r^j(1), \cdots, \omega_r^j(n)])$ present respectively the transmitter and the receiver hardware impairments as depicted in Figure 12.5.

It was demonstrated that $\omega_t^j(\gamma) \sim CN(0, \kappa_t^2 E)$ and $\omega_r^j(\gamma) \sim CN(0, \kappa_r^2 E|h^j(\gamma)|^2)$ [34,35].

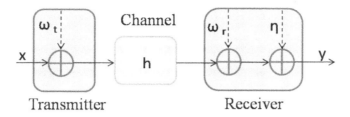

FIGURE 12.5 Transmitter and receiver hardware impairments.

The hardware impairments are defined as the residual impact of impairments after the use of mitigation algorithms for all different distortions types [36,37].

In fact, κ_t^2 and κ_r^2 present transmitter and receiver hardware impairments level in the range of [0, 0.03] [34,38]. The evaluation of the ratio of the average distortion magnitude to the average signal magnitude (EVM) could provide an experimental measurement of system hardware impairment level [39].

Without loss of generality, an aggregate distortion level κ is introduced, such as $\kappa = \sqrt{\kappa_t^2 + \kappa_r^2}$ to simplify the performances analysis.

Hence, Figure 12.5 can be presented using the aggregate model of hardware impairments as a general model in Figure 12.6.

Therefore, (12.9) becomes

$$y = (\sqrt{E}X + \omega)h + \eta, \tag{12.10}$$

where $\omega = \mathrm{diag}([\omega^j(1), \cdots, \omega^j(n)])$ models the hardware impairments which diagonal elements follow the $CN(0, \kappa^2 E)$ distribution. For ideal hardware case $\kappa^2 = 0$.

An optimal maximum likelihood detector (ML) is used at the receiver not only for symbol detection but also to determine active subcarriers indices using the above equation:

$$[\hat{X}]_{ML} = \arg\min_X \|y - \sqrt{E}X\ h\|^2. \tag{12.11}$$

12.4 NUMERICAL RESULTS

First, the WOFDM-IM system is simulated under ideal conditions, $\kappa = 0$, for various mother wavelets over Rayleigh fading channel and compared to OFDM-IM performance according to the simulation parameter shown in Tables 12.1.

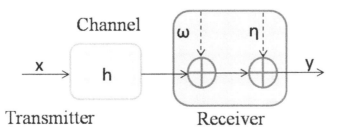

FIGURE 12.6 General model of hardware impairments.

TABLE 12.1 Simulation parameters of WOFDM and OFDM systems comparison over different wavelet family and modulation orders.

SIMULATION PARAMETERS	OFDM-IM	WOFDM-IM
Modulation schema	BPSK\4-QAM\16-QAM\64-QAM	BPSK\4-QAM\16-QAM\64-QAM
Symbol length	10^4	10^4
Number of subcarriers	64	64
Number of active subcarriers	32	32
CP length	16	Nil
Wavelet family	Nil	Symlet, Daubechies, Haar

Figure 12.7 shows that all the wavelets perform better compared to OFDM-IM-based FFT implementation. Moreover, mother wavelet Haar performs better than other wavelets.

As for a standard OFDM system, increasing modulation order from BPSK to 16-QAM for WOFDM-IM as presented in Figure 12.8 leads to an increased error detection probability; however, WOFDM-IM outperform OFDM-IM system for all modulation order.

In addition, the impact of hardware impairments of OFDM-IM and WOFDM-IM is evaluated in Figures 12.9 and 12.10 using Rayleigh channel and with different values of κ and modulation order.

It can be concluded that increasing the hardware impairments value from 0.05 to 0.1 have negative effects on performance degradation of standard OFDM-IM as well as WOFDM-IM systems. Although WOFDM-IM presents significant improvement compared to that with OFDM-IM for all hardware impairments levels using various modulation order, which is shown in Figures 12.9 and 12.10. Simulation

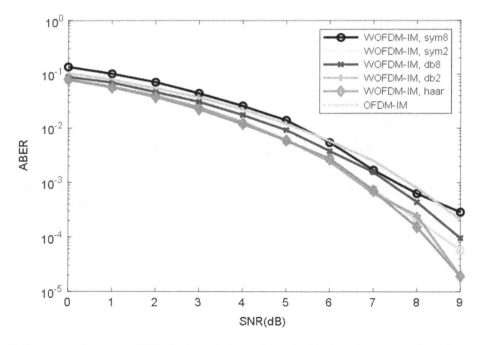

FIGURE 12.7 BER performance of WOFM-IM and OFDM-IM under ideal hardware and with different wavelet families.

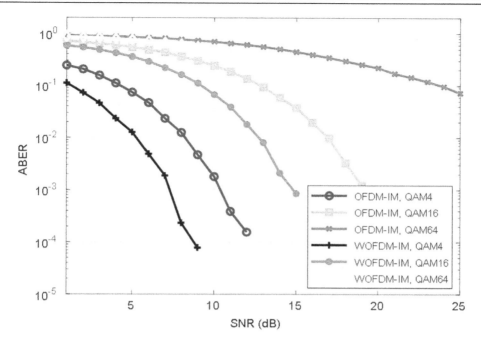

FIGURE 12.8 BER performance of WOFM-IM and OFDM-IM under ideal hardware and with different modulation orders.

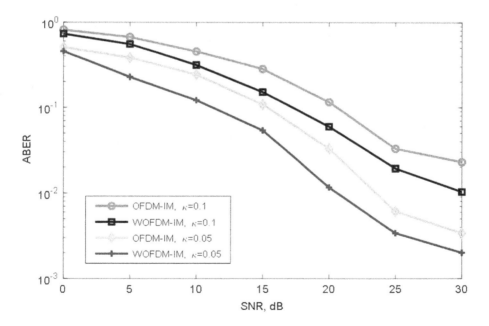

FIGURE 12.9 BER performance of WOFM-IM and OFDM-IM under hardware impairments using BPSK.

results prove that using wavelet transform as an alternative transform for OFDM-IM system could improve the system performance even in the presence of the harmful effects of hardware impairments. Thereby, hardware impairments mitigation algorithms should be applied.

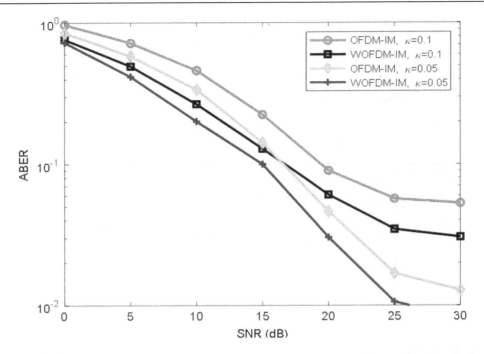

FIGURE 12.10 BER performance of WOFDM-IM and OFDM-IM under hardware impairments using 4-QAM.

12.5 CONCLUSION

The WOFDM-IM performance under hardware impairments is analysed in this chapter and compared with the classical OFDM-IM system. Indeed, inspired by the merits of introducing wavelet transform to overcome Fourier transform drawbacks, this chapter extended the transforms comparison to include the effects of hardware impairments. Simulation results are provided to prove that the proposed WOFDM-IM under hardware impairments improves standard OFDM-IM performance.

REFERENCES

[1]. J. A. Bingham, "Multicarrier modulation for data transmission: An idea whose time has come", *IEEE Commun. Mag.*, vol. 28, no. 5, 5–14, 1990.

[2]. E. G. Larsson, O. Edfors, F. Tufvesson, and T. L. Marzetta, "Massive MIMO for next generation wireless systems", *IEEE Commun. Mag.*, vol. 52, no. 2, 186–195, 2014.

[3]. L. Dai, B. Wang, Y. Yuan, S. Han, I. Chih-Lin, and Z. Wang, "Non-orthogonal multiple access for 5G: Solutions, challenges, opportunities, and future research trends", *IEEE Commun. Mag.*, vol. 53, no. 9, 74–81, 2015.

[4]. R. Y. Mesleh, Spatial modulation: A spatial multiplexing technique for efficient wireless data transmission (Doctoral dissertation, Jacobs University Bremen), 2007.

[5]. R. Mesleh, H. Haas, C. W. Ahn, and S. Yun, "Spatial modulation-a new low complexity spectral efficiency enhancing technique", in *2006 First Int. Conf. Commun. Netw. China*, 2006, (pp. 1–5). IEEE.

[6]. A. Younis, N. Serafimovski, R. Mesleh, and H. Haas, "Generalised spatial modulation", in *2010 Conf. Rec. Forty Fourth Asilomar Conf. Signals systems and Comput.*, 2010, (pp. 1498–1502). IEEE.

[7]. C. C. Cheng, H. Sari, S. Sezginer, and Y. T. Su, "Enhanced spatial modulation with multiple signal constellations", *IEEE Trans. Commun.*, vol. 63, no. 6, 2237–2248, 2015.

[8]. R. Mesleh, S. S. Ikki, and H. M. Aggoune, "Quadrature spatial modulation", *IEEE Trans. Veh. Technol.*, vol. 64, no. 6, 2738–2742, 2014.

[9]. R. Abu-Alhiga and H. Haas, "Subcarrier-index modulation OFDM", in *2009 IEEE 20th IEEE Int. Symp. Pers. Indoor Mob. Radio Commun.*, 2009, (pp. 177–181). IEEE.

[10]. D. Tsonev, S. Sinanovic, and H. Haas, "Enhanced subcarrier index modulation (SIM) OFDM", in *2011 IEEE GLOBECOM Workshops (GC Wkshps)*, 2011, (pp. 728–732). IEEE.

[11]. E. Başar, Ü. Aygölü, E. Panayırcı, and H. V. Poor, "Orthogonal frequency division multiplexing with index modulation", *IEEE Trans. Signal Process.*, vol. 61, no. 22, 5536–5549, 2013.

[12]. A. Bouhlel, S. G. Domouchtsidis, S. S. Ikki, and A. Sakly, "Performance of OFDM-IM under joint hardware impairments and channel estimation errors over correlated fading channels", *IEEE Access*, vol. 5, 25342–25352, 2017.

[13]. M. K. Lakshmanan and H. Nikookar, "A review of wavelets for digital wireless communication", *Wirel. Pers. Commun.*, vol. 37, no. 3, 387–420, 2006.

[14]. M. N. Suma, S. V. Narasimhan, and B. Kanmani, "Analytic discrete cosine harmonic wavelet transform based OFDM system", *Sadhana*, vol. 40, no. 1, 173–181, 2015.

[15]. V. Kumbasar and O. Kucur, "Performance comparison of wavelet based and conventional OFDM systems in multipath Rayleigh fading channels", *Digit. Signal Process.*, vol. 22, no. 5, 841–846, 2012.

[16]. M. I. Doroslovacki and H. Fan, "Wavelet-based linear system modeling and adaptive filtering", *IEEE Trans. Signal Process.*, vol. 44, no. 5, 1156–1167, 1996.

[17]. H. Zhang, H.H. Fan, and A. Lindsey, "A wavelet packet based model for time-varying wireless communication channels", in *2001 IEEE Third Workshop Signal Process. Advances Wirel. Commun. (SPAWC'01). Workshop Proc. (Cat. No. 01EX471)*, 2001, (pp. 50–53). IEEE.

[18]. E. Jaffrot, "Wavelet based channel model for OFDM systems", in *2004 IEEE 15th Int Symp. Pers., Indoor and Mobile Radio Commun. (IEEE Cat. No. 04TH8754)*, 2004, (Vol. 1, pp. 704–708). IEEE.

[19]. T. K. Sarkar, M. Salazar-Palma, and M.C. Wicks, *Wavelet Applications in Engineering Electromagnetics.* Artech House, 2002.

[20]. M. H. C. Dias and G. L. Siqueira, "On the use of wavelet-based denoising to improve power delay profile estimates from 1.8 GHz indoor wideband measurements", *Wirel. Pers. Commun.*, vol. 32, no. 2, 153–175, 2005.

[21]. S.D. Mantis, Localization of wireless communication emitters using Time Difference of Arrival (TDOA) methods in noisy channels. Naval Postgraduate School, Monterey, CA, 2001.

[22]. X. N. Fernando, S. Krishnan, H. Sun, and K. Kazemi-Moud, "Adaptive denoising at infrared wireless receivers", *Infrared Technol. Appl. XXIX*, vol. 5074, pp. 199–207. International Society for Optics and Photonics, 2003.

[23]. B. G. Negash and H. Nikookar, "Wavelet-based multicarrier transmission over multipath wireless channels", *Elect. Lett.*, vol. 36, no. 21, 1787–1788, 2000.

[24]. G. W. Wornell, "Emerging applications of multirate signal processing and wavelets in digital communications", *Proc. IEEE*, vol. 84, no. 4, 586–603, 1996.

[25]. H. Nikookar *Wavelet Radio: Adaptive and Reconfigurable Wireless Systems Based on Wavelets.* Cambridge University Press, 2013.

[26]. M. Gautier, M. Arndt, and J. Lienard, "Efficient wavelet packet modulation for wireless communication", in *Third Adv. Int. Conf. Telecommun. (AICT'07)*, 2007 (pp. 19-19). IEEE.

[27]. B. L. Sturm, Stéphane Mallat: A Wavelet Tour of Signal Processing, 2007.

[28]. S. Bougeard, Modélisation du bruit de phase des oscillateurs hyperfréquences et optimisation des systèmes de communications numériques (Doctoral dissertation, Rennes, INSA), 2001.

[29]. C. Muschallik, "Influence of RF oscillators on an OFDM signal", *IEEE Trans. Cons. Elect.*, vol. 41, no. 3, 592–603, 1995.

[30]. L. Tomba, "Analysis of phase noise effects in OFDM modems", IEEE Trans. Commun, vol. 46, no. 5, 580–583, 1998.

[31]. E. Costa, M. Midrio, and S. Pupolin, "Impact of amplifier nonlinearities on OFDM transmission system performance", *IEEE Commun. Lett.*, vol. 3, no. 2, 37–39, 1999.

[32]. D. Dardari, V. Tralli, and A. Vaccari, "A theoretical characterization of nonlinear distortion effects in OFDM systems", *IEEE Trans. Commun.*, vol. 48, no. 10, 1755–1764, 2000.

[33]. H. Liu, and U. Tureli, "A high-efficiency carrier estimator for OFDM communications", *IEEE Commun. Lett.*, vol. 2, no. 4, 104–106, 1998.

[34]. T. Schenk, RF imperfections in high-rate wireless systems: Impact and digital compensation. Springer Science & Business Media, 2008.

[35]. C. Studer, M. Wenk, and A. Burg, "MIMO transmission with residual transmit-RF impairments", in *2010 Int. ITG workshop Smart Antennas (WSA)* (pp. 189–196). IEEE, 2010.

[36]. E. Björnson, J. Hoydis, M. Kountouris, and M., Debbah, "Hardware impairments in large-scale MISO systems: Energy efficiency, estimation, and capacity limits", in *2013 18th Int. Conf. Dig. Signal Process. (DSP)*, 2013, (pp. 1–6). IEEE.

[37]. M. Matthaiou, A. Papadogiannis, E. Bjornson, and M. Debbah, "Two-way relaying under the presence of relay transceiver hardware impairments", *IEEE Commun. Lett.*, vol. 17, no. 6, 1136–1139, 2013.

[38]. E. Björnson, P. Zetterberg, and M. Bengtsson, "Optimal coordinated beamforming in the multicell downlink with transceiver impairments", in *2012 IEEE Global Commun. Conf. (GLOBECOM)*, 2012 (pp. 4775–4780). IEEE.

[39]. M. D. McKinley, K. A. Remley, M. Myslinski, J. S. Kenney, D. Schreurs, and B. Nauwelaers, "EVM calculation for broadband modulated signals", in *64th ARFTG Conf. Dig* (pp. 45–52), 2004.

A Systematic Review of 5G Opportunities, Architecture and Challenges

Rishiraj Sengupta[1], Dhritiraj Sengupta[2], Digvijay Pandey[3], Binay Kumar Pandey[4], Vinay Kumar Nassa[5], and Pankaj Dadeech[6]

[1]Network Engineer, Facebook connectivity Deployment, Facebook, London, United Kingdom
[2]Post-Doctoral Fellow, State Key Laboratory of Coastal and Estuarine ResearchEast China Normal University. 500, DongChuan road, Shanghai, China
[3]Ph.D. Candidate, IET Lucknow, Lecturer, Department of Technical Education Kanpur, India
[4]Assitant Professor, Dept of IT, College of TechnologyGovind Ballabh Pant University of Agriculture and Technology, U.K, India
[5]Principal/Professor, Department of Computer ScienceScience, EnggSouth Point Group of Institutions, Sonepat-131001
[6]Computer Science & EngineeringEngineering, Swami Keshvanand Institute of Technology, Management & Gramothan (SKIT), Jaipur, India

13.1 INTRODUCTION

5G or 5th generation mobile communication was presented in the release 15 of the 3GPP specifications. 5G is advancing to provide faster throughput and minimal latency. As per **3GPP 21.915 and 3GPP 21.916**, 5G is mainly intended to support numerous facilities with diverse traffic flow outlines such as very high throughput, low latency, and massive connections. The 3GPP specifies the introduction of a new- radio interface known as new radio, which further offers a comprehensive system

DOI: 10.1201/9781003175155-13

to develop a plethora of facilities [1,2]. The key advantage of 5G is that the technology can grow side by side to the existing technology and work side by side with technologies like 4G. A 5G access network integrated to a 5G main backbone network is known as just like a no standalone based architectural design, whereas a 5G access network linked to the a 5G network infrastructure is known as a standalone architecture. We will discuss this in detail in later sections. The new radio type base stations are designed to work together with the LTE base stations to provide combined service to the user equipment (UE) in case of NSA. With the growing demand in cellular network usage, there is congestion in the lower bands due to wireless technology like 4G 3G and 2G. Therefore, 3GPP spaced a variety of frequency bands to deploy 5G and to offer all those advantages that come from 5G. As per **3GPP 38.104,** there are two groups of frequency FR1 (410 MHz to 7125 megahertz) and FR2 (24250–52600) Megahertz. Table 13.1 and Table 13.2 given below illustrated the supported frequencies in both ranges [3].

In contrast with LTE, in which the sub-carrier range (SCS) is fixed at 15 kHz, the new radio has an adaptable sub-carrier spacing, ranging (15 to 120) Kilohertz, with 15 Kilohertz, 30 Kilohertz, and 60 Kilohertz fitting to F1 and 60 Kilohertz and 120 Kilohertz to FR2. The bandwidth ranges from 5 to 100 MHz (FR1) and 50 to 400 MHz (FR2).

As we go higher into the frequency range coverage will be a challenge due to basic wireless fundamentals which state that frequency is inversely proportional to wavelength. Therefore beam forming and MIMO are two important features in both uplink and downlink of NR. Beam-forming is indeed a major characteristic that aims to enhance the networks as a whole efficiency by managing the SINR (signal to noise ratio), especially at higher frequencies like FR2, and multi-user MIMO aims to enhance the network's total spectrum utilization. To accomplish all these, large-scale antenna components would be required in the 5G wireless communication antenna structure.

Regardless of the fact that now the 5G new radio air interface is based on LTE, there are some differences. In new radio, cyclic prefix OFDM is used in the down-link and up-link directions, while discrete Fourier transform, which is analogous to LTE's SC-FDMA, is used to improve uplink coverage at the cell boundary with both the addition of/2 BPSK in areas of weak coverage. [4]

Now, few important differences that exists among LTE and 5G are given below in Table 13.3.

13.2 5G OPPORTUNITIES

The use cases were described by 3GPP through accordance with international telecommunications union's regulations for international mobile telecommunications in 2020.

As per ITU, the following are the main use cases for international mobile telecommunications in 2020[5]. (Figure 13.1)

1. Enhanced mobile broadband
2. Massive machine-type communications
3. Ultra-reliable and low latency communications

Again, all these categories have different types of requirements, like eMBB focuses on high data rates URLLC focuses on latency with low data rates. There are applications that demand high security along with low latency. (Figure 13.2)

3GPP 22.891 further classifieds this uses cases into the following [6]

1. Improved broad-band mobile
    - e.g., mobile broad-band, holo-gram, high degree of mobility, presence virtually

TABLE 13.1 New radio operational frequency bands in FR1

NEW RADIO OPERATIONAL BAND	UP-LINK OPERATIONAL FREQUENCY BAND B-S RECEIVE / U-E TRANSMIT ($F_{UL,LOW} - F_{UL,HIGH}$) IN MEGAHERTZ	DOWN-LINK OPERATIONAL FREQUENCY BAND B-S TRANSMIT / U-E RECEIVE ($F_{DL,LOW} - F_{DL,HIGH}$) IN MEGAHERTZ	DUPLEX MODE
n1	1920–1980	2110–2170	FDD
n2	1850–1910	1930–1990	FDD
n3	1710–1785	1805–1880	FDD
n5	824–849	869–894	FDD
n7	2500–2570	2620–2690	FDD
n8	880–915	925–960	FDD
n12	699–716	729–746	FDD
n13	777–787	746–756	FDD
n14	788–798	758–768	FDD
n18	815–830	860–875	FDD
n20	832–862	791–821	FDD
n25	1850–1915	1930–1995	FDD
n26	814–849	859–894	FDD
n28	703–748	758–803	FDD
n29	N/A	717–728	SDL
n30	2305–2315	2350–2360	FDD
n34	2010–2025	2010–2025	TDD
n38	2570–2620	2570–2620	TDD
n39	1880–1920	1880–1920	TDD
n40	2300–2400	2300–2400	TDD
n41	2496–2690	2496–2690	TDD
n46	5150–5925	5150–5925	TDD[3]
n48	3550–3700	3550–3700	TDD
n50	1432–1517	1432–1517	TDD
n51	1427–1432	1427–1432	TDD
n53	2483.5–2495	2483.5–2495	TDD
n65	1920–2010	2110–2200	FDD
n66	1710–1780	2110–2200	FDD
n70	1695–1710	1995–2020	FDD
n71	663–698	617–652	FDD
n74	1427–1470	1475–1518	FDD
n75	N/A	1432–1517	SDL
n76	N/A	1427–1432	SDL
n77	3300–4200	3300–4200	TDD
n78	3300–3800	3300–3800	TDD
n79	4400–5000	4400–5000	TDD
n80	1710–1785	N/A	SUL
n81	880–915	N/A	SUL
n82	832–862	N/A	SUL
n83	703–748	N/A	SUL
n84	1920–1980	N/A	SUL

(Continued)

TABLE 13.1 (Continued) New radio operational frequency bands in FR1

NEW RADIO OPERATIONAL BAND	UP-LINK OPERATIONAL FREQUENCY BAND B-S RECEIVE / U-E TRANSMIT ($F_{UL,LOW} - F_{UL,HIGH}$) IN MEGAHERTZ	DOWN-LINK OPERATIONAL FREQUENCY BAND B-S TRANSMIT / U-E RECEIVE ($F_{DL,LOW} - F_{DL,HIGH}$) IN MEGAHERTZ	DUPLEX MODE
n86	1710–1780	N/A	SUL
n89	824–849	N/A	SUL
n90	2496–2690	2496–2690	TDD
n91	832–862	1427–1432	FDD[2]
n92	832–862	1432–1517	FDD[2]
n93	880–915	1427–1432	FDD[2]
n94	880–915	1432–1517	FDD[2]
n95[1]	2010–2025	N/A	SUL
n96[4]	5925–7125	5925–7125	TDD[3]
n97[5]	2300–2400	N/A	SUL
n98[5]	1880–1920	N/A	SUL

NOTE 1: One such frequency band is really only acceptable in China.
NOTE 2: Changeable duplex procedure doesn't really allow the network to configure diverse variable duplex, and is used to assist on down link and up link frequency ranges independently in any valid frequency range for the band.
NOTE 3: One such frequency band, as described in, can only be used with capable of sharing spectrum communication links [20].
NOTE 4: This band is applicable in the USA only subject to FCC Report and Order [FCC 20–51].
NOTE 5: The requirements for this band are applicable only where no other NR or E-UTRA TDD operating band(s) are used within the frequency range of this band in the identical terrestrial area. Particular co-occurrence specifications could spread over through circumstances in which other NR or E-UTRA TDD operational frequency bands are used within the frequency range of this band in the same geographical area, which would not be addressed by 3GPP specs.

TABLE 13.2 New radio operational frequency band in the FR2

NEW RADIO OPERATIONAL FREQUENCY BAND	UP-LINK AND DOWN-LINK OPERATIONAL FREQUENCY BAND B-S TRANSMIT/RECEIVE U-E TRANSMIT/RECEIVE ($F_{UL,LOW} - F_{UL,HIGH}$) IN MEGAHERTZ ($F_{DL,LOW} - F_{DL,HIGH}$) MEGAHERTZ	DUPLEX MODE
n257	26500–29500	TDD
n258	24250–27500	TDD
n259	39500–43500	TDD
n260	37000–40000	TDD
n261	27500–28350	TDD

2. Serious infrastructure of communications
   - e.g., collaborative game / sports, Industrial monitor, drone /robot / vehicle, emergency

3. Enormous machine type communications
   - e.g., Sub-way / Stadium based service, e-Health, Wearable, Inventory Control

4. Communication network's operation
   - e.g., slicing of communication network, routing, migration and inter-working, saving of energy

TABLE 13.3 Difference between LTE and 5G bands

Parameters	LTE	5G
Frame	15 KHz	Flexible(15 kHz, 30 kHz, 60 kHz, 120 kHz)
Channel Coding	Turbo Coding (data), Convolution Coding (Control)	Polar Codes(control); LDPC(DATA)
Refference Signals	Cell Specific Refference Signal	PBCH DMRS and PDSCH DMRS. No CRS. Introduction of PTRS
Initial Access	No Beamfoarming	Beamfoarming
Latency(Air interface)	10 ms	1 ms
Waveform	CP-OFDM for DL, SC-FDMA for UL	CP-OFDM and DFT-s-OFDM both for UL
Subframe Length	1 ms	1 ms
Slot Length	14 Symbols in 1 ms	14 symbols and Duration depends on SCS. 2,4 and 7 symbols for mini-slots
Maximum Bandwidth	20 Mhz	50 MHz(@ 15khz), 100 MHz (@ 30 Khz), 200 MHz (@60kHz), 400 MHz
MIMO	8x8	8x8
Duplexing	FDD, Static TDD	Flexible TDD
Beam Management	Wide Beam	Concept of SSB and Beamsweeping
User Plane Protocol Stack	No SDAP Layer	Addition of SDAP Layer
Types of SRB	SRB0, SRB1, SRB2	Addition of SRB3 (EN-DC)
RACH	LTE is only available on FR1	Additional Preamble Formats specific to FR2
Transport Block size	TBS is given by a table as per 3GPP	TBS is calculated
Slot Length indicator	Time domain is fixed	Time domain is determined by SLIV parameter
Modulation scheme	pi/2 BPSK on uplink not supported	pi/2 BPSK on uplink is supported

5. Improvement of vehicle to everything
   - Such as autonomous driving, safety and non-safety-based aspects related to the vehicle

NGNM 5G white paper 1 defines the use cases along with their families and categories [7] (Figure 13.3)

13.3 ENHANCED MOBILE BROADBAND

Enhanced mobile broadband is the next generation of mobile broadband service which involves users accessing internet services including downloads and streaming. LTE offered similar services but with 5G, new radio it will improve the data throughput further resulting enhanced experience for existing applications and will make room for new applications.

13.4 ULTRA-RELIABLE COMMUNICATIONS

Ultra-reliable communications mainly focus on reliable and low latency requirements to achieve the services. This is crucial for services in the health care sector like remote surgery or connected ambulance. We will explain this use case further in this chapter.

All such critical communication services require priority dealing in comparison to eMBBs which can be achieved via a network slice as well as given priority radio programming.

Following are some examples in reference to 3GPP 22.891

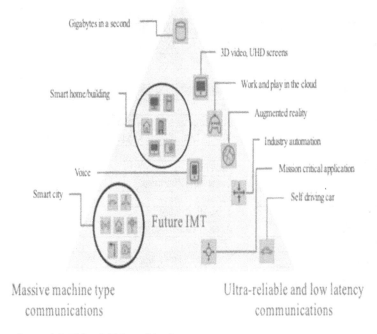

FIGURE 13.1 Procedure of IMT for 2020 and in future.

As given in figure 1, the three use case groups of 5G new radio use cases are as follows:

Enhanced Mobile Broadband (eMBB): data-driven use cases that necessitate high throughput over even a broad service area.

Ultra Reliable Low Latency Communications (URLLC): ultra-low transmission delay as well as dependability prerequisites for mission-critical communication systems like remote surgery, autonomous vehicles, or the Tactile Internet.

Massive Machine Type Communications (mMTC): must endorse a large number of devices in a small area, some of which could just send data infrequently, as in Internet of Things use cases.

1. Industrial control systems
2. Remote monitoring, diagnosis and treatment, linked emergency services and managed rural medi-care.
3. Vehicle-to-vehicle interaction, vehicle-to-everything connectivity, traffic flow, and accident avoidance.
4. Monitoring of smart power grids
5. For public safety situations, crucial information should be communicated.

13.5 SLICING OF NETWORK

Slicing of network is another important feature with 5G. As 5G NR provides flexible network configuration therefore there is a need to build different slices for different services such as some service might need low

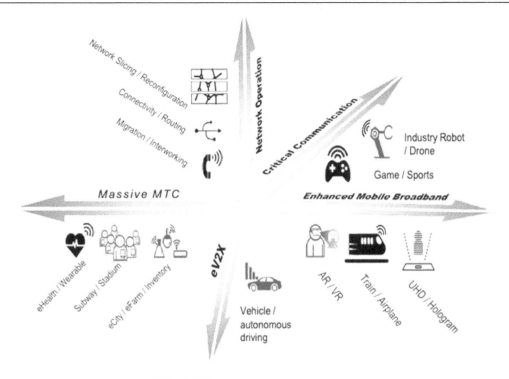

FIGURE 13.2 Use cases as per 3GPP 22.891.

Figure 2 depicts the use cases as described by 3GPP 22.891. The primary dimensions of the this use case are massive MTC, EV2X, Enhanced Mobile Broadband, critical connectivity, and network operation.

latency and low throughput so they can be assigned with a different subcarrier spacing and lower bandwidth so that the TTI reduces and also the time to synchronize to a larger bandwidth. (Figure 13.4)

Some of the other use cases as described in NGMN white paper 1 and white paper 2. [7,8]

13.6 NATURAL DISASTER

5G must've been able to give reliable radio communications in the event of natural hazards such as floods, tidal waves, heavy rains, storms, and some other ecological degradation. Some who have attempted to flee the major disaster will need a variety of basic communication tools based on voice and text-messages. The survivors must also be able to give indication for their existence because then that, they are being apprehended as soon as possible. During an urgent situation, efficient network and user terminal energy consumption is crucial. Some few days of operation should be possible.

13.7 PRODUCT MANUFACTURING INDUSTRY

The worldwide engineering industry needs a very high-quality, time-sensitive, automated, thoughtful, but flexible commercial supervision in order to monitor, improve, as well as govern modules, objects, but

254 Future Trends in 5G and 6G

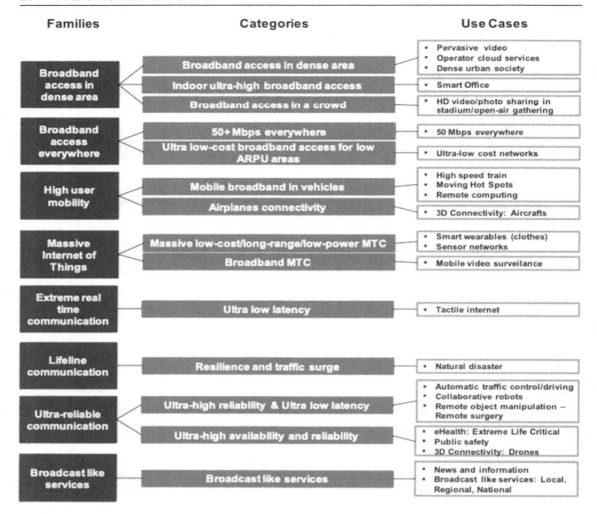

FIGURE 13.3 Use case categories as per NGNM 5G white paper v1.

Figure 3 illustrates the use case categories as described in the NGNM 5G white paper v1: This provides a simplified representation of the relationship that exists among families, categories, and NGNM 5G use cases.

also processes in near real-time. That very many manufacturing facilities presently utilize network connections that also meet scientific requirements and are complicated as well as rigid in terms of reconfiguration. Business sector requirements are defined by 5-G, that will provide more flexible and convenient communication networks. The 5G system will enable monitoring, computerization, but also optimization by including features such as:

- An inter-connection, range-of-motion, but also precise location of equipment like sensors and actuators are of less latency, highly reliable, as well as highly available for both real time watching and regulation of procedures.
- Interconnectedness but also local processing has been needed for live real-time video capture and video-based novelties.
- Massive numbers of device sensors are connected, but instead platforms for storing and analyzing large amounts of information are being developed.

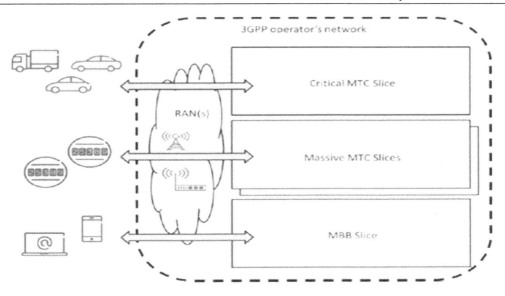

FIGURE 13.4 Network Slicing use cases as per 3GPP 22.891.

Figure 4 describes Slicing of the network is an essential aspect of 5G that is adaptive network configuration, under which there is a need to create various slices for various services, such as certain providers would need low latency and low congestion, so they may be allocated with a different subcarrier distribution and lower bandwidth such that the TTI decreases and the time to synchronize to a greater bandwidth may become possible.

- Sustainability efforts can be optimized but mostly improved using augmented reality.

13.8 CONSTRUCTION INDUSTRY

The construction sector must complete activities on time and under budget whilst also upholding the rigorous quality of workplace safety but mostly in workplace circumstances. As a result, remotely supervision of the automobiles in use, independent procedure, device safety, and automobile synchronization are becoming increasingly important.

The 5G based system will facilitate, an autonomous high end type system for construction projects by providing, among several other things.

- Reduced latency, elevated steadfastness, and enormous readiness connectivity for sensors and controllers in such a specific geographical location.;
- Elevated systems but also human locations, as well as high-quality real-time video transmission.
- a local edge-based computing used for controlling, like movement motor vehicle.

13.9 STREET TRANSPORTATION

A street carriage business is supposed to offer well-organized, harmless, ecologically sustainable, and agreeable carriage facilities, particularly by leveraging machine learning to obtain linked but also fully automated driving via perceived notion, judgment, but also regulate. It will be facilitated by a 5G system that will encourage, among several other things.

- communication of good audiovisual or pictures of street situations as well as side of the street structure to aid mapping, distant but also fully automated driving, but instead classification of blind areas and other automobile security issues.
- In live communications happens between automobiles and available infrastructures of road, collective with exact automobile location and local- edge computation competences, permits the discovery of likely risks that may support in making of any important decision. Processes such as traffic updates, emergency stop, and smart vehicle collision avoidance.
- Sensor interconnection as well as information gathering from vehicles and roadside infrastructure enable the development of external situational practices and business applications like sophisticated pavement chargers.
- Remote monitoring is accomplished via the installation of software upgrades on automobiles and also the distribution of actual vehicle data to automobile producers.

13.10 HEALTH CARE INDUSTRIES

The health-care business necessitates a well-adjusted distribution of resources related to medical aid, and also portable and smart health care related equipment, enhanced handling competences for public health care automobiles, and also the expansion of medical procedures beyond the operating room. The number of healthcare facilities presently utilizes wired based bio-medical instrument, that is considerable trustworthy but stiff. Additionally, a growing demand for remote medical aid, particularly to facilitate the frail and elderly or to manage emergencies. 5G could even enable extra adaptable and convenient wireless connections, which can accommodate a majority of health industry application areas. Now a day's artificial intelligence [9,14] and 5G are highly used in health care industries.

- Ambulance services have broad health coverage, such as the ability to send live video and patient vital signs to a hospital's control room in near real-time.
- Sensors are used to collect vital signs from patients or even older people by using wearable technologies to locate, wherever they are, to enable remote medical staff to decide quickly and to administer medicine remotely.
- Sensors monitor retrieve important data from hospitalized patients in a homogeneous manner, and data organization is performed on a common system.
- The designated, dependable direct communication among, say, health facilities, allowing for remote surgery and guidance.
- maintaining but also supervising hospital instruments, assets of hospital, and employees.

13.11 SMART CITIES AND COMMUNITIES

A smart city supposes to expand the inter-networking space for intelligent urban infrastructures, enhance knowledge compilation but also processing power, enhance economic security management capacities, combine new smart community application areas, as well as improve public quality and efficiency of living. 5G would then make it possible for smart cities and communities to high quality video transmission, positioning and tracking to improve security monitoring efficiency.

- interconnected smart sensor systems, such as video, to support enhanced integrated city management and various citizen services such as traffic and transportation planning, and also resource planning and management.

13.12 EDUCATION

For various reasons, such as inequity in funding, technological progress as well as the increasing complexity of instruction, new educational models and educational materials are being developed together. Approaches which tailor learning to concrete objectives that use a blend of learning techniques (e.g., online and offline). That 5G method will allow the learning industry to achieve its greatest potential of innovation.

- The use of virtual- reality and augmented-reality amenities to enhancement and animates conventional education.
- Virtualized experimentations offer students with virtualization tools and systems so that they can communicate, particularly for experiments involving expensive products or severe perils.
- Remote interactive learning, which provides a simultaneous experience by allowing students and teachers from various schools to interact remotely.
- Supervision of institute personnel, resources, and environmental management, also aids with the protection of the people and assets.

13.13 TOURISM

A tourism sector could even take advantage of new 5G network capabilities to provide further engrossing, informative, as well as exciting experiences, and also a wealth of information about frequented region's sites, places of interest, but rather infrastructure. Touristic attractions are typically happens in much more packed than other locations: thanks to 5G networks' increased information utilization and ability, huge numbers of tourists could even gain from, among many other things, higher bandwidth throughput but rather ability.

- Augmented reality can provide 3-D virtual reconstructions of archaeological sites, thematic based tours in galleries, and glossed city tours that enrich its knowledge of a city with architectural design but also historical documents, among other things.
- Autonomous cloud storage with elevated quality of video and photos.
- Smart crowd flow management.

13.14 AGRICULTURE

Intelligent agriculture is indeed a revolution in agricultural production which has greatly improved productive capacity. A improved 5G system capabilities encourage so many intelligent farmers, such as:

- Agro robotic systems like self-driving tractors, precise seeders, computerized weed controls as well as harvesters allow more and more quality crops and fewer jobs to also be produced.

- Drones or unmanned aerial vehicles include a panoramic view for imaging, sowing, but rather plant spraying to aid throughout land and crop management.;
- Sensor systems for observing agricultural productivity, livestock wellness and site, and environmental and climate conditions in near real-time;
- In the real-time processing, strategic thinking, and stream-lining operations related to land, crop, livestock, logistic support, and equipment, information sharing among all devices and sensor systems, and also with systems, seems to be essential.

13.15 FINANCE

To enable advanced financial services, the finance industry requires a secure, efficient, and widely embraced framework. 5G will provide such a platform and hasten banks' digitalization, allowing for more widespread financial institutions but better customer service.5G does have a range of abilities that can be matched to the requirements of the financial sector, which include:

- Gigabit connection speeds allow for the access to huge amounts of high-frequency stock market data for quick review.
- Utilizing multi-access edge computing technology, ultra-reliable but also Reduced delay in information sharing allows for quick reply but also regulate in mobile internet trading, and makes sure that time-critical buying and selling transactions occur only at cellular network's side.
- Reliable interconnection supported by a large variety of sensors to speed up the ability of devices to share data and provide personalized payments and micropayments over even a range of interconnected devices. These numbers of devices and their interconnection may create a new generation with financial products.;
- Security requirements, that will improve the reliability and trustworthy process in banking operations for both banks and customers, ensuring payment validity as well as detecting abuse.

13.16 VIRTUAL OFFICE

After the global pandemic of COVID-19, virtual offices might be one of the new normal. Employees can be at the office from anywhere in the world with help of virtual reality and augmented reality. It can provide a 360° viewing experience with a glass with an exact office environment as they are in the office. The offices will be equipped with cameras to transmit video feed on 5G network.

13.17 HOLOGRAPHIC COMMUNICATION AND HAPTIC FEEDBACK [10]

Holographic communication and haptic feedback can open up whole new opportunities with 5G. It will provide a virtual touch and feel of an object in front of you the picture below is of a scientific report (mentioned) and portrays first-person interaction with the haptic bio-holographic heart. On the user's hands, tracking indicators were also overlaid. Haptic feedback is activated when the markers intersect

FIGURE 13.5 First person view of the MR headset during interaction with the haptic bio-holographic heart.

Figure 5 depicts a research report and first-person contact with the haptic bio-holographic heart. Tracking markers were often overlaid on the user's arms. Whenever the marker collide with holographic heart, haptic feedback is triggered. Correspondingly, there are numerous options with this use case, and 5G can have the requisite flexibility and delay.

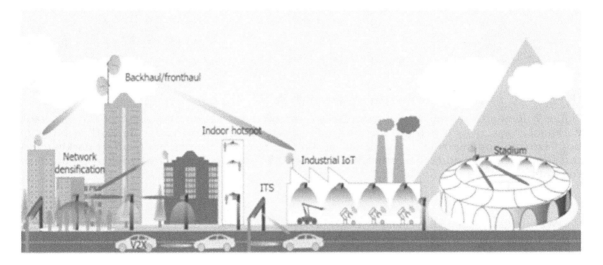

FIGURE 13.6 Use cases for 5G NR above 52.6 GHz.

Figure 6 describe the various use cases of Use cases for 5G NR for above 52.6 GHz that it had researched to achieve as part of its 38.8 GHz release study are analysed in 3GPP 38.807

with the holographic heart. Similarly, there are lots of possibilities with this use case and 5G will provide the reliability and required latency it requires. (Figure 13.5)

3GPP also studied the requirements for the use cases for NR beyond 52.6 GHz as part of release 16, described at **3GPP 38.807**. (Figure 13.6)

13.18 INTEGRATED ACCESS BACKHAUL [11]

One of the use cases that might get popular is integrated access backhaul. As the major challenge to deploy any network is backhaul infrastructure in terms of both cost and deployment challenges. Therefore this feature mainly provides backhaul connectivity using a 5G NR node to another 5G NR node. This feature mainly benefits nodes or devices that operate in LOS (Line of sight). (Figure 13.7)

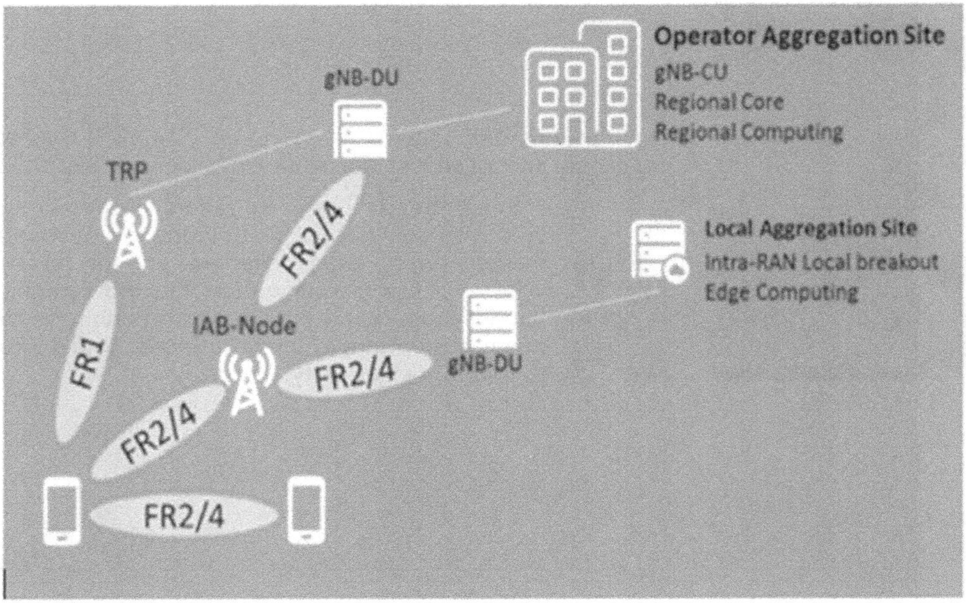

FIGURE 13.7 Integrated Access Backhaul as per 3GPP 38.807.

Figure 7 demonstrates one of the potentially common usage cases: integrated control backhaul. Backhaul connectivity is perhaps the most difficult aspect of deploying any network in terms of both expense and implementation issues. As a result, such function primarily offers backhaul access from one 5G NR node to another 5G NR node. This functionality clearly benefits LOS nodes or computers.

We will talk about gNB-DU and gNB-CU in later sections. Many other use cases can get integrated with 5G which needs a higher data rate, high reliability, and low latency.

13.19 5G ARCHITECTURE

5G architecture are of mainly two types-

1. Standalone Architecture.
2. Non-standalone Architecture.

In addition to the above, there is a functional split of layers mainly known as central unit (CU) and distributed unit (DU). We will discuss this later in the following sections.

Non-Standalone Architecture also called dual connectivity has multiple options described as per 3GPP **37.340**. Non-Standalone means that 5G will work in combination with legacy technology LTE where the signaling will be carried over LTE first and then data will be carried over 5G technology and the transfer of the technology will be based on the communication between the nodes over X2 interface. There are various multi radio deployments which can be possible like, EN-DC(Enodeb-NR), NE-DC(NR-enodeB),NR-DC and LTE-DC. [12]

13.20 NON-STANDALONE ARCHITECTURE, OPTION 3X

This will be the most commonly deployed architecture in practical scenario in terms of 5G non-standalone. In this architecture, existing LTE EPC (core) will be used to connect to 5G nodes. The main key highlight of the architecture is that once the user is move to 5G the user plane data to both LTE and 5G will flow from the 5G node, ie. the PDCP (Packet data convergence protocol) will be split on the SCG(secondary cell group) i.e., the 5G node and towards the LTE node. (Figure 13.8)

13.21 NON-STANDALONE ARCHITECTURE, OPTION 3A

In this architecture as well the LTE EPC (core) will be used to connect to 5G nodes but the main key difference is how data is delivered to the nodes. In this case as you can see from the figure there is no user plane between LTE and 5G node and therefore the user data has to via S-GW which add additional latency to the data. (Figure 13.9)

13.22 NON-STANDALONE ARCHITECTURE, OPTION 3

In this architecture also the 5G nodes will be served by LTE EPC (core) but the main key difference is that the slip of PDCP data will be on LTE node and not on the 5G node ie the split will be on the MCG (master cell group) which is the LTE node. (Figure 13.10)

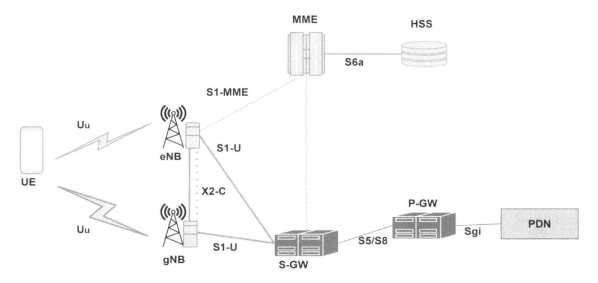

FIGURE 13.8 Non-Standalone Architecture, option 3x.

Figure 8 represents a non-standalone architecture, choice 3x, which will be the most widely implemented architecture in a realistic scenario for 5G non-standalone. Existing LTE EPC (core) will be used to bind to 5G nodes in this architecture. The main point of the architecture is that if the user is switched to 5G, the user plane data to both LTE and 5G will flow from the 5G node, i.e. the PDCP (Packet data convergence protocol) will be split on the SCG (secondary cell group), i.e. the 5G node, and towards the LTE node.

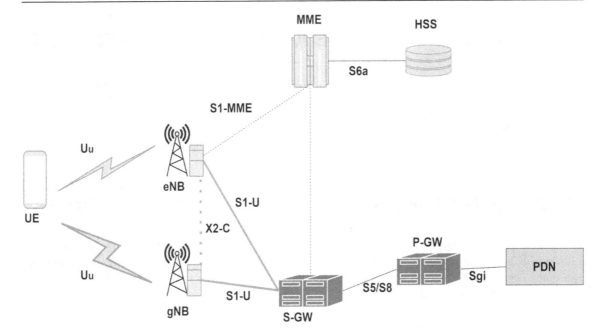

FIGURE 13.9 Non-Standalone Architecture, option 3a.

Figure 9 depicts Non-Standalone Architecture, Option 3a Architectural design, and LTE EPC (core) will be used to link to 5G nodes, but the main difference will be how data is transmitted to the nodes. In this case, since there is no user plane between the LTE and 5G nodes, the user data must pass through S-GW, which brings extra delay to information.

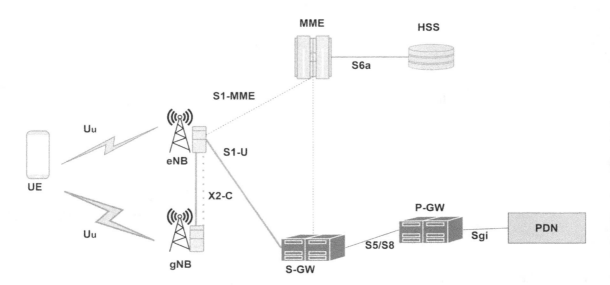

FIGURE 13.10 Non-Standalone Architecture, option 3.

Figure 10 addresses Non-Standalone Architecture, Choice 3 - In which 5G nodes will be served by LTE EPC (core), but the main difference is that the slip of PDCP data will be on the LTE node rather than the 5G node, i.e. the break will be on the MCG (master cell group) which is the LTE node.

13.23 STANDALONE - ARCHITECTURE, OPTION 2

In this architecture, both the control plane and data traffic are carrier over the 5G nodes i.e., there is no dependency on the LTE network and 5G network will be a standalone network. The core network consists of AMF (access and mobility function) which does all the signaling and control functions. UPF (user plane function) takes care of the user plane data, for packet routing and forwarding. (Figure 13.11)

13.24 STANDALONE ARCHITECTURE, OPTION 5

In this architecture, the LTE nodes are served by the next-generation core network (5G core network), i.e., by AMF and UPF. This is connected via NG-C and NG-U interface. (Figure 13.12)

13.25 NON-STANDALONE ARCHITECTURE, OPTION 4

This architecture uses 5G core network and serves the next generation enode B and next generation 5G gnode B. In this architecture the split will be on the 5G node PDCP and will connected via Xn interface to the enode B. The Xn interface carries both user plane and control plane traffic. The 5G acts as a master node (MN) and the LTE acts as a secondary node (SN). (Figure 13.13)

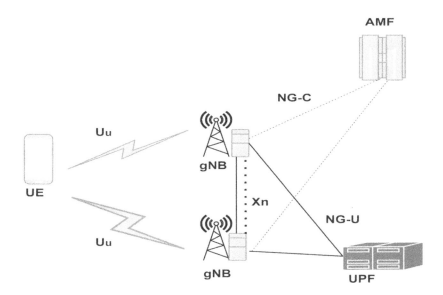

FIGURE 13.11 Standalone - Architecture, option 2.

Figure 11 depicts the Standalone-Architecture, Choice 2- In this architecture, both control plane and data traffic are transported over the 5G nodes, implying that the 5G network is not dependent on the coverage area and would be a standalone network. The AMF (access and mobility function) performs all signalling and control systems in the network infrastructure. The user plane function (UPF) manages user plane data for packet routing and routing.

264 Future Trends in 5G and 6G

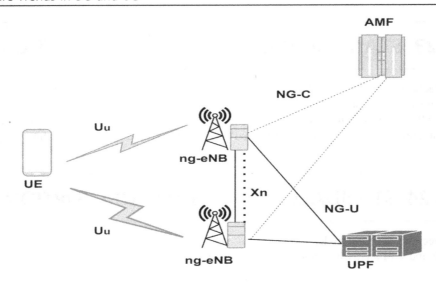

FIGURE 13.12 Standalone architecture, option 5.

Figure 12: Standalone design, choice 5-In this architectural design, the LTE nodes are operated by the next-generation network infrastructure, which includes AMF and UPF. These are linked through the NG-C and NG-U interfaces.

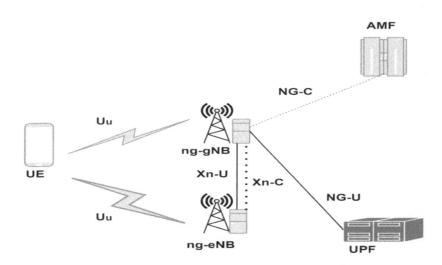

FIGURE 13.13 Non-Standalone Architecture, option 4.

Figure 13, shows a Non-Standalone Architecture, option 4 - which utilizes a 5G network infrastructure to represent next-generation enode B and 5G gnode B. The separation in this architecture will be on the 5G node PDCP, that will be attached to the enode B via the Xn interface. All user plane and control plane traffic passes through the Xn interface. 5G serves as a master node (MN), while LTE serves as a secondary node (SN).

13.26 NON-STANDALONE ARCHITECTURE, OPTION 4A

In this architecture, the next generation enode B needs to support NG-U and the user data flows from or via the UPF. The Xn interface only carries control plane. This option also does not have any connection to the AMF. (Figure 13.14)

13.27 NON-STANDALONE ARCHITECTURE, OPTION 7A

This architecture is similar to option 3a in terms of functionality that the LTE node carries the control signal and 5G node carries the data. Also for option 7a, there is no user plane connectivity over the Xn interface between both the nodes which is also similar. The main difference is that the architecture is connected to a 5G core with AMF and UPF with NG-C and NG-U interfaces respectively. (Figure 13.15)

13.28 NON-STANDALONE ARCHITECTURE, OPTION 7

This architecture is similar to option 3x in terms of access connectivity and data split. The main difference is that it is connected to the 5G core. The LTE node carries the control and the 5G node carries the data with split will be on the LTE PDCP. (Figure 13.16)

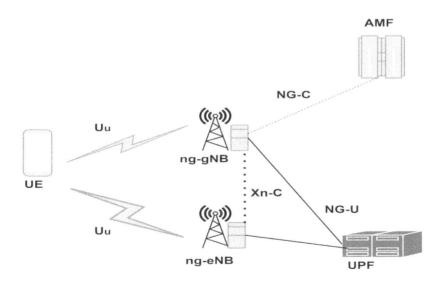

FIGURE 13.14 Non-Standalone Architecture, option 4a.

Figure 14 is a representation of Option 4a: Non-Standalone Architecture, wherein the next-generation enode B must encourage NG-U but also user data must stream from or through the UPF. Only the control plane is carried by the Xn interface. One such option, too, has nothing to do with the AMF.

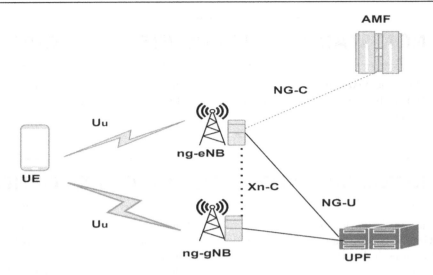

FIGURE 13.15 Non-Standalone Architecture, Option 7a.

Figure 15 depicts Option 7a's Non-Standalone Architecture. In terms of usability, this design is similar to option 3a in that the LTE node carries the control signal and the 5G node carries the data. Option 7a also appears to lack user plane interconnection between the nodes via the Xn interface, which is also similar. The main difference is that the architecture is linked to a 5G core via AMF and UPF with NG-C and NG-U interfaces.

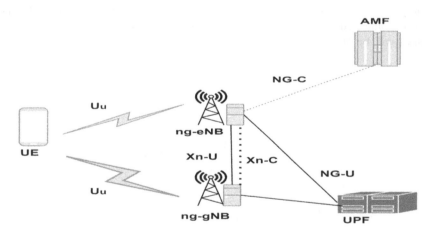

FIGURE 13.16 Non-Standalone Architecture, option 7.

Figure 16 is depicted Non-Standalone Architecture with Option 7. In terms of access connectivity and data split, this architecture is similar to option 3x. The main distinction is that this is linked to the 5G core. The control is carried by the LTE node, and the data is carried by the 5G node, with the split being carried by the LTE PDCP.

13.29 5G CHALLENGES

Though 5G helps to overcome many bottle neck that LTE has like throughput, latency and flexibility it also comes with some challenges which will lead to initial and long term hiccups for its deployment.

1. **Low device penetration** - As with every new technology the initial challenge will be low device penetration with very few devices available the utilization of the network will be a challenge as it will take some time for users to migrate which might also affect the initial revenue growth.
2. **Coverage and Interference** - As a basic engineering concept where the frequency is inversely proportional to wavelength and therefore the major basic challenge is a reduction in coverage leading to an increase in the number of base station deployment and further it will become difficult to manage interference due to an increase in a number of transmitting points.
3. **Geospatial challenges** - As we go higher into the bands of millimeter wave's geospatial challenges kick in and need more diverse planning and optimization are required. Also we have to keep in mind that the main component which suffers the most is transmission on the uplink mainly due to lack of limited power. [9]

Some of the key geospatial challenges are-

1. High propagation loss
2. High vegetation loss
3. Effect of rain attenuation
4. Effect of atmospheric absorption
5. Low signal penetration

The following graph shows the effect of rain at various frequencies. [13] (Figure 13.17)

13.30 COMPLEX HARDWARE DESIGN AND ARCHITECTURE

Design of the physical layer for such technology is a challenge when it comes to meet the demands of such a variety of services with various requirements. As some services might need higher throughput and some services require low throughput but need less latency. As we go higher into frequency bands massive mimo will be a basic necessity to overcome some of the coverage challenges but it creates more challenges in terms of handling the multiple beams and processing information due to frequency changes of beams by devices and services as it becomes narrower with increase in frequency. Algorithms will be introduced to overcome the inter base station interference due to multiple beams. Overall complex hardware implementation and also needs higher process on both the device side and on the base station side.

13.31 CONCLUSION

This study was undertaken to evaluate and provide a systematic overview of key application, primary architecture and main challenges pertaining to implementation of 5G. With complex network, and introduction of higher frequencies effective methods are needed to achieve an optimum capability of the network in reality. In this context, it is critical to understand the network architecture which provides the required capability for beam, resource, and interference management. Upcoming wireless networks need

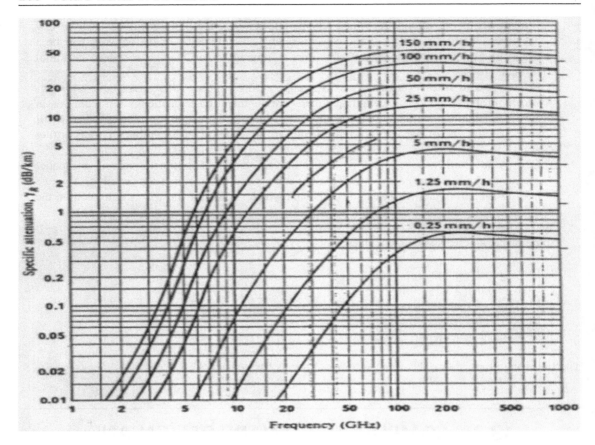

FIGURE 13.17 Attenuation due to Rain. (Reference: Federal Communications Commission Office of Engineering and Technology Bulletin number 70 july, 1997).

Figure 17 shows a attenuation due to the effect of rain in different frequencies, since Raindrops are roughly the same size as radio frequencies, causing the radio transmission to scattered.

to be self-predictive and pre-emptive to manage futuristic applications like holographic communication, haptic feedback, or any latency dependent requests. Radio frequency itself is a dynamic component, and therefore rather than having fixed rules, it must have a vigorous adaptable network [9].

REFERENCES

[1]. Retrieved 21 April 2021, from https://www.3gpp.org/DynaReport/21915.htm
[2]. [SPEC] 3GPP TS 21.916 – Release description; Release 16 – iTecTec. (2021). Retrieved 21 April 2021, from https://itectec.com/archive/3gpp-specification-ts-21–916/
[3]. Retrieved 21 April 2021, from https://www.3gpp.org/DynaReport/38104.htm
[4]. Johnson, C. *5G new radio in bullets*.
[5]. ITU towards "IMT for 2020 and beyond". (2021). Retrieved 21 April 2021, from https://www.itu.int/en/ITU-R/study-groups/rsg5/rwp5d/imt-2020/Pages/default.aspx
[6]. 3GPP 22.891. Retrieved 21 April 2021, from https://www.3gpp.org/dynareport/22891.htm
[7]. https://www.ngmn.org/wp-ontent/uploads/NGMN_5G_White_Paper_V1_0.pdf

[8]. 5G White Paper 2 | NGMN. (2021). Retrieved 21 April 2021, from https://www.ngmn.org/work-programme/5g-white-paper-2.html

[9]. Sengupta, Rishiraj, Sengupta, Dhritiraj, Kamra, Aashish, and Pandey, Digvijay. (2020). Journal of Critical Reviews Artificial Intelligence and Quantum Computing for a Smarter Wireless Network. 10.31838/jcr.07.19.21.

[10]. Frish, Sam, Maksymenko, Mykola, Frier, William, Corenthy, Loic, and Georgiou, Orestis. (2019). Mid-Air Haptic Bio-Holograms in Mixed Reality. 348–352. 10.1109/ISMAR-Adjunct.2019.00-14.

[11]. 3GPP specification series: 38series. (2021). Retrieved 21 April 2021, from https://www.3gpp.org/DynaReport/38-series.htm

[12]. 3GPP specification series: 38series. (2021). Retrieved 21 April 2021, from https://www.3gpp.org/DynaReport/38-series.htm

[13]. (1997). Retrieved 21 April 2021, from https://www.fcc.gov/Bureaus/Engineering_Technology/Documents/bulletins/oet70/oet70.pdf

[14]. Manne, R., and Kantheti, S.C., "Application of Artificial Intelligence in Healthcare: Chances and Challenges", Current Journal of Applied Science and Technology, vol. 40, no. 6, pp. 78–89, 2021. 10.9734/cjast/2021/v40i631320.

The Latest 6G Artificial Intelligence Network Applications

K. R. Padma[1] and K. R. Don[2]

[1]Assistant Professor, Department of BiotechnologySri Padmavati Mahila Visvavidyalayam (Women's) University, Tirupati, AP
[2]Reader, Department of Oral Pathology and Microbiology, Shree Balaji Dental College and Hospital, Bharath Institute of Higher Education and Research (BIHER) Bharath University, Chennai, Tamil Nadu, India

14.1 INTRODUCTION

Despite of the fact that 5G is still in the fundamental stage of peaks in the profitable scale, whether it be connected with technical features which further needs to be augmented. Furthermore, the establishment of Internet of things (IoTs) which is applicable in several industrial purposes needs to be examined thoroughly. However, the investigation for new innovation with latest developments is essential for next generation communication [1]. The start-up enterprises always focus on future generation communication. Although, 5G networks were been employed in all disciplines but the upsurge of 6G communication technology is attracting many researchers due to its prominent features and its promises. The 6G would revolutionize the whole world from 2030 onwards. Although, several important features were been depicted in leadingconferences [2–4].

Even though the 5G networks are being implemented by both industrial persons as well as academicians yet several researchers are also concentrating in further development in communication network[5,6]. The robust advancement in science and technology is anticipated to substitute the 5G completely after 2030 [7]. The 6G is technology potentiality also very advanced where its connectivity is made to all electronic devices. The Figure 14.1 provides a perception about the 6G networking.

Nevertheless, prior to portraying about 6G networking application augmentation, it is essential to understand the capacity of transforming interactions globally. Previously, we have noticed the alterations from first generation to 5th generation network communications but today 6G is considerably upsurging due to some drawbacks in 5G networking communication whichcan be overtaken with 6G implementation [8].

272 Future Trends in 5G and 6G

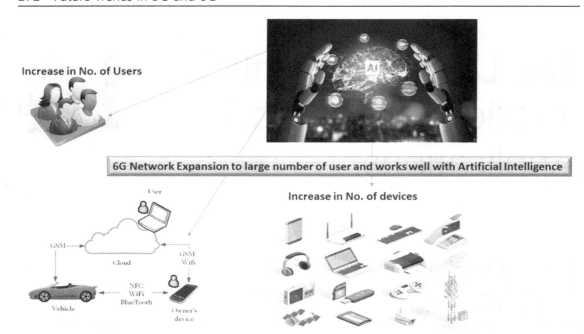

FIGURE 14.1 6G Enabled Artificial Networking Pattern.

The change in networking pattern in each generation time was roughly found to be 10 years. Hence, the deployment of 6G is expected to be in 2030.

The 6G communication system will totally revolutionize the whole world as it covers globally with integration of artificial intelligence, machine learning along with deep learning programmes which has the ability to predict the outcomes and helps in management of automobiles to computer communication in an automated fashion. Our current article provides insight to readers about the transformation from 5G to 6G networking pattern which is built with AI technology and was found to possess ultra-fast internet with high storage data along with massive coverage of networking through satellites [9]. Thus, this article main focus is the applications of 6G networking in several fields and its advantages as well as disadvantages with disclosing its future directions.

14.2 THE 6G SATELLITE COMMUNICATION SYSTEMS

The coordinated function of 6G with 5G communication network is based on wireless networking along with satellite communication systems. The telecommunication between the earth satellites is employed for decoding any kind of voice, any data, Wi-Fi connection is utilized for collecting information regarding climate as well as environmental data. The navigation of satellite connectivity is through the the global positioning system (GPS). Among the four majornations that developed such GPS based satellite systems are USA [10,11]. The principle objective of 6G network is to supply diverse services to mobile users with various multimedia Apps with which association can happen without much interruption and also process the data with very high speed. The Figure 14.2 shows the 6G satellite network communications.

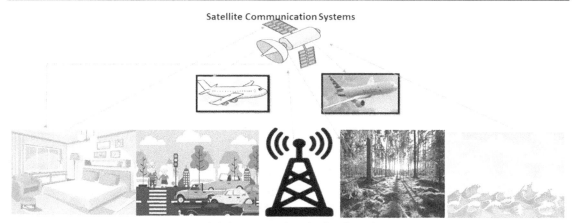

FIGURE 14.2 The 6G Global Satellite Navigation through GPS.

14.3 THE VISION IN COMBINATION OF ARTIFICIAL INTELLIGENCE WITH 6G AND ITS EXPECTATIONS

Our universe in future will be automated with artificial intelligence mode of communication technology. However, the Artificial intelligence has significant impact on 6G communications system [12]. Today most of the gadgets along with electronic appliances are built with latest augmenting artificial intelligence driven machinery and in future days it is obvious to see the expansion of AI driven machinery all round the world.Although, researchers are more keen in creating a new generation communication system for the benefit of human lives [13]. The involvement of AI technology with machine learning plus deep learning program helps in transformation of communications system to synapses junction which interconnects with artificial neural networking pattern which is possible with the 6G communication introduction. The Figure 14.3 shows the vision of 6G networking.

14.4 TRENDS AND TECHNOLOGY APPLICATIONS OF 6G NETWORKS

The reinforcement of digital learning technology constructed with artificial intelligence leads to transformation from the Self-Organizing Networks (SON) to Self-Sustaining Networks (SSN) The SSN is the chief accomplishment marker for performing dynamic actions in complex environmental states and creates revolution in the augmenting AI technology which out powered the 5G [14]. The 6G requires greater bit rates in comparison to 5G.The frequency bands also is greater than 6GHz and further mm wave increases to THz which leads to shrinkage of the size of small cells to tiny cells. However, the involvement of AI networking bought various alterations in machine learning plus deep learning based programs which are reliable to perform several critical tasks in comparison with the 5G communication system [15].

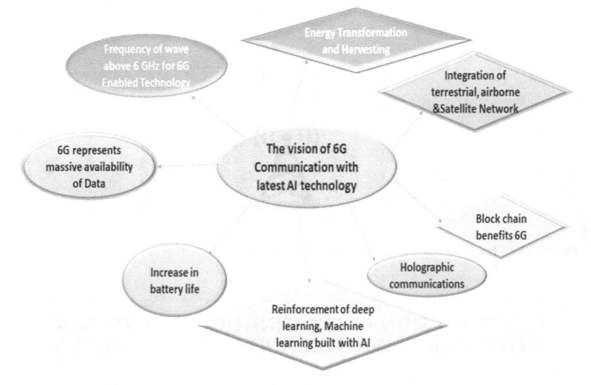

FIGURE 14.3 The role of 6G networking vision with latest AI technology.

14.5 THE NEED FOR 6G NETWORKING SYSTEM

Although, the existence of 5G not even have been vigorously established, the deployment of 6G has been into reinforcement with much vigour. Thus, it is crucial to understand why we need to inculcate 6G networking system. Nevertheless, usage of wireless networks had been augmented extensively along with huge amount of data produced by using of online platforms. However, in future days to come, the holographic communication which in turn requires larger data consumption, where 6G could compensate that in near future. Thus, the tremendous expansion of smart devices with wireless networking coverage initiates the employment of the sixth generation in communication technology [16–18]. The Table 14.1 shows the need for 6G networking system.

14.6 REQUIREMENT OF 6G APPLICATION IN HEALTHCARE SYSTEM

The 6G technology employment in healthcare system will create revolution. The wireless communication technology will enable to reach the needy in time. The advancement of AI built drones such as ambulance drones along with other kinds UAV with inbuilt thermal sensors helps in detection of any

TABLE 14.1 The Transformation of 5G to 6G due to enhanced networking communications

APPLICATIONS	5G NETWORK	6G NETWORK
Types of devices	Smart phonesUAV (Unmanned Aerial Vehicles) Sensors	Smart Implants Computation Oriented Communications (COC) eMBB (enhanced mobile broadband), uRLLC (ultra-reliable and low-latency communications)
The spectral and Energy Efficiency	100x in bps/Hz/m2	1000x in bps/Hz/m3 (volumetric)
Frequency Bands	Sub-6 GHz. MmWave for fixed access at 26 GHz and 28GHz	Sub-6 GHz. Exploration of THz bands (above 140 GHz)
Processing delay	50 ns	10 ns

change in temperature among the populations. The prime focus of 6G communication is the enormous facility it provides to all users, the quality of services plus the distinguishing experiences which helps a lot in all fields. Although, the envision of using 6G technology built with AI intelligence and deep learning program which provides adorable services to the population at prime time. The 6G provides a better wireless network communication through satellite communication and can be deployed in hospitals, remote areas [19]. The high frequency signals generated covers a larger areas withoutthe restriction of electromagnetic interference (EMI) is the advantage of 6G networking systems. With the outbreak of covid-19 globally, the remote healthcare system is promoted which initiates in mitigation of the transmission of the diseases and thereby decrease the patient load in hospitals [20,21]. Therefore, 6Gcommunication system architecture provides envision of its complete infrastructure depicted in Figure 14.4.

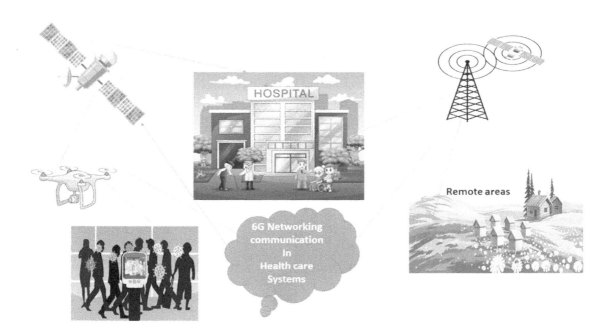

FIGURE 14.4 The 6G Architecture in Healthcare Systems.

14.7 6G INTELLIGENT CONNECTIVITY

Several researchers, entrepreneurs have strong faith that in constructing 6G network built with AI based technology will be an ordainedselection, as well as considered as "Intelligent" intrinsic feature of 6G communication, hence can be also termed as "Intelligent connectivity". Although, the construction of 6G networks have various restrictions due to its complexity nature with diverse kinds of terminals along with massive users. Thus, its complexity as well diverse use in business and all fields these 6G networking connectivity has to meet few requirements to support the massive connecting devices. The development of 6G "Intelligent Connectivity" with few basic characteristics in order to support thechief features of 6G network are Deep Connectivity programed with deep learning language, Holographic Connectivity plus Ubiquitous Connectivity. The 6G intelligent connectivity is activity as neurons which passes signals at a faster rate. Nevertheless, in coming 2030 years there will be complete 6G communication systems with "Deep connectivity". The deep mind helps to sense and communicate robustly with support of deep learning program, it acts like telepathy where mind to mind communication takes place. The higher fidelity of AR/VR, holographic interaction through wireless networking anywhere provides immense pleasure among the population globally.The recognition of 6G communication with holographic vision of connectivity with free accessibility is the major vision of 6G networking pattern [22]. However the surplus connection of anytime, anywhere in future years tends to achieve the actual "Ubiquitous connectivity"worldwide. Thus, the 6G envisage the future connectivity with help of its intelligent brain similar to human brain, where neurons helps in passing signals to generate any sort of action potential.The soul of the "Intelligent Connectivity" is the inbuilt important characteristics i.e"Deep Connectivity", "Holographic Connectivity" and "Ubiquitous Connectivity" which are regarded as significant trunk for 6G network connectivity [23]. (Figure 14.5)

14.8 6G COMMUNICATION SMART SOCIETY

Nevertheless, for building smart societywith fast access certain definite structures of 6G networking is essential which support in smart society enhancement. The artificial intelligence constructed M2M

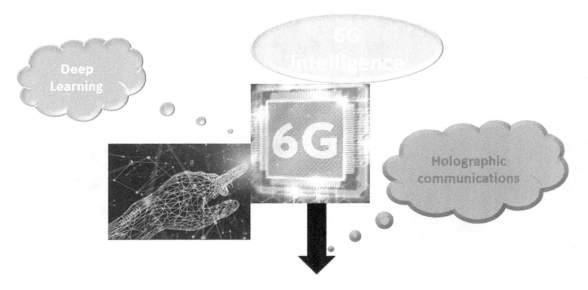

FIGURE 14.5 Intelligence communication and connectivity with 6G.

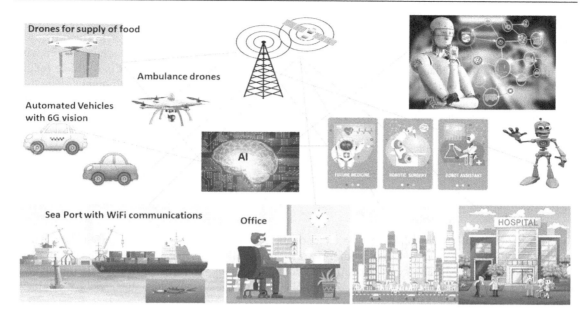

FIGURE 14.6 6G communication networking for smart society.

connectivity assist infast life class progression, employment of6G robotics in medical care system [24]. The 6G connectivity is an extension of augmented reality services with fully automated systems. [25]. The 6G connected robotics with WiFi communications has bought drastic changes in all fields.The automated vehicles i.e cars, bus plus trains which are used for means of transportation constructed with 6G wireless communication cruciallyalter our everyday lives [26]. The whole interaction/connectivity is AI brain which are like wireless brain computer interactions which has ability to receive signals as well as transmit signals promptly. The BCI are helpful to build smart life without any assistance along with sense of touch is use by non-verbal communication. However, the 6G robotics with wireless communication builds smart healthcare with ease in remote monitoring system. Although, the speed with less failure even promotes remote surgery in medical care system. The transportation of huge quantity of medical data to any place with much reliabilityis possible with 6G connectivity. (Figure 14.6)

14.9 6G AUTOMATED COMMUNICATION SYSTEMS

The introduction of 6G technology fully integrated with artificial intelligence utilized for transportation purpose i.e self-driving vehicles which has been made without any human operation. These vehicles are fully automated and constructed with latest 6G connectivity to complete the journey at specified time from point of source to end point. In order to complete its destination the operation the AI enabled machinery has to be fed with large data so as to perform the task in a self-organized manner with high speed and low latency communication [27]. The Brain computing information (BCI) plays a key role incontinuous integration and self-directed with support of internet connection which relays in co-ordination withmassive number of computing elements, sensors, objects and various devices to processes the real data [28,29]. The internet connectivity with Wi-Ficonnections provides massive support for 6G connectivity forintegrationof one frame with another frame. Henceforth, AI brain intelligence will automatically bind all people, data, processes, and physical objects into one system [30]. Therefore, automated communication system with 6G connectivity enabled to introduce smart cars which are fed with required data-pathfor guidance along with traffic rules, high resolution maps so that the vehicle totally depends on that data which is received from satellite communication network and completes its task successfully [31–33]. However, the signals received fromsatellite communications provides any warnings, new updates about traffic information to run the smart cars in smooth manner without any sort of interruption.

14.10 6G IMPLEMENTATION IN INDUSTRIAL SECTOR

The 6G communications are anticipated to enhance the 5G communications with the establishment of improved services from the assessment of network data availability, the cell phone data rate withcontinuousuniversal connection. Additionally, 6G communications will operatebypeculiar communication method to gain recognition to numerous mobile data categories and link them via conventional enhanced radio-frequency networks.

Nevertheless, Apple Inc. telecom Industry launched its first i-Phones with 5G wireless speeds a few months ago. Today, it works with sixth generation communication i.e 6G technology constructed with artificial intelligence which provide services at a robust rate. Today, this is an indicative example for implementation of 6G in industrial area.(Shown in Figure 14.7).

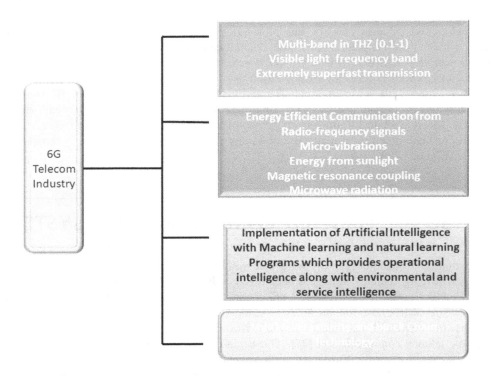

FIGURE 14.7 6G Telecommunication Industry built with AI machinery.

14.11 FUTURE PERSPECTIVES FOR 6G NETWORKING SYSTEMS

The upsurge in AI based devices with dynamic programming and powerful networking performance arose interest in several researchers to comprehend its architecture, efficiency plus prediction strategies. In future, the AI constructed machinery with power of storing large amount of real data and transmit signals to required areas robustly, provides potential solution to many problems. The 6G intelligent connectivity networks has complex architecture with 3D graphics, deep learning programs are the promising procedures forhigh performance. The corresponding hardware development is very challenging when designing 6G networks. Moreover, the 6G connectivity major limitation is higher power consumption and frequency of transmission from mm to THz bands. The future perception majorly rely on the benefit of human population to communicate anywhere fast without any interruption which is possible with the development of 6G connectivity. Therefore researchers must significantly focus on energy management which is most challenging task in 6G technology [34].

14.12 CONCLUSION

In our current article, we have focussed on the latest 6G connectivity technology constructed with AI based intelligence. Although, each generation bought several modifications in communications, the progression from 1G to the 5G communications are regarded as major leap in technology. Nevertheless, the next generation is building a road map for deployment of 6G network communication which is fully automated without human handling. Moreover, in this paper we have highlighted the applications of 6G technology which bought several transformation to bring a smart society with holographic communications.Finally, we have portrayed the 6G networks which cuddle latest spectrum bands connecting advanced network circuits designed with artificial intelligence. Thus, the future generationwill gain benefits with such advanced technological developmentwhich bring forth industrial revolution and for sure will change the style of living.

REFERENCES

[1]. W. Saad, M. Bennis, and M. Chen, "A vision of 6g wireless systems: Applications, trends, technologies, and open research problems", *IEEE Network*, pp. 1–9, 2019.

[2]. G. Gui, M. Liu, F. Tang, N. Kato, and F. Adachi, "6g: Opening new horizons for integration of comfort, security and intelligence", *IEEE Wireless Communications*, pp. 1–7, 2020.

[3]. M. Giordani, M. Polese, M. Mezzavilla, S. Rangan, and M. Zorzi, "Toward 6g networks: Use cases and technologies", *IEEE Communications Magazine*, vol. 58, no. 3, pp. 55–61, March 2020.

[4]. S. Chen, Y. Liang, S. Sun, S. Kang, W. Cheng, and M. Peng, "Vision, requirements, and technology trend of 6g: How to tackle the challenges of system coverage, capacity, user data-rate and movement speed", *IEEE Wireless Communications*, pp. 1–11, 2020.

[5]. K. David and H. Berndt, "6G Vision and Requirement", *IEEE Vehic. Teh. Mag.*, vol. 13, no. 3, pp. 72–80, Sept. 2018.

[6]. P. Yang, Y. Xiao, M. Xiao, and S. Li, "6G Wireless Communications: Vision and Potential Techniques", *IEEE Network*, vol. 33, no. 4, pp. 70–75, Jul. 2019.

[7]. WalidSaad, Mehdi Bennis, and Mingzhe Chen, "A Vision of 6G Wireless Systems: Applications, Trends, Technologies, and Open Research Problems", IEEE accepted for publishing.
[8]. ArockiaPanimalar, S., Monica, J., Amala, S., and Chinmaya, V., "6G Technology", Sep -2017, IJERT.
[9]. 1Rukmani Khutey, Ghankuntla Rana, Vijay Dewangan, Anil Tiwari, and Adarsh Dewamngan, "Future of Wireless Technology 6G & 7G", June 2015.
[10]. Khutey, R., Rana, G., Dewangan, V., Tiwari, A., and Dewamngan, A., "Future of wireless technology 6G & 7G", *International Journal of Electrical and Electronics Research*, vol. 3, no. 2, pp. 583–585, 2015.
[11]. Kalbande, D., Haji, S., and Haji, R., "6G-Next Gen Mobile Wireless Communication Approach", In 2019 3rd International conference on Electronics, Communication and Aerospace Technology (ICECA), 2019, June (pp. 1–6). IEEE.
[12]. Helin Yang, Arokiaswami Alphones, Zehui Xiong, Dusit Niyato, Jun Zhao, and Kaishun Wu, "Artificial Intelligence-Enabled Intelligent 6G Networks".
[13]. Khaled B. Letaief, WeiChen, YuanmingShi, JunZhang, and Ying-Jun AngelaZhang, "The Roadmap to 6G: AI Empowered Wireless Networks", *August*2019, IEEE.
[14]. Y. Xing and T.S. Rappaport, "Propagation measurement system and approach at 140 GHz-moving to 6G and above 100 GHz," arXiv preprint arXiv:1808.07594, 2018.
[15]. A. TalebZadehKasgari, W. Saad, and M. Debbah, "Human-in-the-loop wireless communications: Machine learning and brain-aware resource management," arXiv preprint arXiv:1804.00209, March 2018.
[16]. Dhyey Patel, and Parth Bhalodiya, "3D Holographic and Interactive Artificial Intelligence System", 2019, IEEE.
[17]. Md. JalilPiran, and Doug Young Su, "Learning-Driven Wireless Communications, towards 6G", IEEE.
[18]. Marco Giordani, and Marco Mezzavilla, "Towards 6G Networks: Use Cases and Technologies", *March* 2019.
[19]. Ahmed, I., Karvonen, H., Kumpuniemi, T., and Katz, M., "Wireless communications for the hospital of the future: requirements, challenges and solutions", *Int. J. Wireless Inform. Netw.*, vol. 27, pp. 4–17, 2020. 10.1 007/s10776-019-00468-1.
[20]. Nayak, S., and Patgiri, R., 6G communication technology: a vision on intelligent healthcare. arXiv preprint arXiv:2005.07532, 2020. 10.4108/eai.17-8-2020.166293.
[21]. Ohannessian, R., Duong, T.A., and Odone, A., "Global telemedicine implementation and integration within health systems to fight the COVID19 pandemic: a call to action", *JMIR Public Health Surveill*, vol. 6, p. e18810, 2020. 10.2196/18810.
[22]. Bastug, E., Bennis, M., Médard, M., and Debbah, M., "Toward interconnected virtual reality: Opportunities, challenges, and enablers. *IEEE Communications Magazine*, vol. 55, no. 6, pp. 110–117, 2017.
[23]. Yajun, Z., Guanghui, Y., and Hanqing, X.U., "6G mobile communication networks: vision, challenges, and key technologies", *SCIENTIA SINICA Informationis*, vol. 49, no. 8, pp. 963–987, 2019.
[24]. B. Li, Z. Fei, and Y. Zhang, "—UAV communications for 5G and beyond: recent advances and future trends," *IEEE Internet of Things Journal*, vol. 6, no. 2, pp. 2241–2263, April 2019.
[25]. Chowdhury, M.Z., Shahjalal, M., Ahmed, S., and Jang, Y.M. (2019). 6G Wireless Communication Systems: Applications, Requirements, Technologies, Challenges, and Research Directions. arXiv preprint arXiv:1909.11315.
[26]. W. Xia, M. Polese, M. Mezzavilla, G. Loianno, S. Rangan, and M. Zorzi, Millimeter Wave Remote UAV Control and Communications for Public Safety Scenarios, 2019. 6G. [Online]. Available: http://mmwave.dei.unipd.it/research/6g/.
[27]. K.B. Letaief, W. Chen, Y. Shi, J. Zhang, and Y.A. Zhang, "The roadmapto 6g: Ai empowered wireless networks", *IEEE Communications Magazine*,vol. 57, no. 8, pp. 84–90, August 2019.
[28]. C. Han and Y. Chen, "Propagation modeling for wireless communicationsin the terahertz band", *IEEE Communications Magazine*, vol. 56, no. 6, pp. 96–101, 2018.
[29]. S. Nayak, R. Patgiri, and T.D. Singh, "Big computing: Where are weheading?" *EAI Endorsed Transactions on Scalable Information Systems*, vol. 4, 2020.
[30]. S. Ullah, H. Higgins, B. Braem, B. Latre, C. Blondia, I. Moerman, S. Saleem, Z. Rahman, and K.S. Kwak, "A comprehensive survey ofwireless body area networks", *Journal of medical systems*, vol. 36, no. 3, pp. 1065–1094, 2012.
[31]. Internet of Everything. (2019). Internet of everything (IoE). [Online]. Available: https://ioe.org/
[32]. CISCO. (2019). The internet of everything. [Online]. Available: https://www.cisco.com/c/dam/en_us/about/business-insights/docs/ioevalue-at-stake-public-sector-analysis-faq.pdf.
[33]. K.B. Letaief et al, "The roadmap to 6G - AI empowered wireless networks," arXiv:1904.11686.
[34]. H. Yang, X. Xie, and M. Kadoch, "Intelligent Resource Management Based on Reinforcement Learning for Ultra-Reliable and Low-Latency IoV Communication Networks", *IEEE Trans. Vehic. Teh.*, vol. 68, no. 5, pp. 4157–4169, May 2019.

A Review of Artificial Intelligence Techniques for 6G Communications: Architecture, Security, and Potential Solutions

15

Syed Hauider Abbas[1], Nazish Siddiqui[1], and
Sanjay Kumar Agarwal[2]

[1]*Integral University, Lucknow, India*
[2]*Dolphin PG Institute, Dehradun, India*

15.1 INTRODUCTION

In recent years, communication networks have evolved considerably and have had a significant effect on the way people perceive and communicate. The new communication technology, the fifth generation of mobile technology, has been adopted in different regions worldwide and is expected to link the entire globe in the near future. As son as the 5G technology is deployed, an axiomatic query arises as to what is next. In future communication network, i.e., 6G, scholars have already started their work on the in-coming genesis. It was observed that a new communication network is launched every 10 years. With the complete rollout of 5G in the year 2020, the emphasis is now on the 6th generation communication technology. By the year 2030, the sixth generation network is projected to replace 5G partially or entirely. In comparison, the 6th era will be advanced and would have all the capabilities that were lacking in the 5G technology [1,2]. In comparison to early generations, 6G design is anticipated to be even many revolutionary and will present a worldview move in communication innovation by advancing from "interfacing gadgets" to "interfacing organize insights" and will feature exceptionally minute details that would address the growing requirements [3]. In this chapter, we explain precisely what the forces that come with 6G are going to be and how profoundly artificial intelligence (AI) and 6G will affect every one, and potentially revolutionize the business enterprise.

From 1G network to the recent 5G network, wireless networks have advanced with focus on different aspects such as information rate, end-to-end interval, power consumption, extent and use of orbit. 5G systems have three fundamental sorts of utilization scenarios. In accordance with the guidelines of the International Telecommunication Union (ITU): To support diverse networks, the primary requirements are highly reliable and low latency communication, improved and enhanced mobile broadband and massive use of machine-based communication [3–6]. Various technologies, namely multiple input multiple output (MIMO), millimeter wave, etc., are being utilized in this respect to supply consumers with improved quality of service and enhanced experience and to advanced network efficiency [3,4]. Although 5G systems are being actualized, people from both industry and the research fields are, as of now, focusing on investigating the 6G systems [7], where high-quality facilities, new emerging technologies and unrestricted property for the galactic bit of smart terminals are expected to effectively support 6G networks. Reference [5], for example, addressed the path towards 6G networks together with specifications, strategies and architectures supporting them. Unlike former genesis networks, 6G networks is an approach towards revolutionizing the world by creating smart society by the 2030s, which include [8]: Ultra-fast data speeds, prime data rate of around 1 terra bits per second and at least 1 giga bits per second user-experienced data rate; ultra-low latency and very little end-to-end delay; ultra-high reliability, increased energy efficiency of the order of 1 pJ/b; greater mobility, that may extend up to 1,000 km per hour; extremely massive link, that could connect up to 107 devices per square km and could handle traffic capacity of about 1 Gbs per meter square; larger bands of frequency and most importantly connected intelligence with higher AI potentiality [8].

In addition, the underwater, ground, air and space altogether integrated in a network would be the central future of 6G network architecture [9], which is expected to enable nearby-instant and coherent super-connectivity, as shown in Figure 15.1, where it consists chiefly of the following four levels: Low, medium and geostationary earth orbit satellites [7] are deployed by the Space Network Tier to stipulate orbit services for non-Earth networked topics. Air network tier makes use of different aerial platforms to facilitate versatile and secure wireless communication for remote areas, providing connection to moving base stations with the airships, unmanned aerial vehicles and balloons. The key mixture for bearing diverse networks for a large number of devices is the ground network tier. This region specifically makes use of low frequency waves, visible light and microwaves, along with terahertz bands of frequency for 6G network to offer different services. The underwater network provide access, observation and surveillance facilities for underwater communication for broad-sea and deep-sea activities. Initially, the 6G architecture will basically be backed by the current 5G foundations, such as SDN, NFV and cutting-edge structures, in compliance with the previous history of advancements. However, relative to 5G networks, 6G networks need to adhere to strict specifications as listed above. With high erudition cognition, strong logical thinking ability and intellectual identification quality, AI [10] enables designs of 6G systems to memorize and adjust so as to boost diverse administrations automatically without manual involvement. In [11], AI-enabled strategies were connected to accomplish shrewdly remote communication, intelligence and closed-loop optimization of 6G networks. N. Kato et al. [7] deployed the concepts of deep learning to improve and enhance the efficiency of interconnected space-air-earth networks and demonstrated the way to pick the most effective routes for satellite networks using deep learning. In addition, deep reinforcement learning was assumed by learning the environment dynamics to maintain stable wireless communication for UAV-enabled networks [12,13], and framework showed that DRL importantly accomplish the traditional methods. Hence, an AI-enabled architecture for 6G systems is displayed in this chapter to figure out shrewd asset administration, mechanized arrange adjustment and high-level insights that brilliantly benefit provisioning, where the design comprises the previously mentioned layers: Detecting layer, information mining and analytics layer, control layer and the application layer. In an attempt to handle upgraded physical layer plan, complicated decision-making, arrange administration and asset optimization errands, the proposed engineering is competent of scholarly people extricating valuable and relevant information out of endless information, analyzing, learning and

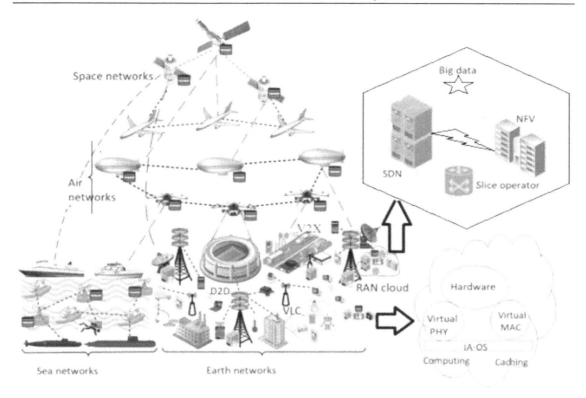

FIGURE 15.1 The proposed 6G network architecture.

proposing different capacities for self-optimization, auto-configuration and auto-healing in 6G network systems. We execute the AI application procedures within the edge computing, which is highly versatile and AI-empowered, shrewdly versatility and handing over administration and shrewd range administration based on AI-enabled brilliant 6G. After that, future inquires about bearings for AI-enabled 6G keen networks are addressed.

15.2 WHAT IS 6G AND ITS REQUIREMENTS?

We need to understand precisely what 6G means, the criteria and vision of 6G with respect to different definitions. 6G is the sixth generation of technology for mobile communication. Past eras, namely 2G, 3G, 4G and currently 5G possess technological capacities with certain boundaries and have been put to use to meet the computational limitations. Each generation evolved in approximately a decade and by 2030, 6G is anticipated to be implemented. It is not possible to define a specific meaning of 6G at the moment. In communication technology, 6G can be defined as the successor to 5G. It will significantly get the better of the impediments of 5G and would have a plenty more focal points support the development of future communications [14]. The global reach of the 6G networking infrastructure will be the convergence of the 5G [15]. It is inferred that 6G, alongside a expansive arrange scope that will be exceptionally steady and vitality effective, will give ultra-fast Web with exceptionally high information speeds and moo idleness.

Even before 5G was fully deployed across the globe, 6G testing began vigorously. Even as 5G implementations are under way, we need to consider that we need 6G in the near future. IP traffic worldwide is projected to exceed three times by 2022, when compared to traffic that was in 2017 and could increase around 400 exabytes/month. It is observed that around 71% of IP activity is produced from portable and remote gadgets [16]. This is due to the increasing requisite for portable devices, smart phones and similar other devices and the tremendous quantity of knowledge created by social networking platforms and applications like YouTube. It has become important to invent a technology that will have the potential to satisfy this need to manage this massive IP traffic and to maintain its standard. In communication technology, however, 5G would be a promising evolution.

The network difficulty together with exponentially growing data stages every day will inevitably exceed beyond the technological capabilities and potentialities of 5G and it may soon fail to be solved. In the future of communication, new applications and innovations will be launched, such as holography [16]. For such upcoming applications, data rates must be around some terabits/second and 5G may fall short of delivering such speed even though if it is anticipating certain advancements in the coming future. However, it is planned and proposed to produce a data rate of around 1 terabyte/second and ensure that these applications can run smoothly in the near future. In addition, both the organize scope and 5G capacity would ought to be improved with the development and exponential advancements in tech gadgets or IoT devices to keep them associated and to communicate successfully [17]. Also, manual network configurations and optimizations will not be sufficient for wireless networks in the future. To fulfill the network requirements, automation would be extremely necessary.

15.3 6G ANTICIPATIONS

Presently that the requirements of 6G versatile network are known, we can take a guess at the 6G characteristics that are supposed to be part of the communication technologies of this future era (Figure 15.2).

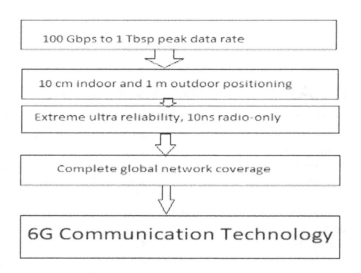

FIGURE 15.2 Expectations and 6G vision.

15.4 INTEGRATING AI WITH 6G

Our planet will soon be witnessing the exponential growth of information technology. In 6G communications, AI plays a significant role [18]. Nowadays, all the new technologies support AI. In the years to come, we should expect AI-powered technologies and devices. The 6G can, in many ways, boost AI. AI is expected to be the key motivational component behind the development of various novel technologies in mobile communication and will be the basis of a whole new wave of machine learning applications [18]. The remote community has appeared a sharp intrigued in manufactured insights in later considers and breakthrough comes about in profound learning in conjunction with the rise within the number of gadgets, particularly keen gadgets and huge information era. It is anticipated that 6G would offer a fully wireless connectivity and will also enable AI to be an integral part of the upcoming technologies. 6G connectivity comprises advanced equipments linked to intelligence, where the intelligence of course will be realized through AI and its subsets, machine learning and deep learning [19] (Figure 15.3).

We may assume that they are all connected to each other by manufactured insights, robotized frameworks and 6G portable communications. AI at the pivot allows for automated machine technology. The dominant force behind automation is numerous machine learning algorithms and deep learning principles. The notion of real-time learning lets an automated system work effectively. When talking about 6G communications technology, automated systems are essential. When associated with the globe it is required for the numerous frameworks which should be mechanized to the limitations for 6G communications. Numerous apps presently running on smart phones or tech devices are fueled by manufactured insights, especially AI and machine learning, hence, acting as a primary element in driving 6G in different aspects, namely as profound neural systems, semantic communications, machine learning, deep learning, caching and various control assets administration. In case we attempt to combine fake insights, mechanized frameworks and 6G innovations, it is observed that totally different ways of 6G will be groundbreaking and remarkable innovations in remote communication. If made to run with human intercession, the long run systems that would be built will be as well complex. These remote and complex systems can moreover turn out to be debilitated by human administrators. A learning-based, AI-empowered organized innovation is required in handling zero-touch communications to realize such uncommon capabilities. For arranging, optimizing, inquiring, disappointment discovery and asset administration, AI became a major capability for empowerment of 6G communications. The combination of the Web of Everything and AI permitted by 6G would build up a solid interface and thus ended up being an awfully effective interface between two innovations where 6G will offer assistance by giving information and AI will dissect and evaluate the information from it [20].

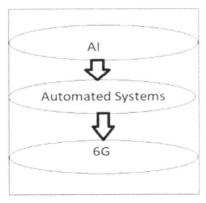

FIGURE 15.3 Integrating AI with 6G.

15.5 SELF-DRIVING VEHICLES: A USE CASE

This is a great combination of the three previously mentioned terminologies. In the deployment of self-driving cars, the convergence of automated systems and 6G communications may be determined. Self-driving cars are fully autonomous and not powered by humans. All these advanced automated vehicles are major applications in the area of 6G communication developments. To complete their journey from source to destination, they use real-time data. This involves an immense amount of knowledge. This can be referred to as low latency and ultra-high speed case [21], which is only feasible with the introduction of 6G communications now, when its capabilities are known. The autonomous vehicle, based on AI, automation and 6G, in itself is an automated device that, although operated by AI, is only functional. To anticipate the following steps, while driving and persistently upgrading the data in their frameworks using AI, these vehicles need real-time data [22]. Presently, it needs continually upgraded data, which produces a enormous sum of information each moment, counting activity overhauls, tall determination maps, live gadget overhauls, alerts, etc. Information is collected through lackey communication, and the arrangement must have adequately tall information rates that comply with the prerequisites of independent vehicles. In the event when the information could not be received and interpreted by the vehicle due to some interference during the communication, these self-driving automated vehicles, when out of control, can become dysfunctional and indeed dangerous for people and could also lead to serious accidents. The communications which are based on quantum machine learning complement independent vehicles that totally depend on self-organization, counting self-configuration, self-optimization and self-healing [23]. Only when networking infrastructure is powerful enough that it can felicitate enormously high data rate with reduced latency, can real-time network decisions be made to facilitate self-organizational ability. At the other side, rapid developments in vehicles during movement in real time are leading to increased road traffic. This makes the concept of these vehicles much more difficult for communication technologies to incorporate and meet the puritan specifications of ultra-high data rates, high reliability and protection and a very low latency. Machine learning on the other hand, has emerged as a propitious and encouraging AI technique towards driving wireless networking even more flexible and adaptable, thus clearing the path for upcoming 6G vehicle network intelligence [24]. 6G is intended to address the rigorous criteria of such principles efficiently and to open up a multitude of possibilities for all of these upcoming innovations [25].

15.6 A 6G VISION

15.6.1 Holographic Communications

In recent years, holographic communications have been buzzing and there are comprehensive studies and work ongoing to change this idea from mere vision to a ground reality. Our motivation is to sympathize how a major supporting factor for these holographic communications would be 6G communication technology. Holographic interactions are nothing but a simulated replication at the same time of a real-time object at two distinct locations. Such encounters will bring about a revolution in human relationships. The existing remote human contact approaches will become redundant in the coming years and holographic communication will become the new normal for communication. The 5D holographic communications and facilities that are supposed to combine all kinds of human sense data can grow all the way together and give a quite unreal but an extremely mesmerizing experience. The advanced cameras that provide multiple-views and modern sensors available with holographic communications are

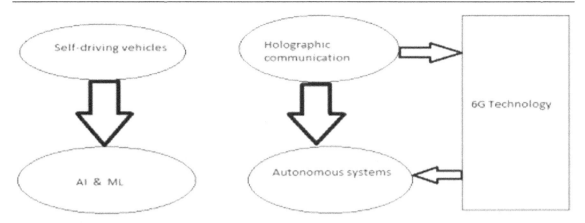

FIGURE 15.4 Relationship between 6G and its applications with respect to AI.

assumed to be requiring higher data rates per second, which is expected to be difficult to be fulfilled in a 5G network age. This demands the need for 6G to assist and enable these forms of communications easily [26] (Figure 15.4).

15.6.2 Tactile Communications

It is conceivable to communicate a visual observation of very close to real vision of people, exercises and situations by means of holographic communication. Without a material web that would empower the real-time transmission of a picture, the cinematic experience would be incomplete [27]. Teleoperation, co-operative autonomous driving and interpersonal communication are some of the anticipated beneficiaries of this technology. For these innovations, using communication networks, a haptic touch may easily be introduced. Experiencing the observations of this advanced technology will drive us towards the eradication of the network paradigm of open networks interconnection along with the introduction of communication system cross-layer architecture. This cross-layer system structure should fulfill these technologies' rigorous requirements. The design of new and modern physical layer systems may be triggered by this situation, which can improve the execution of signaling system designs and multiplexing of waveforms. How to develop processes like buffering, scheduling, queuing, protocols and handover, that could fulfill the requirements of 6G networks, is another factor that requires attention. It is clear that the current wireless network systems are obviously not sufficient to meet these requirements and, thus, it is required to conduct studies in the field of over-the-air fiber communication systems [28].

15.6.3 Communications Based on Human Bond

It is anticipated to be one among the most fundamental drivers towards the evolution of 6G communication. As a result of this innovation, people are anticipated to get to share and express physical highlights or physical wonders. An illustration of the innovation is the communication that occurs through breath, which makes it possible to be carried by examining human bio-profile by utilizing their exhaled breath and the inhaled breath utilizing unstable natural compounds [29]. Such technology, as a result, helps in diagnosis of various diseases, detecting emotions, biological features collection and interaction with the human body remotely. Thus, working to design a communication system which is very much capable of copying the different human senses, requires the multidisciplinary research collaborations. Such inquiries about endeavors would normally lead to cross-breed communication advances able to extricate different physical amounts and after that disseminate these towards the aiming recipient by means of secured channels.

Hence, with the arrangement of future administrations from the viewpoint of arranged information, versatile information rate and consistent ubiquitous connect, 6G communications are anticipated and believed to lift the standard set by 5G communications, as of now. Moreover, 6G communications would make use of an offbeat approach towards organizing endorsements for diverse types of versatile information and transmit them through conventional advanced radiofrequency systems. This strategy will permit the advanced mechanism of transferring of feelings remotely with virtual nearness and association. One of the most recent non-existent cutting-edge remote organizing scenarios of 2030 are anticipated in [30], which comprises holographic calls and the material web. 6G will escort within the same unwavering quality as it was in the existing wired systems with vey little bit-error-rate taking into account the sort of applications that will be verified. The figure abridges the most characteristics of the potential 6G [31]. These developments mark a complete departure from the conventional concepts of designing and standards of execution practised in the wireless mobile telecommunications industry (Figure 15.5).

15.7 CHALLENGES AND THEIR POTENTIAL SOLUTIONS

The communications within a 6G network comprises certain fundamental components: Organize assurance arrangement and information security compliance, achievement of the economical perspective towards fast arrange sending and extension with an accentuation on the inaccessible and standalone ranges, decrease within the cost of versatile communications utilization, approaches to lead versatile

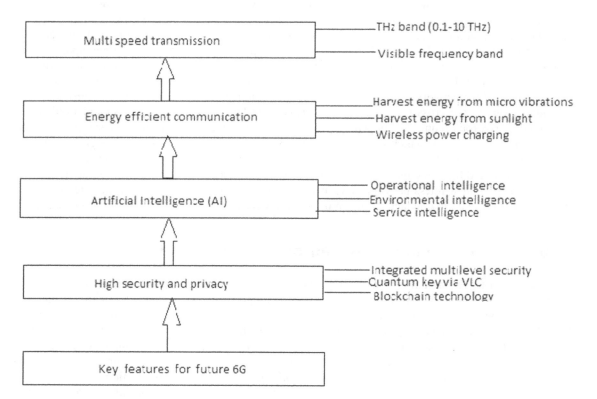

FIGURE 15.5 Features of future 6G.

hardware battery life span and achievement of the higher information rate buoyed with end-to-end, ultra-reliable moo idleness administration. However, it is difficult to address all characteristics, but a method must be devised to sufficiently balance these characteristics. A few of the expected issues and challenges of future 6G communication systems and conceivable arrangements are highlighted in this segment.

15.7.1 High Intelligence versus Complexity

For 6G systems, since they are humanoid systems, the adjust between protection and insights will be critical. In differentiate, AI calculations must be associated with refined private information to improve the usefulness of the organize, alter organize figures and give predominant quality administrations [32]. Therefore, to the detriment of superior intellect, privacy will be foregone. Using a negotiator at the intermediary level in between AI algorithms and the end-user data is a candidate approach. Such a negotiator should, if necessary, be a third party, independent and functioning in a dispersed manner. An intermediary negotiator agent would anonymize all the personal and confidential data. Likewise, the dominating intelligence provided by the algorithms of AI and intelligent nodes, decreases the independence possessed by humans. The user inclination is most likely not continuously tuned towards the enhanced approach created by AI algorithms. When multiple users are considered, the contradictory situation becomes more complex. The mentioned issue becomes a matter of concern and may be carved by tracing the central point between the intelligence and customization in 6G communication systems [33]. The later term customization is certainly attracting enhanced interest for AI algorithms and smart nodes intelligence versus exorbitant routines sub-routines. The most elementary approach for 6G communication networks should define certain routines and sub-routines. Intelligent anthology or portfolio may easily be transferred within the permissible limits by using this method. In terms of network sophistication, higher intelligence is available at a price [34] and could lead to greater network operators and budgets for device manufacturers. Such advances will lead to higher consumer device prices and negate the dream of supplying affordable gadgets. Technological advances in smart structures are important in tackling these issues by ensuring protection and confidentiality.

15.7.2 Security versus Spectral Efficiency

A number of range transfer speed preventive measures are required to attain end-to-end, attack-proof remote information connectivity, resulting in a decrease in the accessible range of information for further transmission [35]. Complex computations would require finding an adequate system to resolve the protection and spectral efficiency issues of wireless communications, but can be solved by three possible computations strategies. First, it is expected that encryption algorithms developed by security experts are highly effective. It may be slightly cumbersome to understand such algorithms with the encryption algorithm approaching the maturity level. Second, a technique to incorporate PHY protection technologies with spectrum spectral loss should be defined by security experts. Third, AI calculations are prepared with the capacity to find organize vulnerabilities professionally and will be valuable in 6G systems to set up an early caution arrangement for security fortress.

15.7.3 Potential Health Issues

In fast communication networks, the huge data rate of the order of THz band and all the bandwidth can also be harnessed. However, the forecast of the experts says that the use and implementation of 6G networks are still elementary and immature. However, 6G could very well get benefitted from the extremely high bands of frequencies. But here, it is relevantly more important to consider human protection, which is certainly going to be compromised by the transmission of the waves of THz order.

In particular, THz radiations coming out of high frequency waves are non-ionizing in nature since the energy of the photon is inadequate. To put it in other way, the energy level possessed by the non-ionizing photon is less than that of an ionizing photon, up to three times [36,37]. The Federal Communications Commission (FCC) [38] along with the International Commission on Guidelines for the Safety of Non-Ionizing Radiation [39] are in place to protect humans from possible exposures to certain parts of the body, specially the eyes and skin tissues that are particularly susceptible to the normal heat, which is caused due to low blood flow levels. Furthermore, careful consideration is needed to observe the molecular and biological effects of these radiations, which is of the order of terrahertz, on the surrounding environment. Another new idea that will be presented in 6G to alleviate health concerns is electromotive force transmission [40].

15.7.4 Fundamentals of 3D Network Latency and Reliability

It is believed that the 6G will bolster 3G applications and execution. 3D base stations will appear for illustration. It is imperative to explore estimation and data-driven displaying of the engendering environment. Organize arranging and 3D recurrence utilization ought to be characterized where, due to the modern elevation measurement and degrees of flexibility, 3D systems are considerably distinctive from conventional 2D systems. Therefore, the basic 3D results of 6G communication systems deals with the rate-reliability-latency tradeoffs. This type of research should determine the demands for spectrum resources and communication and also that 6G must be able to support the driving applications. Current studies in [41,] provide the necessary basic knowledge in this path. Similarly, the Poisson point method carries a significant advantage in the field of mathematical tractability and further is used in the deployment of network modeling and its probability of coverage. However, there is a random distribution of mathematical objects in space. Multi-tier heterogeneous systems include random deliveries in actual applications. Networks having smaller cells generally have user-centric hotspot areas that indicate attraction, while repulsion is observed to display rural and urban deployments at the macro cell level. In very small cell conditions, this distinction would be more emphatic. However, the trusted PPP will not be present in the 6G networking environment since the latter is generally restricted to 2D plane. Emerging implementation scenarios will be performed in 3D in the future. In any case, since the blockages and the exceedingly directional bar designs, such a situation is complicated within the THz recurrence band.

15.8 CONCLUSION

Today's age is the finest one to think and aspire beyond 5G and make some provisions for a quantum jump in terms of advancements in the following decade, which is anticipated to head the vision of technological revolution by the coming 2030, when 5G communications would cease to be used. We tended to the requirement for 6G communication innovation in this chapter and imagined the capabilities it holds. In expansion, we combined 6G communication with counterfeit insights to assess distinctive applications which are going to alter the approach of people communicating with one another using communication devices. We, in this chapter, have addressed two future developments, namely self-driving cars and holographic communications, which certainly will be among the significant applications that would have been derived from the evolution and beginning of smart 6G networks. In terms of networks and communications, this latest wave of communication technologies would lead to a boom in technological growth. It is predicted that 6G will go along with the industry 4.0 or the fourth industrial revolution, which undoubtedly will change the idea of life. 6G technology, apart from the launch of a group of new services, enables very high bit rates, which is expected around 1 Tbps and a very low latency, which is expected to be less than 1 ms. This research began by keeping a vision in mind and some fundamental characteristics with a goal of promoting future 6G network, especially in the following dimensions: Intelligence; energy efficiency; spectral efficiency; protection, privacy and confidentiality; customization and affordability. We also presented many issues and problems presently linked with 6G network technology and the possible potential strategies and approaches for encouraging the future of 6G. At last, this work concludes with universal inquire about exercises pointed at building the vision for 6G with regard to its future.

REFERENCES

[1]. S. Aggarwal and N. Kumar, "Fog computing for 5g-enabled tactile internet: research Issues, challenges, and future research directions", *Mob. Netw. Appl.*, pp. 1–28, 2019. doi:10.1007/s11036-019-01430-4

[2]. M. H. Alsharif, A. H. Kelechi, K. Yahya, and S. A., Chaudhry, "Machine learning algorithms for smart data analysis in internet of things environment: Taxonomies and research trends",*Symmetry*, vol. 12, p. 88, 2020.

[3]. Arockia Panimalar, S., Monica, J., Amala, S., and Chinmaya, V., "6G Technology", *IJERT*, Sep. 2017.

[4]. M. Bennis, M. Debbah, and H. V. Poor, "Ultrareliable and low-latency wireless communication: Tail, risk, and scale", *Proc. IEEE*, vol. 106, pp. 1834–1853, 2018.

[5]. L. Chiaraviglio, A. S. Cacciapuoti, G. di Martino, M. Fiore, M. Montesano, D. Trucchi, and N. B. Melazzi, "Planning 5G networks under EMF constraints: State of the art and vision", *IEEE Access*, vol. 6, pp. 51021–51037, 2018.

[6]. C.H. Liu, Z. Chen, J. Tang, J. Xu, and C. Piao, "Energy-efficient UAV control for effective and fair communication coverage: A deep reinforcement learning approach," *IEEE J. Sel. Areas Commun.*, vol. 36, no. 9, pp. 2059–2070, Sep. 2018.

[7]. S. Dang, O. Amin, B. Shihada, and M.-S. Alouini, "What should 6G be?", *Nat. Electron.*, vol. 3, pp. 20–29, 2020.

[8]. M. Z., Chowdhury, M., Shahjalal, M., Hasan, and Y. M., Jang, "The role of optical wireless communication technologies in 5G/6G and IoT solutions: Prospects, directions, and challenges", *Appl. Sci*, vol. 9, p. 4367, 2019.

[9]. C. Cho, M. Maloy, S. M. Devlin, O. Aras, H. Castro-Malaspina, L. T. Dauer, A. A. Jakubowski, R. J. O'Reilly, E. B. Papadopoulos, M. -A. Perales, et al., "Characterizing ionizing radiation exposure after T-cell depleted allogeneic hematopoietic cell transplantation", *Biol. Blood Marrow Transplant*, vol. 24, pp. 252–253, 2018.

[10]. F. Tang, Y. Kawamoto, and N. Kato, *"Future intelligent and secure vehicular network toward 6G: Machine-learning approaches"*, invited paper, IEEE, 2019.

[11]. H. Gacanin, "Autonomous wireless systems with artificial intelligence", *IEEE Veh. Technol. Mag.*, Sept. 2019.

[12]. E. C. Strinati, S. Barbarossa, J. L. Gonzalez-Jimenez, D. Kténas, N. Cassiau, L. Maret, and C. Dehos, *"6G: The Next Frontier from Holographic Messaging to Artificial Intelligence using Subterahertz and Visible Light Communication"*, IEEE, August 2019.

[13]. F. Boccardi, R. W. Heath, A. Lozano, T. L. Marzetta, and P. Popovski, "Five Disruptive technology directions for 5G," *IEEE Commun. Mag.*, vol. 52, no. 2, pp. pp. 74–80, Feb. 2014.

[14]. H. Yang, A. Alphones, Z. Xiong, D. Niyato, J. Zhao, and K. Wu, "Artificial Intelligence-enabled intelligent 6G networks". *IEEE Netw.*, vol. 34, no. 6, pp. 272–280.

[15]. O. Holland, E. Steinbach, R. Prasad, Q. Liu, Z, Dawy, and A. Aijaz, "The IEEE 1918.1 "Tactile Internet" standards working group and its standards",*Proc. IEEE*, vol. 107, pp. 256–279, 2019.

[16]. I. Tomkos, E. Pikasis, D. Klonidis, and S. Theodoridis, *"Toward the 6G Network Era: Opportunities and Challenges"*, IEEE, 2020.

[17]. J. Andrews *et al.*, "What will 5G be?" *IEEE J. Sel. Areas Commun.*, vol. 32, pp. 1065–1082, Jun. 2014.

[18]. A. T. Z. Kasgari and W. Saad, "Model-free ultra-reliable low latency communication (URLLC): A deep reinforcement learning framework", *Proc. 2019 IEEE Int. Conf. Commun. (ICC)*, Shanghai, China, May 2019, 20–24; pp. 1–6.

[19]. K. B. Letaief, W. Chen, Y. Shi, J. Zhang, and Y. A. Zhang, "The roadmap to 6G: AI empowered wireless networks," *IEEE Commun. Mag.*, vol. 57, no. 8, pp. 84–90, Aug. 2019.

[20]. K. David and H. Berndt, "6G vision and requirement," *IEEE Vehic. Teh. Mag.*, vol. 13, no. 3, pp. 72–80, Sept. 2018.

[21]. K.G. Kibria, K. Nguyen, G.P. Villardi, O. Zhao, K. Ishizu, and F. Kojima, "Big data analytics, machine learning, and artificial intelligence in next-generation wireless networks," *IEEE Access*, vol. 6, pp. 32328–32338, May 2018.

[22]. K. B. Letaief, W. Chen, Y. Shi, J. Zhang, and Y.-J. Angela Zhang, *"The Roadmap to 6G: AI Empowered Wireless Networks"*, IEEE, August 2019.

[23]. M. Khalid, O. Amin, S. Ahmed, B. Shihada, and M.-S., Alouini, "Communication through breath: Aerosol transmission", *IEEE Commun. Mag.* vol. 57, pp. 33–39, 2019.

[24]. T. Kleine-Ostmann, "Health and safety related aspects regarding the operation of THz emitters", *Towards Terahertz Communications Workshop*. European Commission, 2018. Available online: https://ec.europa.eu/digital-single-market/events/cf/towards-terahertz-communications-workshop/item-display.cfm?id=21219 (accessed on 28 March 2020).

[25]. L. Zhang, Y.-C. Liang, and D. Niyato, *"6G visions: Mobile ultrabroadband, super internet of-things, and artificial intelligence"*, IEEE, May 2019.

[26]. K. B. Letaief, W. Chen, Y. Shi, J.Zhang, and Y. -J. A., Zhang, "The roadmap to 6G: AI empowered wireless networks", *IEEE Commun. Mag.* vol. 57, pp. 84–90, 2019.

[27]. M. Giordani and M. Mezzavilla, "Towards 6G Networks: Use Cases and Technologies", *IEEE Commun. Mag.*, vol. 58, no. 3, Mar. 2019.

[28]. Md. A. Rahman, "Network Intelligentizing for Future 6G Wireless Networks How AI will Enable Network Intelligentizing?", Vision for Future Communications Summit, Lisbon, Nov. 2019.

[29]. Md. J. Piran and D. Y. Su, "Learning-driven wireless communications, towards 6G", *2019 Int. Conf. Comp. Elect. Commun. Engineering (iCCECE)*, Aug.2019.

[30]. M. Yao, M. Sohul, V. Marojevic, and J.H. Reed, "Artificial intelligence defined 5G radio access networks", *IEEE Commun. Mag.*, vol. 57, no. 3, pp. 140–147, Mar. 2019.

[31]. N. Kato *et al.*, "Optimizing space-air-ground integrated networks by artificial intelligence", *IEEE Wireless Commun.*, vol. 26, no. 4, pp. 140–147, Aug. 2019.

[32]. P. Yang, Y. Xiao, M. Xiao, and S. Li, "6G wireless communications: Vision and potential techniques", *IEEE Network*, vol. 33, no. 4, pp. 70–75, Jul. 2019.

[33]. T. S. Rappaport, Y. Xing, O. Kanhere, S. Ju, A. Madanayake, S. Mandal, A. Alkhateeb, and G.C. Trichopoulos, "Wireless communications and applications above 100 GHz: Opportunities and challenges for 6G", *IEEE Access*, Jan. 2019.

[34]. R. Khutey, G. Rana, V. Dewangan, A.l Tiwari, and A. Dewamngan, "Future of Wireless Technology 6G & 7G", June 2015.

[35]. S.H. Abbas and Nazish Siddiqui, "A selective Reading on Future Generation of 5G Wireless Mobile Network Framework", *Compliance Eng. J*, vol. 10, no. 12, pp, 621–631, 2019.

[36]. S. J. Nawaz, S. K. Sharma, S. Wyne, M. N. Patwary, and Md. Asaduzzaman, *"Quantum Machine Learning for 6G Communication Networks: State-of-the-Art and Vision for the Future"*, 2019, IEEE

[37]. S. J. Russell and P. Norvig, *"Artificial Intelligence - A Modern Approach"*, 2010. Prentice Hall.

[38]. W. Saad, M. Bennis, and M. Chen, "A Vision of 6G Wireless Systems: Applications, Trends, Technologies, and Open Research Problems", IEEE accepted for publishing.

[39]. T. Wu, T. S. Rappaport, and C. M. Collins, "The human body and millimeter-wave wireless communication systems: Interactions and implications",*Proc. 2015 IEEE Int. Conf. Commun. (ICC)*, London, UK, June 2015, vol. 8–12, pp. 2423–2429.

[40]. A. Yastrebova, R. Kirichek, Y. Koucheryavy, A. Borodin, and A. Koucheryavy, "Future networks 2030: Architecture & requirements", *Proc. 2018 10th Int. Congress on Ultra-Modern Telecommunications and Control Systems and Workshops (ICUMT)*, Moscow, Russia, November 2018, 5–9, pp. 1–8

[41]. Z. Zhang *et al.*, "6G Wireless Networks: Vision, Requirements, Architecture, and Key Technologies", *IEEE Vehic. Teh. Mag.*, vol. 14, no. 3, pp. 28–41, Sept. 2019.

Layered Architecture and Issues in 6G

N. Krishna Chaitanya[1], N.V. Lalitha[2], Gulivindala Suresh[2], and Mangesh M. Ghonge[3]

[1] Professor in Electronics and Communication Engineering, RSR Engineering College, Kavali, Nellore District, Andhra Pradesh, India
[2] Assistant Professor in Electronics and Communication Engineering, GMR Institute of Technology, Rajam – 532 127, Andhra Pradesh, India
[3] Assistant Professor in CSE, Sandip Institute of Technology and Research Center, Nashik, India

INTRODUCTION

Internet seems to be a very trending name irrespective of gender, age, country and qualifications. Most of the people around the world are using the internet for better user services. It has been seen that there is a tremendous change in the wireless networks from the second generation to fifth generation. Right from the basic services of voice calling to most advanced services like video calling facilities brought the attention of so many users towards the usage of wireless networks. Day-by-day the communication industry is growing at a rapid rate, especially in terms of wireless networks.

The future is expected to be based on 6G wireless networks. Let us discuss about the wireless generations [1,2] that have evolved over a period for various application service support. These generations have various standards with different capacities. The wireless network that was introduced in the early 1980s was the first generation, which supported only voice services. This first generation network gives data rates up to 2.5 Kbps. The modulation technique that has been used in this first generation is analog modulation. The major problem in the first generation wireless network is a handoff.

Second generation wireless system was based on digital modulation techniques. CDMA and TDMA are popularly used with data rates of up to 64 Kbps. This would support both voice and messaging service. The quality of voice service was better in this generation as compared to that with the first generation network.

The third generation wireless network was introduced aiming high-speed data transmission with a data rate of 2 Mbps. Third generation offers services like web browsing, video services, live TV, GPS and maps.

Thereafter, the fourth generation wireless network was introduced in the early 2000. This supports incomparable speeds over all other wireless generations such as 500 Mbps. It has high spectral efficiency

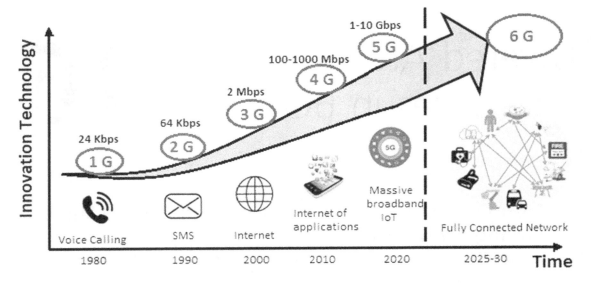

FIGURE 16.1 Change in wireless technologies over the time.

and reduced latency. This supports advanced services like video chats and high definition TVs. This generation provides better terminal mobility. It uses advanced technologies such as Multi Input Multi Output (MIMO) and Orthogonal Frequency Division Multiplexing (OFDM).

The next generation is the fifth generation wireless network, which was introduced in the year 2020 and it is not adopted worldwide so far. It has been designed to support speeds up to 10 Gbps and it uses a microwave band to deliver services. It not only supports high data rates, but also has network reliability, capable of connecting a large number of devices.

The evolution of various technologies with respect to time and the services is shown in Figure 16.1.

Now, researchers are aiming to develop the sixth generation wireless network. This is expected to combine all the various services under single network in a seamless way. The expected traffic increase rates from 2020 to 2030 as per International Telecommunication Union (ITU) are shown in Figure 16.2.

As per ITU the predicted traffic over a month in 2030 is 5016 Exabytes, which is a very high data rate. This means in the near future, wireless networks are expected to make rapid changes in the communication technology and provide best services to the customers.

At present, the 5G network is not so popularly used across nations. The 5G network finds a large number of applications, but developments towards 6G technologies are attracting industries as well as academia. In case of 5G, there are few difficulties, such as performance of communication systems, information speed and another major concerns like the security. These issues can be resolved using the next-generation systems such as 6G. As it is emerging as a new wireless technology, it is capable of supporting the older wireless generations to provide seamless communication between systems. All the previous wireless generations were to fulfill the requirements of customers as well as service providers.

5G WIRELESS NETWORKS

Although the technological evolution is taking place, still, customers are using 4G wireless network whereas the 5G wireless network is used only in few countries. 5G wireless network comprises three bands. Low band corresponds to the frequency range of 600–850 MHz and it provides a speed of

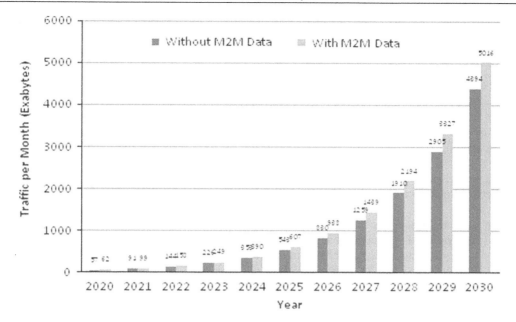

FIGURE 16.2 Predicted mobile data change from 2020 to 2030 as per ITU.

250 Mbps. Mid band uses frequency range of 2.5–3.7 GHz and it provides a speed of up to 900 Mbps. High band uses frequency a range of 25–39 GHz and it provides a speed of up to 1 GHz. It supports the basic smart services in smart homes, smart buildings, smart cities, augmented reality, video services, blockchain, smart grid, autonomous vehicles and high-speed internet.

Some of the important aspects of 5G services are mentioned below:

- Due to interoperability among the operators, it is possible for better revenue for operators
- Possible support for multi-carrier systems and at the same time supports advanced data modulation and coding schemes
- Millimeter wave frequencies are useful for wireless access
- Expected low battery consumption
- High data rates and more coverage area
- Data transfer via multiple paths
- Provides better security than 3G wireless networks
- High spectral efficiency
- Systems that uses smart beam antennas
- Applications are designed with the help of Artificial Intelligence (AI)

The advantages of 5G networks are:

- Capable of faster uploading and downloading of files
- Easy to play online games
- The 5G devices are compatible with 3G and 4G devices
- Provides better Quality of Service (QoS)
- Supports more user services
- Even capable of providing new user services
- 5G supports for older versions of wireless networks
- Capable of connecting heterogenous networks in a seamless way
- User data can be stored and accessed through cloud

BASIC ARCHITECTURE OF 5G NETWORK

The basic architecture of 5G network is shown in Figure 16.3. It clearly shows that it is able to connect with the GPRS/EDGE, 3G, WLAN and LTE technologies. It supports streaming server, data server and the server for real-time communication that provides high-speed data communication. The underlying protocols provide an efficient way of communicating with wireless devices.

The technologies supported by the 5G network [3] are classified into two types:

1. Characteristics of the air interface
2. Network architecture

In case of air interface characteristics that discuss about MIMO support the incorporation of more antennas, thereby providing increased spectral efficiency as well as improved energy efficiency. The air interface supports full duplex mode of operation. But the implementation of full duplex mode of operation is difficult due to self interference. When full duplex mode of operation is used, the complex interference situation, role of analog and digital circuits and how full duplex is combined with a MIMO is a typical scenario.

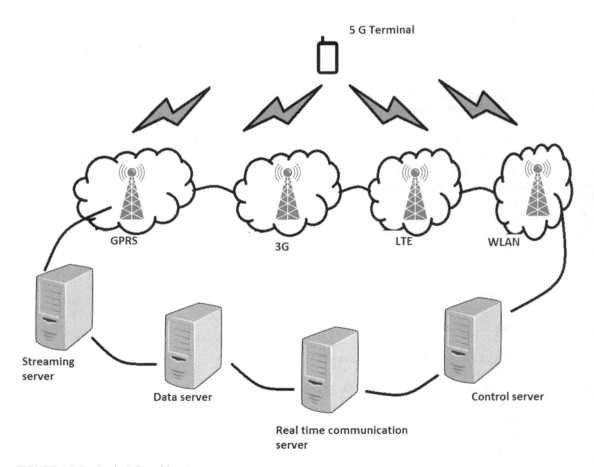

FIGURE 16.3 Basic 5G architecture.

In terms of network architecture, it is very important how the architecture has been designed for effective communication. The role of small cell network architecture is very crucial for the future data growth. With the help of small cell network it is possible to improve the capacity of the system, and also minimise the energy consumption. At the same time, if the nodes in the small cell are in close proximity to each other, then the network topology becomes complicated. Here, device-to-device (d2d) communication provides better data transmission over the network. This improves the spectral efficiency. In d2d, apart from the benefits, there are some challenges. The first is the inter-cell interference due to the frequency reuse. The second is the peer discovery, which is used to identify the neighbors. This causes overload in the network.

6G WIRELESS NETWORKS

The evolutional technology that is going to hit the communication market is 6G. It is expected to provide incomparable services over 5G. The 6G network is said to connect things with intelligence. These 6G network is expected to be the future network featuring high-throughput, low latency and efficient connectivity. This advanced wireless network has to support various networks, capability in connecting numerous devices, direct d2d communication, high-speed data transfer with less delay and high delivery rates. The sixth generation is anticipated to be built based on an intelligent system, where the decisions are taken in the network itself rather than taking them manually.

Figure 16.4 shows various types of networks that are going to support communication along with the technologies expected to be used with it. The features [2] that are expected in 6G are shown below.

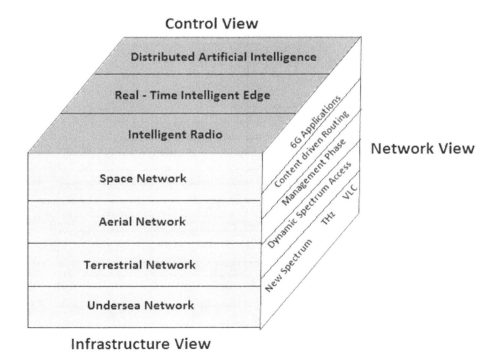

FIGURE 16.4 Architecture of 6G networks.

6G network is finding a number of applications as mentioned in Figure 16.5. The feature wireless communication that is expected to dominate the communication world is definitely 6G because of its applications, speed and security. 6G has been compared with 5G as shown in Table 16.1 [4–6].

From Table 16.1 it is very clear that 6G has more advantages than 5G. 6G is going to support up to 1 Tbps, which means it is possible to download an HD movie within a second. This is going to be an unbelievable communication technology for customers who are dependent on internet. The utilisation of the spectrum is good in 6G as compared with 5G. As it is supporting for higher data rates, the latency is also less in 6G. Though the user is moving with a speed of 1000 kmph still 6G network is capable of

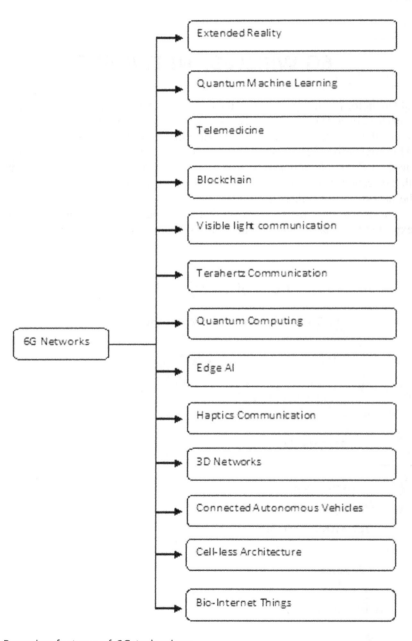

FIGURE 16.5 Powering features of 6G technology.

TABLE 16.1 Comparison of 5G and 6G

PARAMETER	5G	6G
Peak data rate	10–20 Gbps	More than 1 Tbps
Spectrum efficiency	3 to 5 times over 4G	More than 5 times over 5G
Latency	In msec	Less than 1 msec
Mobility	350 kmph	More than 1000 kmph
Traffic density	10 Tbps/km^2	More than 100 Tbps/km^2
Processing delay	100 nsec	10 nsec
Energy efficiency	1000 times over 4G	10 times over 5G
End-to-End reliability	99.999%	99.99999%

providing service. It supports more number of users in the same area. 6G network is highly energy-efficient and a more reliable future network.

Various technologies that are used in the 6G wireless networks are spectrum, physical layer technologies, network architecture and intelligence in the network. The key technologies in the spectrum are terahertz and Visible Light Communication (VLC). Terahertz provides more bandwidth with the help of small size antennas with focused beams. The challenges associated with this technology are complexity in circuit design and high propagation loss. This technology is mainly used in the industries. The next technology is the VLC which is also used because of its low interference and affordable cost. It does not require any license for use. But the problem with the technology is the coverage area, which is restricted. This technology is mainly used in e-health services. In physical layer technologies, various technologies are available such as full duplex, out-of-band channel estimation, sensing and localisation. The full duplex is preferred where continuous transmission and reception of data is required. But the problem in this technology is interference and complexity in scheduling data transfer. This is also preferred in the industries. Out-of-band channel estimation technology has flexible multi-spectrum communication. The problem is lack of reliable frequency mapping. The last technology is sensing and localisation and is capable of providing novel services. This technology requires better multiplexing technique. The technology is mainly used in e-health services and unmanned mobility vehicles.

The technologies used in the network architectures are multi-connectivity and cell-less architecture, 3D network architecture. In the first technology, it provides better mobility and is capable of interconnecting different types of networks. But the problem is that it requires strict scheduling as well as a new network design. In case of 3D network architecture, it provides services in a seamless way. The main challenge in this technology is that proper modeling and topology optimisation is required.

The technologies used in the network to provide intelligence are knowledge sharing and user-centric network architecture. In case of knowledge sharing, the network is capable of adjusting by itself due to intelligence incorporated into the network. But the challenge in this technology is that it requires novel sharing mechanism. In case of user-centric, intelligence is incorporated to the end points in the network. This technology consumes more power. Most of the technologies are used for pervasive connectivity and e-health.

To provide the services, 6G architecture consists of intelligent radio, subnetwork evolution and network intelligence technologies. The decisions are taken on the type of supervisory learning algorithms that are used, communication takes place with protocol, hardware dependent coding, it also looks after the radio condition, traffic characteristics, resource availability and modulation techniques in use.

This 6G network is highly intelligent when compared over 5G networks. Moreover AI plays a crucial role in the sixth generation wireless network. It is important to incorporate intelligence in future networks as it influences advanced and sophisticated wireless communications as well as mobile computing technologies. It enables use of AI-based applications in various supporting devices with restricted computational capability and energy resources. Thereby, end-to-end automated network architecture is realised.

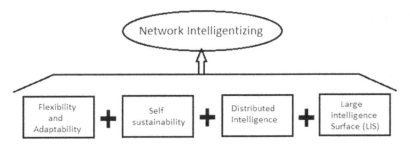

FIGURE 16.6 Intelligentising of network.

Basic network intelligentising in 6G networks is shown in Figure 16.6. Intelligence is expected to cover flexibility and adaptability, self sustainability, distributed intelligence, large intelligence surface. The smart connectivity of 6G network is expected to cover:

- Remote radio heads (RRH)
- Visible light communications (VLC)
- Mounted Network equipments over moving objects, e.g., autonomous vehicles
- Drones
- Base stations(BSs)

The basic features in network architecture of 6G are:

- Ultra-high frequencies in cell-free smart surfaces
- Network in a spray, i.e., air-duct/water-duct
- Using cars as fog/edge devices
- Water duct communications

The 6G network [7–9] was enabled with the help of AI and supports seamless connectivity. It also guarantees Quality of Service (QoS), which is majorly required where more number of devices are connected and are generating large amount of data. Use of AI techniques gives more ability for analysis, learning, optimising and intelligent recognition. It is possible to take intelligent decisions with AI support, updating network environment and organisation of network structure. The basic 6G network with AI [10] is shown in Figure 16.7.

This structure consists of four layers:

- Intelligent sensing layer
- Data mining and analytics layer
- Intelligent control layer
- Smart application layer

Intelligent sensing layer is capable of collecting the data from various sensing devices. It is capable of sensing environmental changes, intrusion detection, identification of frequency utilisation and interference detection.

Data mining and analytics layer is capable of analysing large amount of data received from various sensors. The data is analysed for knowledge discovery. It is also responsible for feature extraction from data, such as images and videos. It also detects if there is any abnormal data received from the network and it automatically filters the data.

Intelligent control layer is responsible for network functionality such as routing, congestion, power control, etc. It also analyses various parameters that are related to network environment. Task scheduling is also performed by this layer for controlling the access of various users through the channel.

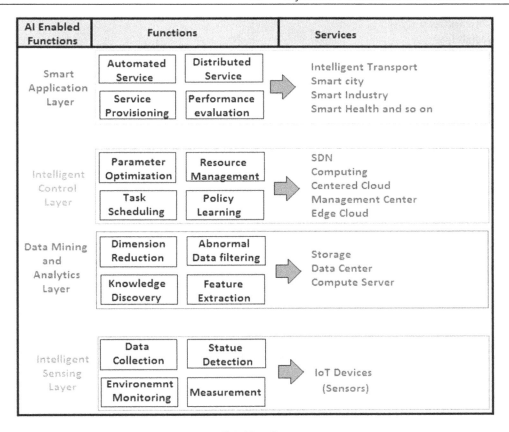

FIGURE 16.7 6G layered architecture with artificial intelligence.

Smart application layer supports various services to the user end.

Figure 16.7 shows the complete functions that are performed by the layers along with their services. *Intelligent sensing layer* basically is responsible for collecting data from the sensors, it also measures the required parameters and monitors the environment. Complete data processing takes place at this layer. In the 6G network, most of the devices are expected to be connected with the internet, where they are able to share real-time sensor data with the server for further processing. In *data mining and analytics layer*, data size is reduced by omitting the unnecessary data. Here, feature extraction of the data is done for further processing, which means it also reduces the data size. *Intelligent control layer* is responsible for resource management like channel allocation and data rates. Task scheduling will also be performed by this layer for scheduling the various tasks performed in the network. The number of parameters that are required for effective communication is also performed here. The *smart application layer* is responsible for automated services by the network, distributing the services, performance evaluation of the network. This layer is mainly supporting the applications of the networks for better communication.

ISSUES IN 6G

Although 6G finds a number of applications, at the same time, there are few issues associated with it. The major issues [11–14] are:

- Network security
- Scalability
- Network coverage
- Network component costs
- Effectiveness of AI algorithms
- Database management
- Protocol integration
- Heterogenous devices mapping
- Computational cost
- Over expectations from 5G

NETWORK SECURITY

The first issue is the lack of network security, which is a very difficult scenario to protect the data from unauthorised access. Network security is the major concern in any wireless networks.

SCALABILITY

Scalability is where the network is able to adopt changes in the future networks. The network bandwidth can be easily modified when the network is a scalable network.

NETWORK COVERAGE AREA

It is also very important to have more coverage area which is also a primary issue in 6G networks. It is expected to have more network coverage area in the advanced network environment.

NETWORK COMPONENT COSTS

To support various services, the devices that are used in the network architecture are costlier when compared to 5G. The reason is that, 6G networks are going to handle high bandwidths and data rates.

ARTIFICIAL INTELLIGENCE

The networks are capable of adapting to the network changes with the help of AI algorithms and are expected to handle large network scenarios in a seamless way.

DATABASE MANAGEMENT

Managing large data base is also very difficult and special care has to be taken for data management.

PROTOCOL INTEGRATION

Whenever the wireless network supports various devices used for various services, it is also very important for protocol integration. All the devices have to support heterogenous protocols for efficient data transfer through the network.

COMPUTATIONAL COST AND OVER EXPECTATION FROM 5G

The operation of the network is based on the advanced AI algorithms that are used. Computational cost of 6G network is high as compared to the existing network. Most of the customers and service providers have huge amount of expectations from 6G. Still the technology has to be tested for its efficiency. If these issues are resolved, then 6G is anticipated to be the revolutionary wireless network for efficient future communication services.

CONCLUSION

This chapter discussed about the 6G wireless communication which is expected to change the communication environment. This technology will dominate the communication world with its high-speed data transfer, security and support for heterogenous devices that are use to connect the network. 6G gives a lot of scope for the industry and academia to work together to enhance the communication services. In this chapter, the basic 6G layered network and the corresponding challenges associated are discussed.

REFERENCES

[1]. K. B. Letaief, et al., "The roadmap to 6G: AI empowered wireless networks", *IEEE Commun. Mag.*, vol. 57, no. 8, pp. 84–90, 2019.
[2]. T. Huang, et al., "A survey on green 6G network: Architecture and technologies", *IEEE Access*, vol. 7, pp. 175758–175768, 2019.
[3]. Y. Wang, J. Xu, and L. Jiang, "Challenges of system-level simulations and performance evaluation for 5G wireless networks", *IEEE Access*, vol. 2, pp. 1553–1561, 2020.
[4]. L. U. Khan, et al., "6G wireless systems: A vision, architectural elements, and future directions", *IEEE Access*, vol. 8, pp. 147029–147044, 2020.

[5]. M. H. Alsharif, et al., "Sixth generation (6G) wireless networks: Vision, research activities, challenges and potential solutions", *Symmetry*, vol. 12, no. 4, p. 676, 2020.

[6]. T. Nakamura, "5G Evolution and 6G", *2020 IEEE Symp. VLSI Technol.*, 2020. IEEE.

[7]. M. Z. Chowdhury, et al., "6G wireless communication systems: Applications, requirements, technologies, challenges, and research directions", *IEEE Open J. Commun. Society*, vol. 1, pp. 957–975, 2020.

[8]. X. You, et al., "Towards 6G wireless communication networks: Vision, enabling technologies, and new paradigm shifts", *Sci. China Inf. Sci.*, vol. 64, no. 1, pp. 1–74, 2021.

[9]. M. Giordani, et al., "Toward 6G networks: Use cases and technologies", *IEEE Commun. Mag.*, vol. 58, no. 3, pp. 55–61, 2020.

[10]. H. Yang, et al., "Artificial-intelligence-enabled intelligent 6G networks", *IEEE Netw.*, vol. 34, no. 6, pp. 272–280, 2020.

[11]. M. Wang, et al., "Security and privacy in 6G networks: New areas and new challenges", *Digital Commun. Netw.*, vol. 6, no. 3, pp. 281–291, 2020.

[12]. I. Tomkos, et al., "Toward the 6G network era: Opportunities and challenges", *IT Professional*, vol. 22, no. 1, pp. 34–38, 2020.

[13]. J. Park, et al., "Wireless network intelligence at the edge", *Proc. IEEE*, vol. 107, no. 11, pp. 2204–2239, 2019.

[14]. S. Dang, et al., "What should 6G be?", *Nat. Electron.*, vol. 3, no. 1, pp. 20–29, 2020.

Artificial Intelligence Techniques for 6G

17

T. Sathis Kumar[1] and N. Kavitha[2]
[1] Assistant Professor, Department of Computer Science and Engineering Technology, Indra Ganesan College of Engineering, Tiruchirappalli, TamilNadu, India
[2] Professor, Department of Information, Saranathan College of Engineering, Tiruchirappalli, TamilNadu, India

17.1 INTRODUCTION

Correspondence innovation has witnessed rapid changes lately and has had a huge impact on communication between people, how they interact with each other as well as on the overall climate. The recent correspondence innovation, i.e., 5G or the fifth epoch, may be a term that refers to the diverse correspondence innovation which is now active in several areas across the world and can in time interface the entire globe. Since 5G has been in use, it also created several challenges. Analysts have just begun their work on the leading edge in correspondence, i.e., 6G. It is been seen that the ages in portable correspondence innovation were sent and view the business sunlight than like clockwork. In the launch of 5G in 2020 and full deployment, the centre is now steadily transitioning to the 6th century. The 6G is required to supplant 5G incompletely or totally constantly 2030. Unique in reference to past age organizations, 6G organizations are going to be needed to vary themselves by acknowledging insight to satisfy more tough necessities and requests for the keen data to the world of 2030, it incorporates huge information data, a pinnacle information pace of in any event 1 Tb/s and a client-based information pace of 2 Gb/s, ultralow idleness, under 2 ms start to end delay, even 20–150 μs, maximum dependability, around 2–20-8, high energy productivity [1], on the request for 2 pJ/b, extremely top versatility, up to 2000 km/h, enormous association, maximum 107 gadgets/km2 and traffic limit of up to 1 Gbs/m2, huge recurrence groups, associated knowledge of AI specification [2].

17.2 WHAT EXACTLY IS 6G?

Before this chapter mentions the capacities, needs and picture 6G regarding different ideas, this chapter like to grasp what precisely 6G methods. 6G is that the sixth era within the versatile correspondence innovation. There are past ages, for instance, 3G, 4G and 5G which have their own algorithms abilities and constraints were conveyed for several timeframes to satisfy the present requirements. All ages has developed generally over like clockwork and 6G is relied upon to be conveyed by 2030. A selected meaning of 6G immediately cannot be resolved because it is an innovation still under examination. 6G are often clarified because the replacement of 5G within the correspondence innovation. 6G will considerably defeat the bounds of 5G and would have tons more points of interest to support the developing necessities to future correspondence [3]. 6G correspondence framework will have a worldwide inclusion which can be a joining of 5G organization and satellite organization frameworks [4]. It is recommended that 6G having super quick web with extremely top information rates and negligible idleness alongside an enormous organization inclusion which can tons of solid and energy proficient (Figure 17.1).

As indicated by the previous advancement rules of organizations, beginning 6G organizations are going to be mostly upheld by the present 5G frameworks, for instance, the models of SDN, NFV and organization cutting (NS). Be that because it may, contrasted and 5G organizations, 6G organizations got to help the previously mentioned tough necessities (e.g., maximum information rates, maximum inactivity, maximum dependability, consistent network). Simultaneously, the advancements in 6G organizations has huge measurement, high intricacy and dynamicity and heterogeneity attributes. All the previously mentioned issues require another engineering that is adaptable, versatile, spry and astute. AI [5], with solid learning capacity, incredible thinking capacity and shrewd acknowledgment capacity, permits the planning of 6G organizations to find out and suits help assorted administrations as needs be without human mediation. AI-empowered procedures were applied to accomplish network intelligent, shut circle improvement and astute remote correspondence for 6G organizations [6]. The profound deciding the way to streamline the exhibition of Space-Air-Ground Integrated organizations, and showed the simplest thanks to utilize profound deciding the way to choose most appropriate ways for satellite organizations. Besides, profound support learning (DRL) was received to safeguard dependable remote availability for UAVs-empowered organizations by learning the climate elements [7], and re-enactment showed that DRL essentially outflanks traditional strategies [6]. Subsequently, it is looking forward to receive AI to 6G organizations to reinforce the organization create and strengthen the organization execution.

FIGURE 17.1 Components of 6G.

17.3 THE ARCHITECTURE OF 6G NETWORKS

In 5G, the "non-radio" angle has gotten increasingly significant, and has been the critical driver behind the new endeavours on "softwarization". All the maximum explicitly, two important 5G innovations are Software-Defined Networking and Network Functions Virtualization, it is been moved current correspondences networks to programming of private organizations. It additionally empower network cutting, it may give a fantastic ability to virtualize to permit numerous virtual organizations to be made available on a collaboration actual framework (Figure 17.2).

The plan of the 6G engineering will follow an "artificial intelligence local" approach where intelligentization will permit the organization to be shrewd, lithe and prepared to find out and adjust as indicated by the changing organization elements. It will advance into an "organization of subnet works", permitting more productive and adaptable updates, and another system hooked in to insightful radio and calculation equipment detachment to adapt to the equipment capacities that are both heterogeneous and upgradeable. Because of those two highlights will abuse AI methods, as an additional delineated within the accompanying subsections.

Artificial intelligence provides two chances and difficulties to 6G organization security and client security insurance. On the one hand, different ML and DL calculations are utilized to upgrade network security, for instance, interruption discovery, irregular traffic location and pernicious client conduct identification. However, the bulk of the present investigations are centered on the common organization side. Few of the jobs are expected to improvise the radio organization authentication by utilizing ML strategies. In important conditions of 6G, for instance, self-governing vehicles, smart assembling, assaults to AI-based control frameworks may prompt obliterating results. Additionally, security dangers may occur during activity [6]. Noxious information infusion could delude AI specialists to make the

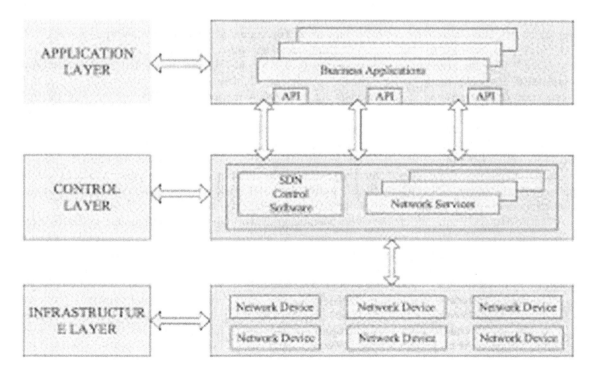

FIGURE 17.2 Different layers of 6G model.

wrong choices. As an example, a pernicious client may communicate counterfeit signs to delude the AI preparing measure so it can get to more assets of the common actual channel while different clients are denied. Additionally, information harming represents another test to the undirected IDS focused on learning and it'd be altered the knowledge utilized for distinguishing malevolent assaults. Also, as long as tons of the original datasets are expected to organize AI models, the way to proficiently distinguish the vindictively dirtied datasets is another test. Consequently, the honesty and protection of the datasets to organize the ML models, the heartiness and secrecy of the ML models need to be all during the AI-powered organization, about instruments improvement [2].

The cloudization, virtualization, softwarization and organization cutting [7] are so far significant highlights of 6G engineering. Nonetheless, the knowledge is expected to be the critical element to empower independent 6G organizations. Empowered by local AI motors, 6G framework can naturally arrange network structure and different assets including cuts, processing, storing, energy, correspondence to satisfy evolving requests. Simulated intelligence-based geography and asset the executives are critical to productively change different assets use hooked in to changing conditions and dynamic client necessities. Given the developing calculation and capacity ability of the client gadgets, AI-based capacities are often conveyed to the organization edge aside from unified knowledge. To beat the restricted calculation, stockpiling, force of the individual gadgets, utilizing the scattered registering assets between network edges and end-devices through various access edge figuring need to be thought about in 6G design. It is imagined that astute administrations in 6G will length between server farms to connect network gadgets with client gadgets [8]. Simulated intelligence put together applications running with reference to cell phones or organization edges can learn and anticipate client conduct, ecological conditions and set about as setting mindful aides to unified AI-based control frameworks. within the interim, to alleviate the concerns on information protection and security of conveyed preparing anxious gadgets, unified learning [9] are often utilized to organize information locally and gain proficiency with the worldwide model through sharing the taking in models from circulated gadgets.

The precise channel models addressing genuine correspondence conditions are essential to the exhibition of the remote correspondence framework. Albeit the present channel models can catch normal highlights of traditional remote channels, they show constraints as far as defects and non-linearities in some intricate situations. But, the expanding number of reception apparatuses in gigantic MIMO correspondence has altered the channel's properties [10], it creates precise distributed demonstrating obscure. Also, the 6G situations, for instance, atomic or submerged acoustic interchanges [11], it is exceptionally difficult to portray the channels by unbending numerical models. During this way, the time has come to research new ways to affect plan a correspondence framework without unequivocally characterized channel models. to the present end, the knowledge-driven worldview need not bother with a particular model to work out the difficulty on the grounds that the arrangement are often straightforwardly gotten from the knowledge produced by networks[12,13]. By exploiting the knowledge-based model, DL can improve the radio connection execution by utilizing enormous preparing dataset with no numerically manageable distributed model. Additionally, DL makes it conceivable to find out the distributed models and by preparing for new distributed model conditions later[12], [13].

The arising programming characterized metamaterial worldview and configurable defective wave radio wires can extraordinarily increase the smart 6G air user interface[17]. Customizable reception apparatuses can result in an enormous framework gains from powerfully adjusting pillar designs hooked in to accessible CSI information on every receiving wire state. The longer term radio wires and metamaterial are often distantly controlled through ML calculations as delicate product. Additionally, the merchandize characterized intellectual radios can furnish solid remote correspondences with proficient utilization of range re-sources through smart tasks hooked in to information gained from general conditions. ML are often normally coordinated into all activities of intellectual radios, for instance, range detecting, impedance examination and dynamic range asset the executives, power control [14]. Furthermore, ML strategies open up the possibilities of together streamlining the beginning to end usefulness the chain of particular layer. Start to put an end to learning plans to deal with the entire correspondence framework including the request, remote channel and collector with a solitary learning structure. The epic strategies make it conceivable to together improve transmitter and collectors hooked in to the beginning to end data recuperation. Improving the knowledge of radio organizations by ML to accomplish self-made administration, self-made-insurance, self-made-mending and self-streamlining is crucial for 6G organizations [15].

17.4 AI-BASED TECHNOLOGIES FOR 6G

The remarkable change of remote organizations will form 6G generously not the same as the past ages, as it will be described by a serious level of heterogeneity in different viewpoints, for example, network foundations, radio access advancements, registering and capacity assets, application models, etc. Likewise, the wide scope of new applications will order a savvy utilization of interchanges, registering, control and capacity assets from the organization edge deeply. To wrap things up, the volume and assortment of information produced in remote organizations are developing essentially.

The entire world is destined to be associated with the fast development in correspondence innovation. Man-made reasoning impacts 6G correspondences [12]. All the most recent innovations these days uphold AI. It can anticipate that the world should be an AI driven really soon in upcoming arena. The 6G have improve man-made reasoning from numerous points of view. Man-made intelligence specialists are without a doubt bound to assume a fundamental part later on. Man-made consciousness will be the primary main thrust in the portable correspondence innovation and will turn into a source to make a totally new age of uses for AI [12]. The new explores and advancement discoveries in profound learning alongside the increment in the no of gadgets particularly brilliant gadgets and age of huge information, the remote local area has demonstrated an unmistakable fascination towards man-made reasoning. The 6G correspondences are relied upon to bring a totally remote and computerized insight and would consequently require man-made reasoning to be a fundamental segment of the innovation. The 6G correspondence will change from associated things to associated insight where the last must be acknowledged with a contribution of computerized reasoning alongside its subsets, AI and profound learning [2,16,17].

The improvement of 6G organizations are going to be enormous scope, multi-layered, high intricate, dynamic and heterogeneous. What is more, 6G organizations got to help consistent network and assurance assorted QoS prerequisites of the big number of gadgets, even as interaction huge measure of data created from actual conditions. AI strategies with incredible examination capacity, learning capacity, advancing capacity and keen acknowledgment capacity, which may be utilized into 6G organizations to insightfully complete execution improvement, information disclosure, refined learning, structure association and convoluted dynamic. With the utilization of AI, this chapter introduce an AI-empowered insightful engineering for 6G organizations it is chiefly isolated into four different layers: Clever detecting layer, information mining and investigation layering, keen control layering and application layering, as demonstrated in Figure 17.3 [16].

At the purpose once this chapter plan to incorporate man-made brainpower, mechanized frameworks and 6G correspondence innovation it tends to be derived that 6G are going to be a ground breaking and upset age of the remote correspondence innovation in several angles. The longer term organizations which will be created are going to be too perplexing to possibly be able to work physically or with the obstruction. The administrators can even find yourself being danger to those remote and sophisticated organizations. To accomplish such outstanding abilities, an AI engaged and the learning-based organization innovation is predicted to affect the interchanges utilizing zero-contact [11]. Computer-based intelligence finishes up being a serious pattern towards 6G correspondences for network arranging, enhancement, investigation, disappointment recognition and asset the executives. The conjunction of 6G empowered web of all entities and AI will create a solid association and grow into a gainful collaboration of two innovations and the 6G will add to the supply information and AI will dissect and choose information resulting from it [14].

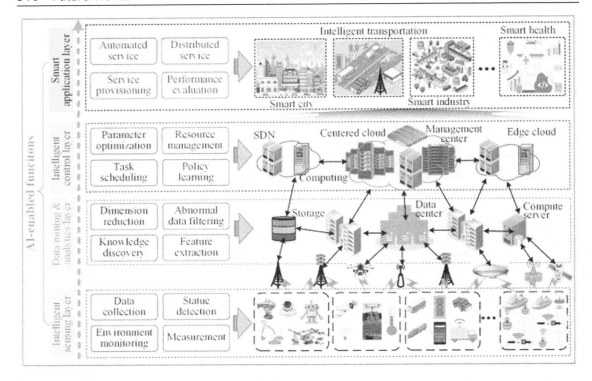

FIGURE 17.3 Architecture diagram for AI-enabled 6G technologies.

This chapter accepts that building 6G organisation hooked in to AI innovation are going to be an inescapable decision, and "Insightful" are going to be the innate component of 6G organization, specifically the alleged "Insightful availability". 6G organizations will confront numerous challenges: more amazing and massive organizations, more sorts of endpoints and organization gadgets and more amazing and various types of businesses. "Wise Connectivity" will have two meetings necessities simultaneously: from one viewpoint, all the related associated gadgets within the actual organization are insightful, and therefore the related administrations are wise; but, the intricate and colossal organization itself needs smart the executives. "Insightful Connectivity" are going to be the essential qualities supporting the opposite three significant highlights of 6G network: Deep Interconnectedness, Holographic Interconnectedness and Omnipresent Interconnectedness. This chapter anticipate that within the following 6–10 years (2030~) of 6G frameworks, access prerequisites will develop from profound inclusion to "Profound availability". Its attributes are often summed up as follows: Deep Sensing: Internet with a physical feel, Profound Learning/AI: Deep data processing, Deep Mind: Telepathy, Brain-to-Mind Communication. It alright could also be normal that in 10 years an extended time (2030 ~), the media correspondence are going to be for the foremost part planar sight and sound, higher devotion AR/VR association, even holographic data association, additionally remote holographic correspondence will become a validity. High constancy AR/VR are going to be universal, even as holographic correspondence and show can likewise be completed at whenever furthermore, anyplace, in order that individuals can appreciate completely inundated holographic intuitive involvement with any time and spot, that is, to know the correspondence vision of supposed "holographic network" [18]. "Whenever, anyplace" association necessity 10 years after the very fact (2030 ~), that is, to accomplish genuine "Omnipresent availability", a tremendous" world will become increasingly available. "Shrewd Connectivity" is the cerebrum of the 6G network at the moment, while the other characteristics of "Profound Network", "Holographic Connectivity" and "Omnipresent Network" make up the 6G network's storage compartment. These four attributes together make the longer term 6G organization an entire natural entire with "soul". Later on, the correspondence framework are going to be additionally evolved and improved supported the present 5G. The info will get through the control of reality, the organization will close the space all things considered and the consistent mixture of human and everything are going to be acknowledged [12].

To settle the perplexing radio correspondence issues, the gap and overcome approach is usually utilized. As needs be, the present correspondence frameworks are planned with a progression of characterized blocks, for instance, channel coding and deciphering, regulation and demodulation, and so on. The radio correspondence issues are normally tackled by upgrading each capacity block freely. Albeit numerous specialists have endeavoured to enhance every particular square and accomplished certain additions by and by, the perfect presentation of the entire framework cannot be ensured [6]. The reason for this is often that the elemental issue of correspondence relies upon the fruitful message recuperation by the collector after the message is sent via the remote channel by the transmitter [5]. The interaction does not actually need a counterfeit square construction. Along these lines, the time has come to reinforce the framework execution from the beginning to end viewpoint as against improving each square freely. In such manner, the support a learning technique such as auto encoder makes it possible conceivable to comprehensively enhance the 6G radio interface comes from a start to end point of view.

To beat the weakness of joint equipment calculation plan and receive the reward of the calculation equipment partition design, this chapter presents a working framework (OS) between the gadget equipment and therefore the handset calculations, where the chapter described here will see a handset calculation as a product running on the operating system. The operating system is fit not just assessing the skills of nearby RF chains, stage shifters, ADCs and receiving wires, then forth, yet additionally estimating their simple boundaries consequently. In sight of the equipment data and AI strategies, the OS will at that time be equipped for designing its own handset calculations through an interface language. This chapter will allude to the present structure as astute radio (IR). Instead of the training based savvy PHY layer overviewed and IR may be a lot more extensive idea counting on the calculation equipment detachment design. In Table I, we expect about key highlights of IR, programming characterized radio (SDR) and intellectual radio. Inferable from Mitola's achievement works [7], IR are often viewed as an extra augmentation, during which the battlefront AI strategies are profoundly included. The regular adjustment/coding modules are supplanted by profound neural networks, which may in an astute manner suits the climate and equipment. IR additionally considers the conventions over layer 3, which can be self-upgraded to support various AI applications.

The primary common use of AI is large information analytics. There are four different sorts of examination which will be applied to 6G frameworks, to be specific clear investigation, symptomatic analytics, predictive investigation and prescriptive examination. Expressive investigation mine recorded information to urge experiences on network execution, profile of traffic, conditions of the channel, client perspectives then forth it enormously upgrades the situational familiarity with network administrators and specialist organizations. Demonstrative examination empower self-ruling discovery of organization blames and administration impedances, recognize the underlying drivers of organization typicalities and eventually enhance the unwavering quality and security of 6G remote frameworks. Prescient examination use information to foresee future occasions, for instance, traffic designs, client areas, client conduct and inclination, content fame and asset accessibility. Prescriptive investigation exploit the forecasts to recommend choice alternatives for asset assignment, network cutting and virtualization, store position, edge computing, autonomous driving, then forth as an example, by foreseeing, expecting and deriving future client requests through enormous information examination, the thought of proactive storing has as lately arose to fundamentally calm pinnacle traffic loads from the remote centre organization.

Machine learning (supervised learning, unsupervised learning and reinforcement learning), deep learning, optimization theory and game theory are all examples of AI techniques.

17.4.1 Supervised Learning

Supervised learning utilizes a bunch of elite named information to fabricate the training model (likewise called preparing), which is comprehensively isolated into grouping and relapse subfields.

Characterization examination intends to allocate an unmitigated mark to every info test, which chiefly incorporates choice trees (DT), support vector machine (SVM) and K-nearest neighbors (KNN). Relapse investigation contains support vector regression (SVR) and Gaussian process regression (DPR) calculations and it appraises or predicts nonstop qualities hooked in to the knowledge measurable highlights.

- **K-Nearest Neighbor:** KNN may be a grouping calculation hooked in to estimating the space between various element esteems. The grouping of an information test is resolved hooked in to the category of K closest neighbors. Within the event that an outsized portion of the K closest neighbors within the element space have an area with a selected class, at that time the instance is sorted into an identical classification. The calculation is not difficult to acknowledge, heartless towards exceptions and appropriate for multi class orders. In any case, it is extremely tedious when utilized for huge datasets.
- **Decision tree**: Each hub of the selection tree addresses a component of an information, each branch addresses the mixture of highlights that cause grouping and every leaf hub addresses a particular category. The selection tree is worked to spice up the info gain of each factor split, which brings a few characteristic variable positioning. The order of the unlabelled example are often accomplished by contrasting its component worth and hubs of the selection tree, which is ready by the named informational collection. The many benefits incorporate high arrangement precision, basic execution and natural articulation. Be that because it may, it experiences information incorporating all out factors with an alternate number of levels since data acquires are going to be one-sided to highlights with more levels.
- **Random forest:** An arbitrary woodland ordinarily comprises various choice trees. To alleviate the ill-fitting problems with the selection tree, the technique haphazardly chooses a subset of highlights to create every choice tree. Another informational index is arranged by every choice tree, at that time the knowledge test is assessed into a category that is settled most trees depend on.
- **Neural network**: The cranial organization are often utilized to require in experiential information from authentic information by a huge number of preparing units, which add equal. Actuation capacities, for instance, sigmoid and therefore the exaggerated digression capacities are typically applied to those units to acknowledge nonlinear calculations. A neural organization ordinarily has one info layering, one yield layering and a minimum of one covered up sanctum layers. By tuning the number of covered up layers and therefore the quantity of units in each layer, various models are often prepared to settle characterization or relapse issues. The neural organizations model are often prepared by regulated learning or unaided learning.
- **Naive Bayes:** It is an easy probabilistic group approach hooked in to Bayes hypothesis. The classifiers can viably affect a huge number of nonstop or straight out highlights that are free since it can change a high-resolution thickness assessment errand to a just has one dimension part thickness assessment task dependent with the understanding that highlights are autonomous. The gullible Bayes classifier are often utilized as a web calculation since it tends to be prepared in direct time [15].

17.4.2 Unsupervised Learning

The assignment of unsupervised learning is to seek out shrouded designs also as concentrate the helpful highlights from unlabelled information, and it is for the foremost part partitioned into bunching and measurement decrease [15]. Bunching looks to aggregate a bunch of tests into various groups as per their likenesses, and it principally incorporates K-implies bunching and progressive bunching calculations. Measurement decrease changes a high-dimensional information space into a coffee without sacrificing three-dimensional room tons of valuable data. Isometric analysis and principal component analysis (PCA) mapping (ISOMAP) are two exemplary measurement decrease calculations.

- **K-means:** It generally would not be possible to order a bunch of unlabelled information into various groups. K addresses the number of wanted bunches. The target capacity of k-implies addresses the space among information and related centroids. K-implies will generally appoint every information to a gaggle with the centroid that is closest to the knowledge. The way towards refreshing centroids hooked in to an allotted information point are going to be rehashed before neither the point nor the centroid shifts. The selection of K incredibly affects the exhibition of the calculation.
- **Self-organizing map:** SOM is usually would not be possible to acknowledge dimensionedness decrease and knowledge grouping. SOM contains one information layer and one guide layer. Each layer incorporates numerous neurons and each individual neuron features a weight vector. During the preparation interaction, SOM can assemble and redesign the guide. Not in the least like customary NNs that apply mistake revision learning, have SOMs utilized an unaided serious learning approach [83]. Within the wake of preparing, another info vector is arranged into a gaggle hooked in to the triumphant neuron on the guide. The procedures are effectively utilized in several example acknowledgment undertakings.
- **Hidden Markov model**: The methodology is often familiar with demonstrating a framework by a Markov interaction with obscure boundaries. The first test is to make a decision the concealed boundaries from known boundaries. This models are broadly utilized in haphazardly powerful situations with memory less property, that is to say the contingent likelihood dispersion in potential future states just relies upon this status. The boundaries of HMM are often prepared during a regulated or unaided manner. The HMM are often utilized to deal with non-fixed arrangements, which allows the framework to vary over the end of the day with various yield likelihood appropriations of every state.

17.4.3 Reinforcement Learning (RL)

In RL, every specialist figures out the way to plan circumstances to activities and makes appropriate choices on what the moves to form through associating with the climate, to amplify a drawn out remuneration. Exemplary RL calculations incorporate MDP, Q-learning, strategy learning, actor critic (AC), DRL and multi-armed bandit are all examples of Markov decision processes (MRB).

The fundamental thought of support learning (RL) is to mimic the training interaction of the cerebrum experimentally [6]. Rather than learning the content, planning of the preparation dataset, RL attempts to research the simplest activities during a strong interaction. The capacity to grasp the climate through activities and input settles thereon reasonable for tackling dynamic issues. The RL are often ordered into two kinds: model-driven and sans model. The model-driven RL system incorporates a specialist, a state-space diagram and an activity space diagram. By way of communicating with the climate, the specialist attempts to deal with the model of the climate and obtain conversant in the simplest activity to reinforce its drawn out remuneration, which is an aggregate limited prize and identifies with both the present prizes and potential compensations. At each progression, the specialist screens a state and makes a move from activity space, at that time it gets a fast prize demonstrating the impact of the activity, at that time the framework moves to a different state. In without model based methodologies, the specialist attempts to realize proficiency with an approach. During the state progress measure, the specialist learns the simplest strategy which may be a guide from state to the activity space to reinforce the drawn out remuneration. To make a decision the drawn out remuneration of activity during a state space, the price capacity is applied. The foremost generally utilized worth capacity is Q-work, which is employed by the Q-learning technique to realize proficiency with a table to stay up state-activity matches and partner end of the day rewards. Without model based RL is more reasonable for portable organizations due to the complexity of construction a particular dynamical model organizations [6]. Contrasted with other learning strategies, a profit of support learning is that it does not depend on an accurate numerical model of the climate. Furthermore, the methodology tends to the drawn out remunerations remembering both the prompt prizes and people for the longer term, which empowers end of the day improvement result. Instructions to plan the framework state, activity, compensation in various situations to hitch the perfect presentation is that the primary test of applying RL to the remote correspondence framework. Lately, the RL has been broadly utilized to settle the dynamic problems with remote correspondences, for instance, client booking [3], range sharing [1] and radio access innovation hand-over [7]. Be that because it may, the tactic faces a couple of difficulties in taking care of issues with an enormous state space or activity space since it is hard to point out each state-activity pair straightforwardly. Thus, RL is seldom utilized practically speaking.

17.5 HARDWARE-ALGORITHM CO-DESIGN

The ability to connect at ever-increasing the cost of data will never go anywhere. It is unavoidable to work at bands of higher and higher frequencies to achieve terabytes per second data rates. To overcome increased path loss and other propagation phenomena, antenna arrays on a large scale are needed. Which necessitate the use of a variety of components of hardware such as signal mixers, ADCs/DACs and power amplifiers. Traditional transceiver architectures are difficult to implement in the mmWave and THz bands due to the high cost and use of energy of these devices, which has an effect on the design of signal processing algorithms. Collaboration between the domains of hardware and algorithms would be needed to design such complex systems effectively, so co-design of hardware and algorithms should be encouraged. The goal is to create structures of transceivers that are both hardware and amiable algorithm. Although previous generations of cellular networks have used a hardware-algorithm co-design approach. AI-based methods can help it play a bigger role in 6G.

As far as new radio go, connectivity advancements IoT devices are becoming more popular as new technologies evolve. Inescapable, equipment imperatives will assume basic parts while preparing 6G organizations. On one side, as radio correspondence is advancing towards millimetre-wave (mm Wave) Terahertz classes, the considerable cost and force use of equipment segments will fundamentally affect the handset engineering and calculation plan [19]. Then again, IoT gadgets have restricted capacity, fuel source and registering power. Such asset compelled stages require an all-encompassing plan of correspondence, detecting and derivation. In this segment, this chapter present another plan worldview for 6G, to be specific equipment mindful interchanges and talk about three promising new plan standards.

Consider mm Wave half breed beam forming for instance, which is a practical approach for giving compelling beam forming gains. It requires few RF chains, and in this manner can altogether lessen equipment cost and force utilization. In any case, countless stage shifters are as yet required for the current equipment structure. Stage shifters at mm Wave groups are as yet costly, and subsequently their number should be diminished. As shown in Figure 17.4 [16], another equipment efficient crossbreed structure was recently proposed in [10]. It only takes a few stage shifters, each with a predetermined step. The unreasonable effort required to change the framework for different equipment settings is one of the challenges posed by equipment heterogeneity. For instance, extraordinary handset models have been proposed for mm Wave frameworks, including simple beam forming, half breed beam forming and 1-cycle computerized beam forming. The traditional methodology depends on a hand-made plan for every one of them, which is wasteful. Since these different types of handsets would be confronted with a similar actual structure, a calculation designed for one will also disclose information about the strategy as an example. Move learning is a promising method for assisting in the transmission of a concept proposal to others (Figure 17.5).

Remote organizations are becoming with different types of access points and portable terminals, becoming increasingly heterogeneous. This heterogeneity started with 4G LTE organizations, and with the implementation of cutting-edge procedures like huge MIMO, the situation will continue to grow across 5G and into 6G. This pattern will confound the correspondence convention and calculation plan.

17 • Artificial Intelligence Techniques for 6G 315

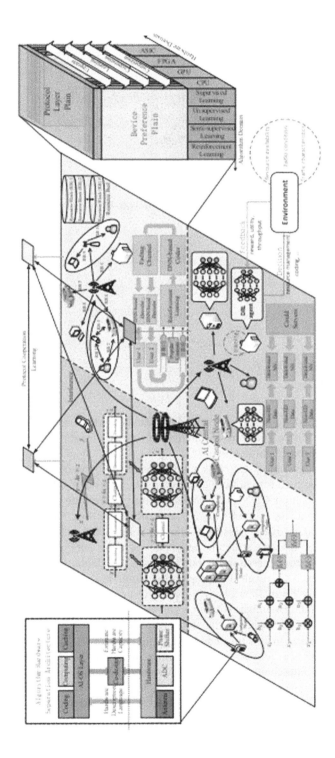

FIGURE 17.4 Architecture diagram of 6G.

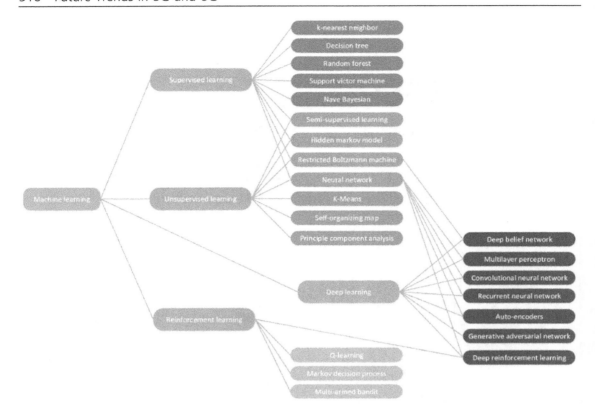

FIGURE 17.5 Different ML methods.

As of late, receiving AI methods to create correspondence frameworks has shown its adequacy, and such approaches have the capability of prompting universally useful insightful correspondences that can adjust to heterogeneous equipment limitations.

Convolutional neural organization (CNN) is a new type of profound neural organization (DNN) that has evolved from fully related feed forward organization to keep a strategic distance from rapid boundary growth. The basic concept behind CNN is to present before addressing contribution to a completely linked network, convolutional and pooling layers must be addressed. Only halfway neurons in the previous layer are connected to every neuron in the convolutional layer. To answer involve guides, these neurons are organized in a lattice structure, and neurons in the same guide share similar loads. The neurons in the element maps are gathered in the pooling layer to calculate the mean weight worth or the most extreme weight esteem. As a result, the preparation limits are greatly reduced prior to the use of fully connected networks. CNN's main strength is its ability to extract important element chains from a large amount of unlabelled data. Boundary sharing, equivariant portrayal and inadequate associations are three approaches that CNN uses to enhance the standard MLP. Because of these interesting highlights, CNN shows awesome execution in imaging handling applications.

Future radio correspondence networks are becoming more unpredictable as well as the new 6G conditions, like sub-atomic interchanges, and new developments, UM-MIMO, mm Wave and other technologies and THz correspondence, emerge. As a result, it is more difficult to accurately assess the channel's standing. Created a DNN-based model assessment method for the symmetrical recurrence OFDM (overlay frequency division multiplexing) system to achieve high-goal channel assessment. After preparing the model under various channel conditions, the DNN yield will recover the information images without the need for explicit channel identification. The methodology outperforms traditional methods in terms of recreation. CNN has also been considered for MIMO channel evaluation. By using the construction of the MIMO model. Devized an approximated, light-weight most extreme probability assessor. It is shown that CNN-based techniques outperform ordinary assessors in terms of computing difficulty. In addition, [10] advocated for a strategy based on the DNN scheme to determine the bearing of appearances (DOA) and channel evaluation issues of massive MIMO. After disconnected DNN preparation, the technique In terms of Bit Error Rate, it outperforms other plans. Furthermore, [2] suggested the intriguing idea of using DL for channel assessment of monstrous MIMO. A DNN device consisting it is to be suggested that an encoder and a decoder be used together. to achieve excellent channel evaluation execution with lower CSI overhead. The encoder uses compressive detecting to modify the vector in code words, and the decoder uses CNN and Refine Net to recover the CSI. It is assured that the methodology can execute the majority of channel assessments with minimal complexity. Furthermore, since typical iterative identification methods have an adverse effects on constant execution, many scientists use DL techniques to unfurl explicit iterative recognition calculations for channel position. The DL-based identification approach can achieve a good balance between discovery precision and computational complexity by using adaptable layer structures. Samuel et al. proposed DetNet, a DL-based identifier that uses got sign and channel network as contributions to layered construction to recreate communicated signals and upgrade the most extreme likelihood by unfurling the expected slope drop measurement. DetNet achieves comparable precision to standard measurements with significantly less time, according to recreation results. It also investigated the use of DL techniques such as fully related RNN, CNN and DNN for direct recognition in sub-atomic correspondence to deal with new situations without numerically manageable channel models. These DL-based finders outsmart the conventional locator, according to the results of the tests. Particularly, the LSTM-based locator shows a remarkable exhibition in the atomic correspondence situations even with entomb image impedance.

The profound support learning (DRL) emerges and demonstrates promising abilities to tackle muddled issues in specific radio conditions with a large state-activity space by using the impressive knowledge portrayal ability of the profound neural organization. DRL refers to strategies that estimate esteem capacities or DNN strategy works, making it possible to deal with complex issues with a high dimensional state and a large activity space. For approximating the best arrangement, DRL relies on DNN. Deep Q-Networks [7], profound approach inclination [11] and other popular DRL strategies. These techniques are well-executed in the areas of autonomous driving, mechanical technology and gaming. AlphaGo, as one of the best of some DRL examples, uses a DNN built using guided learning, reinforcement learning and conventional heuristic calculations. Many portable organization issues can be described as a Markov pattern option interaction, where DRL utilized to make the best long-term selections, like steering, MIMO following power, flexible cell turn on/off for energy conservation and so on (Figure 17.6).

Despite the fact that machine learning techniques have been used to enhance the display of some radio interface handling squares, the global ideal correspondence structure cannot be guaranteed by freely streamlining each square. Recently, a few experts have re-examined the correspondence process as a start-to-finish leisure errand to improve framework execution [2]. The complete correspondence device that includes the transmitter, collector and AWGN model is addressed using the AE structure in [5]. From start to finish, the framework's implementation has been improved. The transmitter and collector, in particular, are shown to be fully associated DNNs, and the AWGN channel is addressed as a commotion sheet. As a result, the communication system can be used. Figure 17.4 shows a diagram of how AI can be used to plan for the future. G. S. Zhang and D. Zhu PC Networks 183 (2020) 107556 11as an auto encoder platform that can be prepared to boost start-to-finish execution, similar to BER. The AE-based methodology can achieve preferred execution over the standard BPSK with hamming code, according to re-enactment results. Furthermore, Kim et al. suggested an AE-based remote correspondence in which the data information at the transmitter side is encoded to a signal using a DNN, and the beneficiary decodes the signal received using the DNN [4]. MIMO channels are protected by the AE-based approach. Both open-circle frameworks without the CSI criticism and completely closed-circle frameworks with CSI input are to be considered. The AE scheme outflanks ordinary open-circle MIMO plans, as shown by the reproduction result of a particular 2 x 2 model of MIMO structure. In terms of SNR, the MIMO model AE with excellent CSI better outperforms the traditional schemes for precoding.

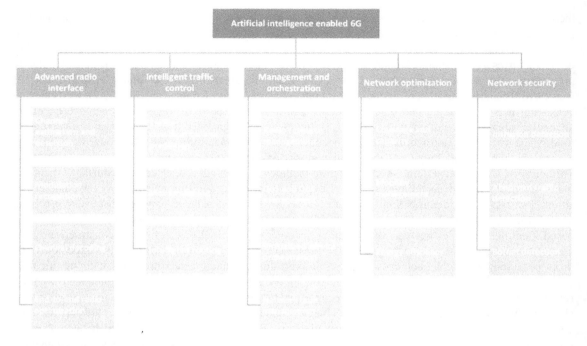

FIGURE 17.6 Components of AI to 6G.

Huang et al., for example, used the DNN to approximate the non-symmetrical many entrance (NOMA) structure, which is considered a black box. The entire NOMA model system is represented as a new AE, in which the methods for encoding and decoding are combined. The suggested arrangement achieves lower BLER than normal hard option strategies, according to the simulation results. Furthermore, DNN and GAN are combined in this process, a start-to-finish correspondence system was suggested. Learning the correspondence model yield utilizing a contingent ingenious ill-disposed organization, in particular, creates a channel-skeptic learning framework. So the results show that the technique works on Rayleigh blurring models as well as additional substance white Gaussian motion models. These researches pave the way for the development of information-driven 6G correspondence frameworks.

One entanglement brought by equipment heterogeneity is the unnecessary exertion to update the framework for various equipment settings. For instance, extraordinary handset designs have been proposed for mm Wave frameworks, including simple beam forming, half breed beam forming and 1-bit computerized beam forming. The traditional methodology depends on a hand-created plan for every one of them, which is exceptionally wasteful [2]. Since these different types of handsets will be confronted with a similar physical structure, a calculation built for one may also provide insight into another's strategy. For disseminating one engineer's strategy to others, move learning is a promising method.

17.6 CONCLUSION

As 5G connections are being set up, now it is a good span of time to look beyond 5G that too create a path map for making a quantum leap advancement towards the cutting edge, which would be expected to witness modern sun by before 2030. In this chapter, the requirement for 6G correspondence innovation is discussed, as well as the capabilities that it is expected to have. This article is a modest attempt to include a 6G examination guide for the future. The 6G has some new functionality revolution were discovered, and empowering inventions were investigated. Furthermore, synchronized 6G connectivity and artificial intelligence to dissect numerous applications that change in every way people collaborate with one another and advanced gadgets. Self-driving vehicles and holographic correspondences are two choices advancements that could be one of the major utilizations resulting as a consequence of the implementation of 6G intelligent organizations. This new era of correspondence innovation would result in a wave of mechanical advancements when it comes to organizations and exchanges. 6G is predicted to arrive in the near future correspond with the upcoming fourth modern transition, known as industry 4.0, which will undoubtedly change people's lifestyles.

REFERENCES

[1]. H. Gacanin, "Autonomous wireless systems with artificial intelligence", *IEEE Veh. Technol. Mag.*, vol. 14, no. 3, Sept. 2019.
[2]. K. Chen et al., "Ultra-low latency mobile networking," *IEEE Netw.*, vol. 33, pp. 181–187, Mar. 2018.
[3]. S. Arockia Panimalar, J. Monica, S. Amala, and V. Chinmaya, "6G Technology", *IJERT*, Sep. 2017.
[4]. S. Zhang and D. Zhu, "Towards artificial intelligence enabled 6G: State of the art, challenges, and opportunities", *Comp. Netw.*, vol. 183, p. 24, 2020.
[5]. J. Andrews et al., "What will 5G be?," *IEEE JSAC*, vol. 32, pp. 1065–1082, June 2014.
[6]. T. S. Rappaport et al., "Wireless communications and applications above 100 GHz: Opportunities and challenges for 6G and beyond," *IEEE Access*, vol. 7, pp. 78729–78757, July 2019.
[7]. B. McMahan et al., "Communication-efficient learning of deep networks from decentralized data," *Proc. Int'l. Conf. Artificial Intell. Stat. (AISTATS)*, vol. 54, pp. 1273–1282, 2017.
[8]. Y. Mao et al., "A survey on mobile edge computing: The communication perspective," *IEEE Commun. Surveys Tuts.*, vol. 19, pp. 2322–2358, Fourth Qtr., 2017.
[9]. R.-A. Stoica and G.T.F. de Abreu, "6G: The wireless communications network for collaborative and AI applications", April 2019.
[10]. X. Yu, J. Zhang, and K. B. Letaief, "A hardware-efficient analog network structure for hybrid precoding in millimeter wave systems," *IEEE J. Sel. Topics Signal Process.*, vol. 12, May 2018, pp. 282–297.
[11]. N. C. Luong et al., "Applications of deep reinforcement learning in communications and networking: A survey," *IEEE Commun. Surveys Tuts.*, May 2019.
[12]. H. Yang, A. Alphones, Z. Xiong, D. Niyato, J. Zhao, and K. Wu, "Artificial intelligence-enabled intelligent 6G networks". *IEEE Net.*, vol. 34, no. 6. doi: 10.1109/MNET.011.2000195
[13]. L. Zhang, Y.-C. Liang, and D. Niyato, *"6G visions: mobile ultrabroadband, super internetof-things, and artificial intelligence"*, IEEE, May 2019.
[14]. K. David and H. Berndt, "6G vision and requirements: Is there any need for beyond 5G?," *IEEE Veh. Technol. Mag.*, vol. 13, pp. 72–80, Sept. 2018.
[15]. S. J. Nawaz, S. K. Sharma, S. Wyne, M. N. Patwary, and M. D. Asaduzzaman, *"Quantum Machine Learning for 6G Communication Networks: State-of-the-Art and Vision for the Future"*, IEEE, 2019.
[16]. K. B. Letaief, W. Chen, Y. Shi, J. Zhang, and Y.-J. Angela Zhang, "The Roadmap to 6G: AI Empowered Wireless Networks", IEEE, August 2019.
[17]. E. C. Strinati, S. Barbarossa, J. L. Gonzalez-Jimenez, D. Kténas, N. Cassiau, L. Maret, and C. Dehos, *"6G: The Next Frontier from Holographic Messaging to Artificial Intelligence Using Subterahertz And Visible Light Communication"*, IEEE, August 2019.
[18]. K. Yang, Y. Shi, and Z. Ding, "Data shuffling in wireless distributed computing via low-rank optimization," *IEEE Trans. Signal Process.*, vol. 67, pp. 3087–3099, June 2019.
[19]. H. Yang, A. Alphones, Z. Xiong, D. Niyato, J. Zhao, and K. Wu, "Artificial Intelligence-Enabled Intelligent 6G Networks", *IEEE Netw.*, vol. 34, p. 6, 2020.

Antenna Array Design for Massive MIMO System in 5G Application

18

Mehaboob Mujawar
Goa College of Engineering

18.1 INTRODUCTION

Conventional multiple input multiple output (MIMO) technologies have already been used in 4G applications. Multi-user MIMO (MU-MIMO) has many advantages over conventional MIMO. Figure 18.1 shows a typical MU-MIMO system, which works with cheap single antenna terminals and creates a rich scattering environment. Conventional MIMO includes resource allocation, which is a simplified process. This is because every active terminal utilizes all the frequency bits. In MU-MIMO, we have equal number of service antennas and terminals. Therefore, we can say that MU-MIMO is not a scalable technology. Figure 18.2 shows a typical Massive MIMO system, which needs extra antennas to focus energy on smaller regions of space. It is important to focus energy on smaller regions of space so as to improve the throughput and radiated energy efficiency. With this MIMO technology, we can extensively use the inexpensive low power consuming components, reduced latency and simplification of MAC layer. The basic building blocks of MIMO technology mainly comprise spatial diversity, spatial multiplexing and beamforming techniques. Spatial diversity enables the transmission of the same data over multiple propagation paths by strengthening reliability of the system. In spatial multiplexing, messages are separated in space, which allows transmission of several message signals continuously at the same time without getting interfered with other message signals. To understand the concept of spatial multiplexing in detail, we can consider the example of a pipeline carrying data between base station and the mobile. If there is single antenna on mobile and single antenna at the base station, then data transmission is limited between the base station and mobile. But if we consider the case wherein there are multiple antennas located both at the base station and on the mobile, then there will be many virtual pipelines in space for the transmission of data between base station and mobile. In this technique there is need for continuous co-ordination between base station and user for the transmission of data, which can be

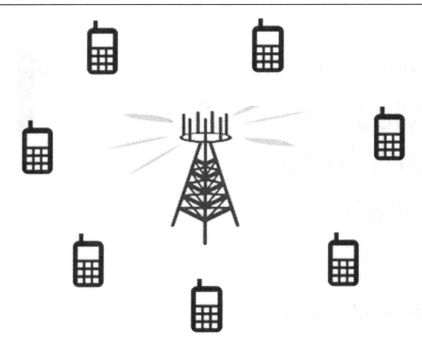

FIGURE 18.1 Typical MU-MIMO system.

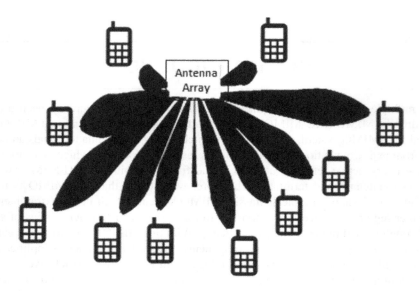

FIGURE 18.2 Typical Massive MIMO system.

improved with the introduction of more advanced techniques. In beamforming technique, the signals are concentrated in a specific direction, rather than focusing radiations in random directions. To understand the concept of beamforming, we can consider the example of a flashlight and a laser pointer. When we use a flashlight, the beam of light is emitted in random directions irrespective of its usage in the desired direction. Similarly, when we consider the example of laser pointer, it will be directed only in the desired direction. Hence, saving the resources is possible with the use of beamforming techniques. The major difference between conventional MU-MIMO and Massive MIMO can be discussed based on the number of antennas deployed, mode of duplexing, acquisition of channels, quality of link after precoding, allocation of resources and performance of the cell edge. In conventional MIMO the number of base station antennas is approximately equal to the number of users. It has limitations in both the number of users and the antennas to be deployed; not more than 10 users and base station antennas can be employed in conventional MIMO. It has been designed to work with both Time Division Duplex (TDD) and Frequency Division Duplex (FDD). Channel acquisition is carried out with predefined angular beams, which are totally based on code books. The quality of the link varies with time and frequency due to small-scale and frequency-selective fading. Since the variations in the channel quality are rapid in the conventional MU-MIMO, the allocation of resources should also be rapid to compensate for the same. Performance of the cell-edges will be good only if the antennas at the base station co-operate. In case of Massive MIMO, the number of base station antennas will be more than the number of users. This system does not have any kind of limitation on the number of users and antennas to be deployed; as many as hundreds of antennas can be employed to provide support to more than 20 users. It has been designed to work with TDD to exploit channel reciprocity. Channel acquisition is carried out by sending up-link pilots and exploiting channel reciprocity. The quality of the link does not vary with time and frequency due to channel hardening. Since the variations in the channel quality are slow in massive-MIMO, the allocation of resources can be planned out easily well in advance. The performance of the cell edges totally depends on the number of antennas, SNR of cell-edge increases with the number of antennas avoiding inter-cell interference.

There are many challenges associated with Massive MIMO, such as making of low cost low precision components, synchronization of terminals and reduction in internal power consumption. The evolution of MIMO, from MU-MIMO to Massive MIMO has led to improvements in the antenna performance. Massive MIMO technology has scaled up MIMO by possible orders of magnitude. The concept of antenna array plays a very important role in improving the antenna performance in communication systems specifically in MIMO systems. Antenna arrays with few hundreds of antennas can serve simultaneously, many terminals in same frequency resource. Massive MIMO has many advantages such as energy efficiency, spectrum efficiency and it can be used with future wireless technologies.

18.2 JOURNEY OF MASSIVE MIMO AROUND THE GLOBE

Massive MIMO will be the key element and an essential building block of 5G. The existing Wi-Fi and 4G standards have already been making use of Massive MIMO principles. The advantages of Massive MIMO will be utilized to their full potential with advancements in 5G communication networks. In future there will be huge requirement of data around the globe to compensate for the increasing demand for data. Massive MIMO and 5G will play a very important role in increasing the data rates. According to the survey conducted by Cisco, there will be approximately 5.5 billion mobile users around the globe by 2022. The approximate data consumed by each mobile user per month will be 20 GB. Massive MIMO enables the communication between several users and multiple devices simultaneously while supporting faster data rates and providing stable performance, which makes it an excellent technology in 5G communication. In Japan and China, Massive MIMO has already been deployed for commercial usage under 4G LTE conditions. Ericsson has already launched a new FDD radio, which could overcome the gap between 4G and 5G, stimulating the ability of the present 4G LTE, while establishing the base for 5G. Many experimentations have been carried out by different companies all over the world to study the effects of Massive MIMO and 5G and to observe the improvements that can be made to overcome the hazardous effects of Massive MIMO and 5G in the real world after their deployment. Vodafone and Huawei had conducted a Massive MIMO technology test under real-time experience in 2017. This test had obtained data transfer rate of 717 Mbps across ten devices. This Massive MIMO test was performed using Vodafone's frequency spectrum along with Huawei's 5G active antenna. In 2018, Nokia took the initiative to develop Massive MIMO technology with the introduction of Reefshark chipset. With the experimentation of Reefshark chipset, the company had claimed that Reefshark would decrease the size of the Massive MIMO antenna by half and energy saving in terms of baseband units was possible. Samsung has also contributed to the research in Massive MIMO by establishing "5G city" or "5G stadium" in Korea, to analyse its effects before its deployment in the market. Under this test, it was possible for thousands of users to stream HD videos without any delay, which proved the effectiveness of this technology. Present smartphones are not Massive MIMO-ready and also the Massive MIMO-equipped network required to operate such smartphones is not yet available. However, many smartphones are presently operating using the MU-MIMO technology for better data speed.

18.3 IMPORTANCE OF A GOOD ANTENNA DESIGN

The antenna is considered to be one of the major hardware components of a mobile phone. Its production is not expensive but it is considered to be important for the design and performance of a mobile phone. A mobile phone reception is solely dependent on how good the antenna functions. That is, an improperly designed antenna will not be able to receive a clear signal from the base station and, therefore, would impact the use of a mobile phone. Furthermore, since mobile phones usually turn off the screen during voice conversations, the component draining most of the energy from the battery is the mobile phone transmitter. Thereby, energy consumption will be heavily dependent on the mobile phone antenna efficiency. When cell phones were invented in the late 1990s, they were totally dependent on the antenna as they were mounted external to the design. But nowadays with antennas being incorporated inside the mobile phone their visual design dependence is less obvious but still important. Finally, there is yet another important aspect where antenna design plays a major role, which is radiation safety. Mobile phones radiate microwaves in frequencies similar to those of microwave ovens and will likewise have a heating effect in the head while using the phone. Design of the mobile phone antenna is the most important factor to consider when trying to keep inside of the allowed radiation standard limits put up by organizations such as IEEE and FCC. Single band antennas support only one or two frequencies of wireless services and these days more and more wireless standards are being supported by the devices. So they employ several antennas for each standard. This leads to large space requirement. One foreseen problem associated with antenna design for such devices is to cover LTE bands while still covering WiMAX and WLAN/Bluetooth bands. Thus, due to space constraints in mobile devices, covering multiple bands with a single antenna structure is the need of the hour. The long distance communication is possible only because of wireless communications and antenna being the main element of the system, which converts electrical power into radio waves, and vice versa. Wireless communication has witnessed an increase in the number of users and there have been restrictions on available bandwidth, so commercial operators have large capacity networks with good quality coverage. There are many merits which are associated with the use of MSA, for example, it is possible to achieve an antenna design which will provide more gain, compact design, narrow bandwidth and low profile. The main requirement of an antenna to be used for commercial applications includes impedance matching and bandwidth enhancement. There is a direct relation between antenna size and resonant frequency of an antenna. As the frequency increases size of the antenna becomes smaller. While designing MSA, we have to choose the shape of the patch and feeding method based on the desired applications. The performance of an antenna can be affected in many ways, due to different shapes of the antenna. Different dielectric substrates have varying dielectric constants that influence the antenna design parameters as well as antenna performance. There are different feeding techniques available to feed the antenna to allow it to radiate. The main motivation of this chapter is in the use of MSA, which provides a huge range of advantages in communication systems that have ultimately led to more demand of antennas for commercial purposes with more enhanced features like multiband, wider bandwidth and low profile. Conventionally, there was a need of specific antenna for specific communication application, since the antenna used to operate on one or two frequencies. This has been a problem in implementation of the antenna on the devices, since these antennas had occupied a lot of space on the device. To get rid of this problem, there was a need of a single antenna which could operate at wide bands of frequencies, which can be obtained using multiband antennas. One of the methods to obtain such an antenna is by using defective ground plane and creating slots on the patch. Defected ground structure is a technique which helps in improving the functioning of an antenna by purposely changing the ground plane element of MSA.

18.4 LITERATURE REVIEW

With steady developments, antennas are a significant advancement to the RF world. The new gadgets restrain the physical space utilized by wiping out the requirement for various antennas. This is particularly indispensable in cell phones, for example, mobile phones that get numerous frequencies like cell tower, Wi-Fi and GPS. The other option in contrast to frequency reconfigurable reception apparatuses is a wideband antenna; notwithstanding, wideband antennas get huge frequency ranges acquainting commotion with the framework. Frequency reconfigurable radio antennas slender the transfer speed to explicit frequencies, ordinarily reducing the measure of noise for the signal. The main aim of this chapter is to design an antenna array using microstrip patch for wireless applications like WiFi, WiMax, WLAN and satellite communication. Paper [1] presents Massive MIMO antenna design for 5G applications. This antenna design involves antenna arrays, which yield high gain and wideband. This antenna had resonated at a frequency range of 3.5 to 5.2 GHz. The antenna array was designed so as to obtain 16 individual antennas on the same patch, basically an array of 4 × 4 antennas using open scale modular technique. The number of antennas on the patch has been increased to 64 units, to form perfect Massive MIMO system. With the increment of antennas on the system, the gain increased to 26.8 dB. Paper [2,13,14] presents dual-polarized broadband antenna, which has an antenna array of 2 × 2 and is in the capped bow-tie shape. This antenna has a bandwidth of 24 GHz, reflection coefficient of 10 dB and the area covered by the antenna is stable. The directivity of this antenna varies from 8 dB to 12 dB. The overall structure of the antenna is quite simple, compact and antenna elements can be easily extended to form large antenna array to provide better performance for 5G applications. Paper [3,15] presents dual dipole massive MIMO antenna, which has been dual-polarized with low coupling. Microstrip lines have been used to connect folded dipole antennas, this structure basically consists of a pair of these antennas. This antenna had a return loss of greater than 15 dB. The 2 × 2 array antenna has been simulated to obtain low mutual coupling antenna. Paper [4] presents massive MIMO antenna design for 5G applications using meta-material. Meta-material has been used for the construction of this antenna, which in turn enhances the performance of the antenna for large antenna arrays. This antenna resonates at a frequency of 11 GHz. In large MIMO antenna arrays, there is need to decrease the diversity factor, which can be achieved by mutual coupling of antennas. Paper [5] presents massive MIMO smartphone antenna array for 5G application. The proposed antenna has a multiband, which makes it suitable for operation in multi-mode. The overall structure of the antenna contains eight planar inverted F antennas (PIFA). Due to the proposed orientation of PIFA elements on the smartphone board, vertical as well as horizontal polarizations are supported. This antenna has good efficiency and S-parameters. Paper [6] presents Full Dimensional Massive MIMO (FD-MIMO) technology. This antenna employs an active antenna system, which allows multiple antennas to be placed on the base station adapting to the beamforming techniques. In FD-MIMO, the structure of the antenna being compact utilizes huge planar antenna arrays to raise spatial correlation. Paper [7] presents the lens antenna design for 5G applications, utilizing microwave metasurfaces. It is possible to obtain lens antennas having higher gains at microwave frequency bands using metasurfaces. Lens antennas offer a wide range of advantages, mainly feeding structures being simple, power focusing in the desired direction and sharing of the aperture is also possible. Paper [8] presents the dual band massive MIMO antenna array for 5G communication. This antenna resonates at two frequencies, that is, at 28 GHz and 38 GHz. The overall structure of the antenna is compact with dimensional size of $13 \times 20mm^2$. This antenna had offered low reflection coefficient and higher gains of 12.07 dB and 13.42 dB at 28 GHz and 38 GHz, respectively. Paper [9] describes a planar inverted F antenna, which will operate on multiple frequency bands and also to gain better results, a slot has been created on the ground plane. The software used for the optimization is HFSS. It was possible to obtain acceptable return loss over multiple frequency bands. This antenna was built using the substrate material of FR4. It basically describes PIFA antenna, whose parameters have been varied to obtain a suitable antenna for various applications. Paper [10] describes Dual Band Microstrip Patch Antenna. This antenna has been constructed using a microstrip patch which is square in shape and operating on dual frequencies. It is operating on C and X bands. The simulation software used was HFSS. This software makes it possible to analyse various antenna parameters. It helped obtain acceptable reflection coefficient for both the frequency bands and also voltage standing wave ratio (VSWR) within the acceptable range for an ideal antenna, that is, between 1 and 2. The frequencies at which the antenna resonated at both the bands are 6.7, 6.4 and 7.3. The papers [11[15]] describe the patch antenna, which is operating at a frequency of 2.4GHz, overall structure of the antenna is C shaped. The software used for the antenna design is Computer Simulation Technology Microwave Studio. The substrate material used for the construction of antenna is FR-4. After obtaining the simulation results, it was analysed that return loss was −10 dB and VSWR was within the range of 1–2. Paper [16] describes the MSA, which is rectangular in shape and can be operated on wide range of frequencies. The optimization of the antenna has been carried out for a wide range of frequencies, which shows improvements in the various parameters of the antenna. The gain of the antenna was increased along with the surface current by making four slots of L shape on the patch. The feeding technique used in this project was inset feed line and the substrate used was RT Duroid having a dielectric constant of 2.2. The simulation software used for the project was CST. Paper [17] deals with MSA, which is compact and has slots that improve the performance of the antenna for wide range of applications including X band and LTE. This antenna operates on a wide range of frequencies having the substrate material FR-4 with 4.4 dielectric constant. In this project, U- and Y-shaped slots were created on the patch for the antenna which was under test. The feeding technique used in this project was inset feed line. The simulation software used for the project was CST. The antenna which was under test had resonated at 7.98 GHz with −20 dB, 4.1 GHz with −13.7 dB and 2.4 GHz with −21.3 dB return loss with frequencies, respectively. Paper [18] describes an antenna, which is specifically designed to operate in C-band and S-band. The main aim of this paper was to reduce the size of the antenna in comparison to other multiband antennas and it was achieved. To operate on multiple frequency bands, this antenna utilized the technique of stubs. It had T- and E-shaped stubs, which helped it operate on multiple frequency bands and also reduced the size of the antenna. The software used in the project was HFSS. The main aim of this antenna was achieved, as it resonated at desired frequencies. Paper [19] deals with an antenna having fractal slots, which is operated at a frequency of 4.1 GHz. E-shape has been mounted on the patch which is resonating at the center. FR4 has been selected as a substrate, having a height of 2 mm. This antenna has been miniaturized with the slots, specifically of H and L shape, to about 60%. It has wide range of applications such as in C-band, S-band, GPRS, GSM and 4G. Paper [20] presents the design of textile antenna, which mainly focuses on the Bluetooth application. This textile antenna was designed to provide better performance without affecting the textile properties. This textile antenna has been provided with microstrip line feed. Paper [21] presents a planar antenna, which has been mainly used in telemetry applications. The feeding technique used in this planar antenna is coaxial cable. Paper [22] presents the UWB antenna design that mainly focuses on the antenna design by employing the Liquid Crystal Polymer (LCP) substrate. LCP substrate helps improve the antenna performance by allowing the antenna to operate in the frequency range of 2.9 to 10.2 GHz. Paper [23] presents the effects and measurement of radiations from the wearable antennas. The human body tissues absorb electromagnetic radiations that are emitted through wearable antennas. Therefore, it is necessary to protect the human body from exposure to high radiations from the antenna. As a solution of this high radiation exposure issue, it is necessary to measure Specific Absorption Rate (SAR) of all the wearable devices before manufacturing them. The SAR has been decided by the different agencies around the globe within their jurisdiction. According to Federal Communications Commission in the U.S., SAR value is 1.6 W/kg for 1 gram of tissue and in Europe, SAR value is 2.0 W/kg for 10 gram of tissue [24]. With advancements and research in the field of wireless communications, the demand for antenna design in millimeter wave applications has also been increasing. Paper [25] presents the additive manufacturing techniques for antenna designs; this technique has a large number of advantages such as materials required for the fabrication process are less expensive, less wastage of materials, etc. Paper [26,27] present UWB antenna for wearable 5G applications, many researchers have contributed in the development of wearable devices for 5G applications and the main point of attraction for antenna engineers is the use of flexible antennas in wearable applications. Flexible antennas can be easily implanted on human body and can be used for a wide range of applications, specifically in monitoring the patient's health. Paper [28] presents conventional design of antennas that is by making use of printed circuit boards, which is not acceptable by few devices because of various antenna performance parameters. For efficient working of antenna and to protect the tissues from damage, due to on body devices, antennas should be implanted on the outer layers of clothing. It is also comfortable for the user, when wearable antennas are placed on the textiles rather than body implant. Paper [29] presents flexible textile antenna, which proves to be more effective compared to traditional metallic antennas. Planar Inverted F antenna has been used to obtain wider antenna band operations. A wearable PIFA was designed using conductive plate and was placed on FlexPIFA substrate at 2.45 GHz. Paper [30] presents meta-material based Left Handed structures, these structures along with LH modes help in the miniaturization of the antenna. This antenna is operating on dual frequencies. This antenna mainly consisted of right T-handed resonator and left T-handed resonator. Paper [31] presents the multiband antennas, these antennas are widely used in wearable devices depending on the specific application. Flexible antennas also find a wide range of application in RFID and sensing. Paper [32] presents multi-frequency antennas, recent development of multi-frequency antennas have increased their demand in market. These antennas are designed to operate on various operation modes over different wireless services. In order to increase the performance of this antenna, it will be effective to have two radiation patterns mainly dipolar and monopolar [32]. Design presented in paper [32] was compact and low profile. Paper [33] presents dual-band textile antennas; it is the most preferred antenna for wireless applications. However, this antenna has a larger surface area, which sets a limitation for its use directly on body. Specifically on chest or back, but still this antenna has proven to be effective in performance. The fabrication complexity of this antenna has been increased due to the presence of vias on the designed antenna. Paper [34] presents wearable device used for 5G, it is simulated at 28 GHz and 38 GHz. This planar inverted F antenna has been placed on the substrate. The substrate used in this antenna design is jeans. Detailed analysis has been done for measuring the SAR values for both the antenna and smart watch. Paper [35] presents Meta material based dual band wearable antenna, this antenna operates at 2.4 GHz and 5.2 GHz. Jeans is used as substrate. The performance is analysed in all possible orientations.

18.5 WEARABLE ANTENNA IN MIMO TECHNOLOGY

Wearable devices have a large number of applications in the field of health Care, military and tracking. The health of the patients as well as the performance of the athletics can be easily monitored with the help of different wearable devices. The main aim of research in wearable devices is to obtain a device, which is compact, good in performance, free from harmful effects, highly reliable and a device which would consume less power. Microstrip patch antennas are widely used antenna for wearable applications because of its special features like miniaturization, low cost and complexity involved in the design and fabrication of the antenna. These antennas can be easily implanted through human wearable like clothing, watch, pendent etc. The use of microstrip antenna as wearable antenna has been widely studied area of research as the advancement in the antenna design with different methods like addition of

Meta-materials, flexible substrates and Formation of EBG structures, antenna Arrays. These have improved the performance of the antenna. The antenna design considerations have to be taken into account in order to enhance the working of wearable antenna. It is desirable to have antenna with low profile. It is preferable to use antennas orthogonal to the body than the parallel antennas. When wearable antenna is implanted on the human body, it results in the detuning of the antenna due to loading effect. Therefore, the designer needs to supervise the changes occurring in the resonating frequency band because of this effect. Ground plane creates a shield around the human body to protect it from radiations. These wearable antennas need to be portable and compact, since they will be used for human body wearable applications. These devices also need to be economical, so that they can be used by all sections of society. These antennas should also be flexible and comfortable to use on any part of the human body. The antenna parameters such as reflection coefficient, resonant frequency and bandwidth will be affected by crumpling of the wearable device. Crumpling creates detuning effect on the antenna. These antennas need to have higher radiation efficiency, when implanted on human body, to reduce the losses. Patch antenna is mainly used because of its low profile, higher directivity and many other advantages. These can be easily fabricated on the printed circuit boards and it tends to be one of the reasons for its large usage in the construction of portable devices. There are many challenges, which are associated with wearable devices. The first challenge associated with wearable device is the location, where the antenna has to be placed on the human body. When the wearable device starts to radiate, it emits electromagnetic radiations. The efficiency will decrease due to the placement of the antenna on human body. For example, if we consider the case of an antenna which initially gives efficiency of 50%. But the same antenna placed on human body will give efficiency of 5%. This indicates that wearable devices have to be designed properly, since it shows large variations in its working depending on the placement of antenna on the human body. The second challenge of wearable device is its compactness. The main aim of the antenna designers is to reduce the size of antenna on the wearable device. While designing a smart watch, we cannot randomly consider big dipole antenna to implant on it. Hence the selection of the type of antenna and its dimensions are of very much importance during the design stage. The performance of the antenna has to be stable under different atmospheric conditions. It is also desirable to have wearable devices which operate on multiple frequency bands, hence enabling the operation of the devices under different applications. It is of great concern for the design engineers to reduce the dimensions of the wearable device keeping constant all its operational features.

18.6 SPECIFIC ABSORPTION RATE

SAR is defined as the amount of power, which has been absorbed by human tissue per unit mass. SAR Values of all the devices must be checked before fabrication process and also before manufacturing the product in market. There is a need to study the effect of human body interaction in wearable antennas. In wearable devices, the resonant frequency of the antenna changes with the change in the permittivity of the body tissues. The permittivity of the body tissues is high, which ultimately leads to detuning of the resonant frequency to a lower value. As the frequency of the wearable device increases, there will be a major effect on the conductivity and relative permittivity of the skin. Therefore, there will be reduction in the relative permittivity and increment in the conductivity of the skin as frequency increases. When the antenna radiates, few radiations will be absorbed by the human body due to its lossy nature and the gain of the antenna will reduce. SAR values have been restricted by regulatory authorities based on the region of application. According to Federal Communications Commission in US, SAR value is 1.6 W / kg for 1 gram of tissue and in Europe, SAR value is 2.0 W/kg for 10 gram of tissue. The SAR values of the wearable device have been determined by using CST microwave studio. While determining the SAR values of the wearable device by locating it on various parts of the human body, the antenna is powered with 0.5 W. A specific standard IEEE C95.1 has been followed for calculating the SAR values.

18.7 SIGNIFICANCE OF ANTENNA ARRAY

When a large number of antenna elements are placed close to each other by maintaining a constant separation, it forms an antenna array. Induction field of each antenna lies within the induction field of its neighboring antenna. The radiation pattern obtained from the performance of antenna arrays will yield better results. The radiation beam formed by the antenna arrays will result in higher gain, high directivity and low losses with better performance. The strength of the signal increases and minor lobes are reduced with the introduction of antenna arrays. There are some disadvantages of antenna arrays which need to be improved with recent advancements in antenna design engineering. Resistive losses of the antenna increases with the introduction of antenna arrays, also there is need for more space for the deployment of antenna elements. Antenna arrays have wide range of applications due to its capability in yielding higher gain and directivity. Its main applications include satellite and wireless communications, military radar communications and so on.

In this chapter, we have designed a Massive MIMO antenna for 5G application. This antenna will support high speed data rates for 5G communication and also Internet of Things applications. We have made use of 12 × 12 antenna array for analysing the performance of the antenna with respect to all the antenna parameters, such as gain, VSWR, S-parameter and other radiation characteristics using the CST software.

18.8 ANTENNA DESIGN

The antenna design is a multi-stage process, which includes initially several mathematical calculations to be carried out to determine the dimensions of the antenna according to the required specifications and then finally obtaining the simulation results using various software. The dimensions of the antenna can be calculated by using the following equations [36].

$$W = \frac{c}{2f} \sqrt{\frac{2}{\varepsilon_r + 1}} \tag{18.1}$$

$$L = L_{eff} - 2\Delta L \tag{18.2}$$

Where W is the width of the patch, c is the velocity of light, f is the resonating frequency, L is the length of the patch, ε_r is the dielectric constant of substrate, L_{eff} is the effective length of patch and is given by

$$L_{eff} = \frac{c}{2f \sqrt{\varepsilon_{reff}}} \tag{18.3}$$

The normalized extension in length is given by

$$\Delta L = 0.412h \frac{(\varepsilon_{reff} + 0.3)\left(\frac{w}{h} + 0.264\right)}{(\varepsilon_{reff} - 0.258)\left(\frac{w}{h} + 0.8\right)} \tag{18.4}$$

Where ε_{reff} is the effective dielectric constant and is given by

$$\varepsilon_{reff} = \frac{\varepsilon_r + 1}{2} + \frac{\varepsilon_r - 1}{2}\left[1 + 12\frac{h}{w}\right]^{-\frac{1}{2}} \quad (18.5)$$

The width and length of the substrate can be determined using the formulas given below

$$L_g = L + 6h \quad (18.6)$$

$$W_g = W + 6h \quad (18.7)$$

W_g is the width and L_g is the length of substrate and h is given by

$$h = \frac{0.0606\lambda}{\sqrt{\varepsilon_r}} \quad (18.8)$$

The below-mentioned equation can be used to determine the length of the feed line

$$Feed\,length\,(L_f) = \frac{\lambda_g}{4} \quad (18.9)$$

Where λ_g is the guided wavelength and is given by

$$\lambda_g = \frac{\lambda}{\sqrt{\varepsilon_{reff}}} \quad (18.10)$$

Efficiency of the antenna is calculated by

$$\eta = \frac{Gain}{Directivity} \times 100\% \quad (18.11)$$

Where η is the efficiency of the antenna. The equations for the radiation box are mentioned below

$$Axis\,position = \frac{-\lambda_g}{6} + \frac{-\lambda_g}{6} + \frac{-\lambda_g}{6} \quad (18.12)$$

$$Length = \frac{\lambda_g}{6} + \frac{\lambda_g}{6} + L_g \quad (18.13)$$

$$Width = \frac{\lambda_g}{6} + \frac{\lambda_g}{6} + W_g \quad (18.14)$$

$$Height = \frac{\lambda_g}{6} + \frac{\lambda_g}{6} + h \quad (18.15)$$

The dimensions of the antenna are calculated by using the abovementioned equations and can be used to design antenna in a software to obtain simulation results. These software results can be utilized for fabrication of the antenna and its practical implementation. The proposed antenna resonates at a frequency of 28 GHz, which is required for practical implementation. The antenna design is considered to be desirable, if it provides VSWR value less than 2 and reflection coefficient less than −10 dB.

Once dimensions of the antenna are calculated, the next step is to select the substrate material. Figure 18.3 shows the structure of the proposed single antenna with the dimensions. The proposed antenna uses Rogers as the substrate material. Then the design of the antenna in the software is carried out to obtain better simulation results. Figure 18.4 shows 12 × 12 antenna array with 144 antenna elements.

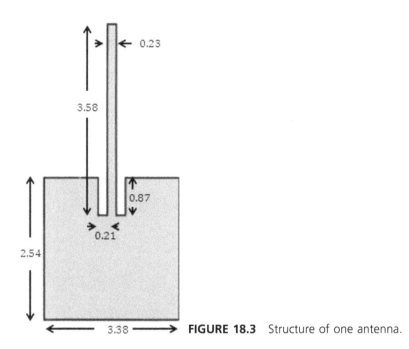

FIGURE 18.3 Structure of one antenna.

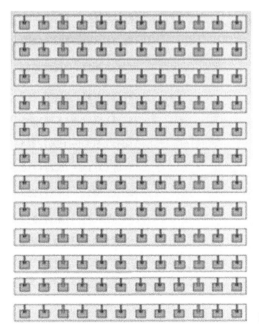

FIGURE 18.4 Antenna array (12 × 12).

18.9 RESULTS

The proposed antenna was designed using CST software and the antenna performance parameters were analysed. The parameters which decide the performance of the antenna mainly include gain, reflection coefficient, VSWR, bandwidth and so on.

Return loss: Return loss is the ratio of the radiowaves transmitted to the radiowaves received by the particular antenna. Figure 18.5 shows a return loss of single antenna element, which is –18.03d B and Figure 18.6 shows return loss values of antenna arrays, which consist of 144 antenna elements in 12 × 12

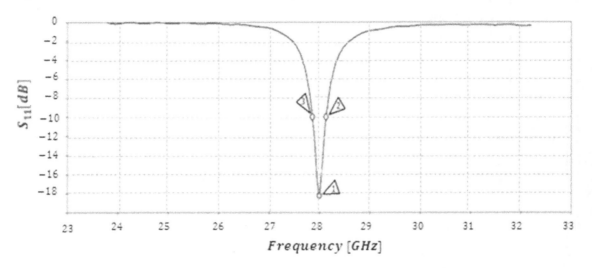

FIGURE 18.5 Graph of return loss for one antenna.

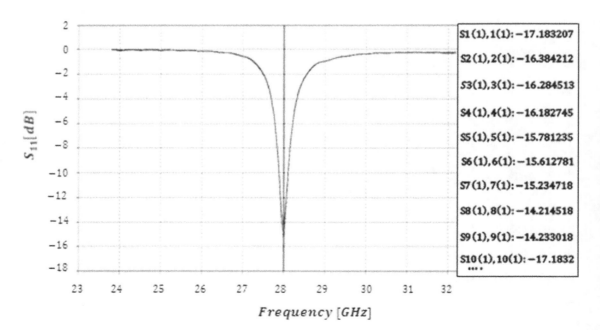

FIGURE 18.6 Graph of return loss for 144 antenna elements.

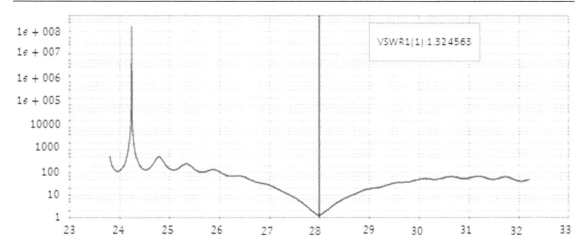

FIGURE 18.7 Graph of VSWR for one antenna.

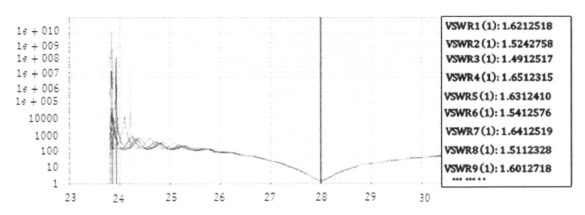

FIGURE 18.8 Graph of VSWR for 144 antenna elements.

array. As can be seen in Figure 18.6, at a frequency of 28 GHz, each antenna element may get different return loss value. These return loss values range from −18 dB to −14 dB, hence the average return loss obtained for 12 × 12 antenna array is −17.18 dB.

Voltage Standing Wave Ratio (VSWR): VSWR represents how effectively RF power can reach the destination when transmitted from source via transmission line. Figure 18.7 shows VSWR value of single antenna, which is 1.32 and Figure 18.8 shows VSWR values of antenna arrays, which consists of 144 antenna elements in 12 × 12 array. As can be seen in Figure 18.8, at a frequency of 28 GHz, each antenna element may get different VSWR value. These VSWR values are ranging from 1.65 to 1.49. Hence, the average VSWR of 1.62 is obtained for 12 × 12 antenna array. Figure 18.9 shows an antenna gain of 144 antenna elements, which is 29 dB.

18.10 CONCLUSION

The design and analysis of 12 × 12 antenna arrays for Massive MIMO system in 5G applications has been explained in the chapter. The antenna design and simulation was done by using CST software. This

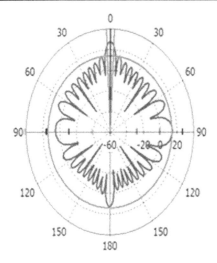

FIGURE 18.9 Gain of 144 antenna elements.

antenna had a resonant frequency of 28 GHz. The average antenna gain of 144 antenna elements was obtained to be 29 dB, average return loss was -17 dB and average VSWR of 1.62 was obtained. The simulated antenna performance parameters indicate that this antenna is suitable for 5G applications.

REFERENCES

[1]. H. Yuan, C. Wang, Y. Li, N. Liu, and G. Cui, "The design of array antennas used for Massive MIMO system in the fifth generation mobile communication", *2016 11th Int. Symp. Antennas, Propag. and EM Theory (ISAPE)*, Guilin, China, 2016, pp. 75–78, doi: 10.1109/ISAPE.2016.7833881.

[2]. R. Wu, Z. Liu, Y. Zhou, X. Wang, J. Yin, and J. Yang, "A broadband dual-polarized capped bow-tie 2 × 2 antenna array for 28 GHz band in 5G systems", *2018 IEEE Int. Symp. Antennas Propag. USNC-URSI Radio Sci. Meet.*, Boston, MA, 2018, pp. 825–826, doi: 10.1109/APUSNCURSINRSM.2018.8608922.

[3]. Y. Qin, Y. Cui, and R. Li, "Dual-polarized dual-dipole antenna with low mutual coupling for massive MIMO", *2019 Int. Conf. Microw. Millimeter Wave Technol. (ICMMT)*, Guangzhou, China, 2019, pp. 1–3, doi: 10.1109/ICMMT45702.2019.8992875.

[4]. K. N. Poudel and W. Robertson, "Metamaterial inspired antenna design for massive MIMO, 5G communications system", *2017 USNC-URSI Radio Sci. Meeting (Joint with AP-S Symposium)*, San Diego, CA, 2017, pp. 103–104, doi: 10.1109/USNC-URSI.2017.8074918.

[5]. N. O. Parchin, H. J. Basherlou, I. A. Yasir Al-Yasir, M. Sajedin, J. Rodriguez, and R.A. Abd-Alhameed, "Multimode smartphone antenna array for 5G massive MIMO applications", *2020 14th European Conf. Antennas Propag. (EuCAP)*, Copenhagen, Denmark, 2020, pp. 1–4, doi: 10.23919/EuCAP48036.2020.9135754.

[6]. Q. Nadeem, A. Kammoun, M. Debbah, and M. Alouini, "Design of 5G Full Dimension Massive MIMO Systems", in *IEEE Trans. Commun.*, vol. 66, no. 2, pp. 726–740, Feb. 2018, doi: 10.1109/TCOMM.2017.2762685.

[7]. Z. N. Chen, T. Li, and W. E. I. Liu, "Microwave Metasurface-based Lens Antennas for 5G and Beyond", *2020 14th European Conf. Antennas Propag. (EuCAP)*, Copenhagen, Denmark, 2020, pp. 1–4, doi: 10.23919/EuCAP48036.2020.9135285.

[8]. M.M.M. Ali and A. Sebak, "Design of compact millimeter wave massive MIMO dual-band (28/38 GHz) antenna array for future 5G communication systems", *2016 17th Int. Symp. Antenna Technol. Appl. Electromagn. (ANTEM)*, Montreal, QC, Canada, 2016, pp. 1–2, doi: 10.1109/ANTEM.2016.7550213.

[9]. S. E.Hosseini, A. R. Attari, and Pourzadi A., "A Multiband PIFA with a Slot on the Ground Plane for Wireless Applications", *Int. J. Inf. Technol. Electr. Eng.*, vol. 3, no. 4, July 2013.

[10]. A. Shamim Banu, M. E. R. Kavitha, R. Aayisha Siddika, M. Elakkiya, and S. Kaovyaa, "Dual Band Microstrip Patch Antenna", *Int. J. Eng. Res. Technol.*, 2018.

[11]. D. Dutta, A. Hira, F. Asjad, and T. I. Haider, "Compact triple C shaped microstrip patch antenna for WLAN, WiMAX & Wi-Fi application at 2.5 GHz", *IEEE Trans.*, Dec. 2014.

[12]. E. G. Larsson, O. Edfors, F. Tufvesson, and T. L. Marzetta, "Massive MIMO for next generation wireless systems", *IEEE Commun. Mag.*, vol. 52, no. 2, pp. 186–195, Feb. 2014.

[13]. Y. H. Cui, X. N. Gao, H. Z. Fu, Q. X. Chu, and R. L. Li, "Broadband dual-polarized dual-dipole planar antennas: Analysis, design, and application for base stations", *IEEE Antennas Propag. Mag.*, vol. 59, no. 6, pp. 77–87, Dec. 2017.

[14]. Y. H. Cui, R. L. Li, and H. Z. Fu, "A broadband dual-polarized planar antenna for 2G/3G/LTE base stations", *IEEE Trans. Antennas Propag.*, vol. 62, no. 9, pp. 4836–4840, Sep. 2014.

[15]. H. Y. Qi, L. L. Liu, X. X. Yin, H. X. Zhao, and W. J. Kulesza, "Mutual coupling suppression between two closely spaced microstrip antennas with an asymmetrical coplanar strip wall", *IEEE Antennas Wirel. Propag. Lett.*, vol. 15, pp. 191–194, 2016.

[16]. N. Saxena, "Design and Analysis of Multi Band antenna for S and C Band", *IEEE Trans. Adv. Comp. Commun. Control Network.*, Oct. 2018. doi:10.1109/ICACCCN.2018.8748427

[17]. A. Dadhich, J. K. Deegwal, and M. M. Sharma, "Multiband slotted microstrip patch antenna for TD-LTE, ITU and X-band applications", *IEEE Trans. Signal Process. Integr. Netw.*, Feb. 2018.

[18]. I. Sai Sampreeth and M. B. Abhishikth, "A multiband slot antenna for wireless communication", *IEEE Trans. Comput. Commun. Netw. Technologies*, July 2018.

[19]. K. Mehr-e-Munir and S . H. K. Mahmood, "E-shape Multiband Patch Antenna for 4G, C-band and S-band Applications", *Int J Adv Comput Sci Appl*, vol. 9, no. 5, pp. 233–237, 2018.

[20]. I. Locher, M. Klemm, T. Kirstein, and G. Troster, "Design and Characterization of Purely Textile Patch Antennas", *IEEE Trans. Adv. Packag.*, vol. 29, no. 4, pp. 777–788, Nov. 2016.

[21]. Y -J Chi and F -C Chen, "On-Body Adhesive-Bandage-Like Antenna for Wireless Medical Telemetry Service", *IEEE Trans. Antennas Propag.*, vol. 62, no. 5, pp. 2472–2480, May 2020.

[22]. M. Ur-Rehman, Q. H. Abbasi, M. Akram, and C. Parini, "Design of Band-notched Ultra Wideband Antenna for Indoor and Wearable Wireless Communications", *IET Microw. Antennas Propag.*, vol. 9, no. 3, pp. 243–251, 2015.

[23]. A. Hirata, O. Fujiwara, T. Nagaoka, and S. Watanabe, "Estimation of Whole-Body Average SAR in Human Models Due to Plane-Wave Exposure at Resonance Frequency", *IEEE Trans. Electromagn. Compat.*, vol. 59, pp. 41–48, 2010.

[24]. K. Zhao, S. Zhang, C. Chiu, Z. Ying, and S. He, "SAR Study for SmartWatch Applications", *Antennas Propag. Soc. Int. l Symp.* (APSURSI), Memphis, U.S, 2014.

[25]. S. Jun, B. Sanz-Izquierdo, and M. Summerfield, "UWB antenna on 3D printed flexible substrate and foot phantom", in *Antennas Propag. Conf. (LAPC), 2015 Loughborough*, 2015, pp. 1–5

[26]. A. Sabban, "Comprehensive Study of Printed Antennas on Human Body for Medical Applications", *Int. J. Adv.*, vol. 1, pp. 1–10, February 2013

[27]. T. Arakawa, K. Asanuma, and T. Maeda, "A Study on the Structural Parameters of a Ring Antenna Including the Effects of a Human Finger", *IEICE Nat. Conf.*, B-1-76, 2018.

[28]. S. Yan, P. J. Soh, and G.A.E. Vandenbosch, "Made to be worn", *Electron. Lett.*, vol. 50, no.6, pp. 420, Mar. 2014.

[29]. P. Salonen, L. Sydanheimo, M. Keskilammi, and M. Kivikoski, "A small planar inverted-F antenna for wearable applications", in *3rd Int. Symp. on Wearable Computers Digest*, 1999, pp. 95–100.

[30]. C. Caloz and T. Itoh, *Electromagnetic Metamaterials: Transmission Line Theory and Microwave Applications.* New York: Wiley, 2006

[31]. T. Leng, "Graphene nanoflakes printed flexible meandered-line dipole antenna on paper substrate for low-cost RFID and sensing applications", *IEEE Ant. Wirel. Propag. Lett.*, vol. 15, pp. 1565–1568, 2016.

[32]. J. Ahzad, "Special issue on multifunction antennas and antenna systems", *IEEE Trans. Antennas Propag.*, vol. 54, no. 2, pp. 314–316, Feb. 2006.

[33]. S. Yan, P. J. Soh, and G. A. E. Vandenbosch, "Wearable dual-band composite right/left-handed waveguide textile antenna for WLAN applications", Electron. Lett., vol. 50, no. 6, pp. 424–426, March 2014.

[34]. A. M. Jajere, "Millimeter wave patch antenna design antenna for future 5g applications", *Int. J. Eng. Res. Technol.*, vol. 6, no. 2, pp. 298-291, 2017.

[35]. S. Agneessens and H. Rogier, "Compact half diamond dual-band textile HMSIW on-body antenna", *IEEE Trans. Antennas Propag.*, vol. 62, no.5, pp. 2374–2381, May 2014

[36]. Y. Huan and K. Boyle, *Antenna From Theory to Practice.* New York: Wiley, 2008.

Index

Page numbers in **bold** represent figures and in *italics* represent tables respectively.

abstraction, 187, 200
access nodes, 82, 83, 93
accessibility, 183, 196
accessible, 187
account, 241
activated subcarrier indices, 238
additive white gaussian, 240
administration, 163–164, 169–172, 174, 176–177, 179–180, 184, 186, 191, 194, 198–199
advantages, 170–172, 177, 180
AF (application function), 29
aggregate distortion level, 243
agriculture, 257
AI (artificial intelligence), 25, 58, 150, 152, 153, 154, 156, 157, 161, 162
AI in wireless communication, 150, 155
AI-enabled wireless communication, 150, 155–156
alternative base, 236
AM RCL entity, 46
AMF (access and mobility function), 28, 29, 31
antenna array, 321, 323, 324, 325, 328, 329, 331, 332, 333
application, 82, 83, 85, 163–165, 168, 170–172, 174–175, 178–182, 183, 187, 189, 194–196, 198, 236
application layer with ML, 159
application of ML in wireless communication, 159
applications deep learning, 15–23
applications of 5G, 92–93
applications of machine learning, 15
architecture, 25, 26, 27, 82, 83, 85, 86, 87, 89, 90, 92, 93, 296–297, 299–302
architecture 5G, 260
Artificial Intelligence (AI), 71, 171, 295–296, 299–300, 302
ASF (authentication server function), 28
assembling, 164
association, 165–166, 168, 171–174, 177
attenuation due to rain, **268**
authentication, 87, 90, 91
average distortion magnitude, 243

backhaul networks, 87
band pass filter, 267, 268, 270, 274, 275
bandwidth, 323, 324, 332
bandwidth efficiency, 240
base station, 126–127, 184, 186, 189, 202, 300, 321, 323, 325
beamforming, 158, 161, 321, 322, 325
ber performance, 244
Big Data Analytics, 155, 156, 161
binary phase-shift keying, 236
BPSK, 236, 239, 244
brief classification of ML for AI, **157**
broadband, 186
building complex and dense network, 50

caching, 127
capacity, 122, 184, 188, 193, 197, 199, 201
CAPEX (capital expenditures), 26
carrier aggregation, 52
centralized, 82, 83, 85, 89, 90
centralized unit–distributed unit (CU-DU), 85
challenges, 25, 50, 82, 86, 87, 88, 89, 266
channel, 235, 241, 243
channel capacity, 182, 183
channel coding and modulation, 158
channel estimation, 158, 159, 161
classification of bands, 68–70
classification of congestion, **2**
Cloud, 82, 83, 85, 87, 89, 93
Cloud computing, 26, 163–173, 175, 177–182
Cloud-RAN (C-RAN), 86, 184
Code Division Multiple Access (CDMA), 293
cognitive radio, 53
combine coaxial cavity filter, 267
commercial usage, 323
Communication, 82, 86, 87, 90, 91, 93, 293–294, 296–301, 303
communication system, 241, 323, 324
complex hardware design and architecture, 267
components imperfections, 241
concepts of network slicing, 134–135
Conclusion, 19, 78, 155, 160–161, 267, 279
configuration, 32, 47, 58
congestion alertness phase, 6
congestion alleviation phase, 6
congestion awareness phase, 4
congestion mitigation phases in WSN, 4
conjugate mirror filters, 240
connected mode procedure, 39
connection density, 129–131
constellation diagram, 237, 238
constellation space, 235
construction industry, 255
controller, 186, 189, 194, 196–197, 199
conventional MIMO, 321, 323
conveying, 241
cooperative communication, 72
core network architecture, 86
corporate social responsibility (CSR), 170
correlation coefficient, 253, 259
coupling screw, 267, 269, 270, 271, 275
cross-coupling, 268, 269, 270, 276
CS (circuit switched), 26
cyber physical system, 178

D2D communication mode, 71
D2D control, 70
D2D coverage, 70

D2D integrated with millimeter wave (mmWave), 71
D2D main classification, 70–71
data, 81, 82, 83, 89, 90, 91, 92
data center, 163–165, 167, 169–181
decoupled, 186, 188, 193
deep learning, 137–138, 153, 154, 160, 161, 162
defected ground structure, 325
dematerialization, 169
deployment, 178
design, 321, 323, 325, 328, 329, 331, 332, 333, 334
detail coefficients, 240, 241
device-to-device (D2D) communication, 297
discrete wavelet transform, 240, 241
distortions types, 243
distributed, 82, 83, 85
diversity gain, 259, 262
diversity performance, 260
DL, 153, 158, 160
DL channel mapping, 49
DL logical channel, 49
DL physical channel, 49
DL transport channel, 49
DoS, 87, 89, 90, 91
dual patch, 8, 253, 259, 260, 261
dynamic ranges, 241

education, 257
efficient, 164, 167, 172, 176, 180–182, 189, 194
electric field, 258, 259, 271
electromagnetic radiations, 326, 328
empowering, 198–199
eNB, 26, 27
end-to-end (E2E) service operations, 84, 93
energy, 164, 167, 169–182, 183–184, 200–202
energy efficiency, 122, 125–128, 139
energy management, 51
enhanced mobile broadband, 251
establishment, 32, 34, 36
evaluation, 241, 243
evolution of cellular network technologies, 150–152, **151**
e-waste, 167
exploration, 177, 180

fading, 50
farms, 164, 167–170, 172, 176–180
feature extraction, 157
feeding, 325, 326
Fifth Generation (5G), 149, 150, 151, 152, 153, 154, 160
5G: at a glance, 152
5G enabling technologies, 119–121
Fifth Generation (5G) enhanced overall system architecture, 83
Fifth Generation (5G) networks, 87
5G key enabling technologies, 63
Fifth Generation (5G) security aspects, 89
Fifth Generation (5G) security threats, **88**
5G specifications, 118–119
5G system, 117–121
Fifth Generation (5G) technology overall architecture, 82
Fifth Generation (5G) threat landscape, 87
figuring, 164, 166, 171, 177, 180
filter, 240
First Generation (1G), 149, 150, 151, 152
flat fading channels, 235
flexibility, 236

fluctuations, 241
fog radio access networks, 137
Fourier basis, 236
Fourier exponential functions, 236
Fourth Generation (4G), 81, 87, 88, 90, 149, 150, 151–152
fractional bandwidth, 267, 276
framework, 187, 190, 192–193, 195–197, 199–201
frequencies, 323, 324, 325, 327
frequency domain, 240
frequency reconfigurable, 325
frequency spectrum, 323
full dimensional massive MIMO [FD-MIMO], 325
future, 164, 170–172, 178–181
future directions, 19
future perspectives for 6G networking systems, 279

gain, 325, 326, 328, 329, 330, 332
green cloud computing, 166–168, 170, 172, 178
green communication, 183–185, 199, 201, 203
green computing, 164, 166, 175

handover, 72
hardware, 169, 174, 177, 184, 187, 189, 241
hardware components, 324
hardware impairments, 236, 240, 244
harmful effects, 244
health care industries, 256
healthcare, 26, 55, 59
heterogeneous, 178, 181, 194, 197–200, 202
heterogeneous access & mobility, 25, 54, 60
high data rate, 121
high frequencies, 241
high quality, 235
high speed data, 293, 296–297
high speed wireless, 236
HO handover, 27, 38, 48
holographic communication and haptic feedback, 258
how to realize 5G?, 152–153
HSS (Home Subscriber Server), 26, 28, 53
Hybrid Automatic Repeat Request Operation (HARQ), 72
hybrid cloud, 171–172, 181

ideal hardware, 243, 244
idle mode procedure, 38, 39
IDWT/DWT, 240
imperfections extremely affect, 241
inactive mode procedure, 39
inband D2D communication, 68–69
infrastructure, 189, 199–201, 203
inspector gartner, 165
integrated, 184, 199
integrated access backhaul, 259
integrated access backhaul as per 3GPP 38.807, **260**
integrated features, 71–72
intelligence communication and connectivity with 6G, **276**
intelligence in wireless communication, 156
intelligent connectivity for 6G, **154**
inter element spacing, 255, 258
interference, 51, 53, 184, 194
interference management (IM), 76–77
international telecommunication union (ITU), 294–295
Internet of Things (IoT), 25, 26, 57, 58, 71, 88, 100, 119, 124, 129, 148, 149, 254, 316, 396

Index 329

introduction, 1, 2–3, 26, 44, 59, 63–68, 81–82, 115–117, 149–150, 155, 161, 247, 248, 271–272
introduction of RLC entity, 44, 45
introduction to 5G system architecture, 26, 28
introduction to AI and ML, 155
IoT data and cloud infrastructure, 57, 57
IoT device management, 57
IoT (Internet of Things), 25, 26, 57, 58, 71, 88, 100, 119, 124, 129, 148, 149, 254, 316, 396
IoT protocols and communication standards, 57
IP, 57
Iris, 267, 271, 274, 275

key areas of 5G security, **91**
key steps for ML, 150, 156–157

latency, 121, 122–125, 138
layer, 299–301
layer 1, 26, 45, 48
layer 2, 26, 45
layer 3, 26, 45
learning function and prediction, 157
lens antenna, 325
linear operating, 241
liquid crystal polymer (LCP) substrate, 326
list of abbreviations, 20
local oscillator frequency, 241
lower layer shares the following services with PDCP, 43
lower layer shares the following services with SDAP, 40
low-pass filter, 240
LTE, 323, 324, 326

M2M, 31, 50, 58
MAC Functions, 47
MAC Procedures, 47
machine learning, 7, 128
machine learning process, 7–10
machine-to-machine, 58
maintenance, 27, 43, 54
management and orchestration (MANO), 84
manufacturing, 26
mapped symbols, 236, 237
massive MIMO, 86, 87, 321, 322, 323, 324, 325, 329, 333
massive multiple input, 235
medium access control, 47
medium access control layer with ML, 158–159
message signals, 321
microstrip line feed, 326
microstrip patch, 325, 325, 326
MIMO, 52, 53
ML, 150, 152, 153, 155, 156, 157, 158, 159, 160, 161
ML for wireless communication, 158–159
ML model for wireless communication, 156–159
Ml techniques, 10–15
ML training model, **157**
MM (mobility management), 28, 32, 33
MME, 26, 27
mobile core and external IP networks, 26
Mobile Edge Computing, *125*
Mobile Edge Network (MEN), 84
mobility, 27, 28, 51, 122
mode selection, 73–76

modulation, 235, 236, 244
modulation order, 243, 244
monitoring, 57, 58
mother wavelet HAAR, 244
motivation, 3–4
moving from 5G to 6G, 150, 154
multi input multi output (MIMO), 294, 296
multi-access edge computing (MEC), 85, 89
Multiband, 324, 325, 326
multicarrier modulation, 235, 236
MU-MIMO, 321, 322, 323, 324
mutual coupling, 254, 255, 257, 258, 259, 325

NAS, 26, 28, 31, 32, 43
natural disaster, 253
need for 6G networking system, 274
NEF (network exposure function), 28, 29
network, 169, 172–174, 177–179, 181–182
network capacity, 128–131, 139
network function virtualization, 133–134, 195, 201–202
network functions (NFs), 84, 85, 86, 91
network layer design, 125
network optimization parameters, 121–122
network optimization state-of-art, 131–135
network slice selection function, 5
network slicing, 83, 84, 89, 93, 134, 139
network slicing enabling technology, 135–138
network slicing in 5G, 26, 28, 29, 30
network slicing use cases as per 3GPP 22.891, **255**
new radio operational frequency band in the FR2, **250**
new radio operational frequency bands in FR1, 249, *250*
New-Generation Radio Access Network, 27
NF repository function, 29
NFV (network function virtualization), 25, 26, 30, 54
noise variance, 240
non-3GPP, 26
nonlinear components, 241
non-orthogonal multicarrier communications, 235
Non-Stadnalone Architecture,Option 3, **262**
Non- Stadnalone Architecture,Option 3A, **262**
Non- Stadnalone Architecture,Option 3X, **261**
Non- Stadnalone Architecture,Option 4, **264**
Non- Stadnalone Architecture,Option 4A, **265**
Non- Stadnalone Architecture,Option 7, **266**
Non- Stadnalone Architecture,Option 7A, **266**
novel transmission systems, 235
NSA (non-stand alone), 31
NSSF (network slice selection function), 28, 29

OFDM bandwidth efficiency, 236
OFDM -IM modulator, 239
OFDM -IM, 236, 237, 238, 239, 240
OFDM modulators, 238
OFDM system, 236, 244
open challenges, 138–139
OPEX (operational expenditure), 51
opportunities 5G communication, 82, 90
optimal maximum likelihood, 243
orchestration, 83, 84, 85, 189, 194, 198
orchestrator, 194, 196, 198–199
organization, 163–170, 173, 177, 179–180, 184, 186–187, 192–194, 196–202

orthogonal frequency division multiplexing (OFDM), 294
orthogonality loss, 236
outband D2D communication, 69–70
overall 5G Architecture, **82**
overall RAN Architecture, **85**
overview of 5G technology, 117–118

Packet Data Convergence Protocol, 42, 43, 44
parameters, 325, 325, 328, 329, 331, 332
PCF (policy control function), 28
PCRF, 26, 28
PDCP, 39, 41, 42
PDCP functions, 43, 44
PDCP procedures, 43, 44
peak average power, 241
peer discovery, 72–73
performance analysis, 241
performance degradation, 241, 244
performance enhancement, 240
performance metrics, *17*
Performance Metrics: moving from 5G to 6G wireless systems, *154*
performances analysis, 243
P-GW PDN gateway, 26
phase noise, 236, 241
physical layer, 45, 47, 48, 50
physical layer design, 123–124
physical layer with ML, 158
planar inverted f antennas, 325
popular standards, 235
positioning, 151, 154, 158
power, 164, 167–170, 174, 176–181
power amplifier, 241
power control, 77–78
priority-based congestion control, **16**, *16*
privacy, 81, 87, 89, 92, 93
privacy perspective, 92, 93
product manufacturing industry, 253
programming, 184, 186–187, 189, 192–193, 195–196, 198, 200
protocol, 296, 302
provisioning and authentication, 58
PS (packet switched), 26

QoS, 84, 173, 178–179, 196, 198, 201, 203, 235
quadrature loss, 241
quadruple-mode filter, 267
quality factor, 267, 268, 274, 276
quality of service (QoS), 295, 300
queue-assisted congestion control, *14*

R-15 (release-15), 25
radiation efficiency, 328
radio, 25, 27, 38
radio access node (RAN), 82, 84, 85, 86, 92, 93
radio amplifiers, 241
radio frequency, 236
radio link control, 39, 42, 43
radio resource controller, 38
RAN slicing, 136–137
Rayleigh channel, 243
real implementation, 236
receiver hardware impairments, 242, 243

rectangular patch antenna, 255
References, 54, 78–80, 94, 140–147, 161–162
reflection coefficient, 325, 326, 328, 331, 332
reinforcement learning, 157, 158, 160
related review study, *4*
release, 27, 37, 59
remote radio heads (RRH), 300
renewable, 170, 173, 184, 203
requirement of 6G application in healthcare system, 274
research challenges, 78
research challenges for 6G, 150, 153–154
resource allocation (RA), 73
resource control-based congestion control, **10**, *11*
resource management, 51, 52, 58, 60, 127
return loss, 332, 333, 334
RL, 157, 158, 160
RLC, 39, 42
RLC functions, 38, 39, 40, 41, 42
robotics, 25
role of 6G networking vision with latest AI technology, **274**
RRC, 38, 39
RRC Functions, 39
RRC Services, 39
RRC states and procedures, 39
RT Duroid, 326

SA (stand–alone), 53, 59
satisfy, 235
scale functions, 240, 241
scattering parameter, 258, 260
SDAP (Service Data Adaptation Protocol), 39, 40
SDAP functions, 40
SDAP procedures, 40
SDN (Software Defined Network), 25, 26, 30
SDN/NFV, 135–136
SDN/SDMN, 89
Second Generation (2G), 149, 151
security, 81, 87, 88, 89, 90, 91, 93, 165, 178, 180, 196, 198–200
security challenges of 5G communication, 87, 89, 90
security issues and potential targets in 5G networks, 26, 53
Service Data Adaptation Protocol (SDAP), 39, 40, 41, 85
session multiplexing, 27
S-GW (serving gateway), 26, 27
signal distortion, 241
signal Dwt, 240
signal representation, 236
simulation, 326, 329, 330, 331
Sixth Generation (6G), 150, 151, 152, 153, 154, 156, 159, 160, 161, 162
The 6G Architecture in Healthcare Systems, **275**
6G automated communication systems, 277
6G communication smart society, 276
6G enabled artificial networking pattern, **272**
The 6G Global Satellite Navigation Through GPS, **273**
6G intelligent connectivity, 276
The 6G Satellite Communication Systems, 271–272
6G telecommunication industry built with AI machinery, **278**
6G: vision, 153–154
slice, 83, 84
slicing of network, 252
slot, 325, 326
SM (session management), 37

Index

small cell, 52, 53, 60
small-cell deployment, 52, 53, 60
smart cities and communities, 256
SMF (session management function), 28, 31, 35
software updates, 57, 58
software-defined networking, 132–133
spatial diversity, 321
spatial multiplexing, 321
specific absorption rate, 326, 328
spectrum efficiency, 128–129
spectrum extension, 129
spectrum management, 25, 52
Standalone Architecture, Option 2, **263**
Standalone Architecture, Option 5, **264**
standard MIMO systems, 236
standard OFDM, 236, 237, 244
storage, 83, 89
street transportation, 255
subblock, 238, 241
subcarriers, 236, 237, 238
summary, 23
supervised learning, 157, 159, 161
sustainable, 170, 174
synchronization, 158

TCP, 27, 57
technique, 184, 187, 198–201, 203
technology, 294, 296–299, 303
textile antenna, 326, 327
Third Generation (3G), 149, 150, 151
Threat Landscape of Internet of Things (IOT), 88
Transport Layer Design, 124
three different phases in congestion processes, **5**
three technical direction for 5G, **153**
throughput, 321
time division multiple access (TDMA), 293
TL, 160
TM RLC entity, 43, 45
tourism, 257
traffic rate control-based congestion control, **6**, *7*
training data, 156–157, 160
transfer learning, 157, 160
transform, 236, 243, 244
transformation of 5G To 6G due to enhanced networking communications, 275
transforms comparison, 244
transmission techniques, 235
transport architecture, 85, 93
transport network, 82, 83, 84, 86, 87
transport network infrastructure, 83, 86
trends and technology applications of 6G networks, 273
tuning screws, 269, 274, 275

types of applications, 2
types of congestion, **2**
types of ML categories for wireless communication, 150, 159–160

UDF (unified data management), 27, 52
UDN (ultra-dense networks), 25, 51, 52, 53
UE (user equipment), 53, 54, 57, 60
UL Channel Mapping, 50
UL Logical Channel, 50
UL Physical Channel, 50
UL Transport Channel, 50
ultra-dense network (UDN), 72
unsupervised learning, 157, 160
ultra-reliable communications, 251
UM RLC entity, 44, 45
unloaded Q, 267
UPF (user plane function), 28, 29
use case categories as per NGNM 5G white paper v1, **254**
use cases as per 3GPP 22.891, **253**
use cases for 5G NR above 52.6 GHz, **273**
user equipment's, 53, 54, 57, 60

V2X (vehicle-to-X), 25, 51, 52, 56
vehicle to cloud, 56
vehicle to devices, 55, 56
vehicle to grid, 55, 56
vehicle to infrastructure, 56
vehicle to machine, 55, 58
vehicle to network, 55, 56
vehicle to pedestrian, 55, 56
vehicle to vehicle, 55, 56
vehicle-to-everything, 55, 61
Virtual Network Functions (VNFs), 83, 85
virtual office, 258
virtualization, 167–168, 172, 174, 177–178, 180–181, 183, 186, 191–193, 195–197, 199–203
visible light communication (VLC), 299–300
vision in combination of artificial intelligence with 6G and its expectations, 273
voltage standing wave ratio, 326, 333
VR (virtual reality), 56

wavelet, 236, 241, 244
wavelet families, 242
wavelet functions, 236
wavelet transform, 236, 243, 244
wearable antenna, 326, 327, 329
wireless communication, 86, 87, 88
wireless networks, 293–295, 297, 299, 302–303
WOFDM-IM systems, 242, 244